Conventional to Emerging Advances in Plant Disease Diagnosis and Pathogen Detection

Diseases in crops cause significant economic losses by decreasing both the yield and quality of produce. Early and accurate diagnosis of these diseases is crucial, as it enhances the effectiveness of management strategies and helps prevent long-term damage to infected plants. The need for precise, swift, and efficient methods for diagnosing diseases and detecting pathogens remains a top priority. *Conventional to Emerging Advances in Plant Disease Diagnosis and Pathogen Detection* comprehensively addresses key diagnostic methods and pathogen detection techniques, from their origins to the present, guiding us toward the future of plant diagnostics. It serves as a valuable resource for academics, extension workers, and growers, offering the latest information on developments in disease diagnosis and pathogen detection.

The chapters within this volume offer a comprehensive journey through the historical perspectives and groundbreaking innovations that have shaped disease diagnosis. We traverse the conventional methods that have laid the foundation for our understanding, and then embark on a journey through emerging technologies that promise to revolutionize the field. This compilation aims to connect traditional knowledge with the latest technological advancements, offering a comprehensive view of the trajectory of plant disease diagnosis. Created through collaborative efforts of field experts sharing insights and experiences, this resource is not only informative but also forward-thinking.

Features

- Comprehensive coverage of both traditional and contemporary diagnostic techniques for identifying plant diseases and detecting pathogens.
- Detailed exploration of the latest advancements and innovations in disease diagnosis and pathogen detection, providing up-to-date knowledge.
- In-depth discussion of cutting-edge diagnostic strategies, including artificial intelligence, machine learning, biosensors, next-generation sequencing, point-of-care assays, and CRISPR-Cas systems.
- Abundantly illustrated with diagrams, flowcharts, and representations to support field identification and precise diagnosis of plant diseases.

Conventional to Emerging Advances in Plant Disease Diagnosis and Pathogen Detection

Edited by

Sajad Un Nabi and Sheikh Mansoor

CRC Press
Taylor & Francis Group
Boca Raton London New York

CRC Press is an imprint of the
Taylor & Francis Group, an informa business

Designed cover image: Getty Images

First edition published 2025
by CRC Press
2385 NW Executive Center Drive, Suite 320, Boca Raton FL 33431

and by CRC Press
4 Park Square, Milton Park, Abingdon, Oxon, OX14 4RN

CRC Press is an imprint of Taylor & Francis Group, LLC

ISBN: 978-1-032-75450-5 (hbk)
ISBN: 978-1-032-75451-2 (pbk)
ISBN: 978-1-003-47403-6 (ebk)

DOI: 10.1201/9781003474036

Typeset in Times
by Newgen Publishing UK

Contents

Foreword

Many factors (biotic/abiotic) are responsible for reduced productivity and quality of crops. Among the biotic stresses, plant pathogens such as fungi, bacteria, phytoplasmas, viruses, and viroids are economically important. Plant pathogens are of great concern because of their systemic nature and their safe transfer to succeeding generations through the vegetative propagules causing decline in orchard health. The propagation of exotic varieties is not regulated, causing the introduction of new pathogens, especially viruses and virus-like pathogens. In a time where being accurate and fast is super important, the use of artificial intelligence, machine learning, and high-throughput phenomics is bringing a new age of insights driven by data. As we explore the complex fields of DNA microarrays, isothermal amplification assays, and next-generation sequencing, it's becoming clearer how we can completely change the way we find and fight plant diseases. The chapters in this book tell a story of scientific discovery. It starts with the basic tools like microscopes, moves through the game-changing phase of molecular diagnostics, and ends up in the advanced world of technologies such as Droplet Digital PCR, Lab-on-Chip Platforms, Biosensors, and the revolutionary CRISPR technology.

As the writer of the foreword, I wish for this book to spark curiosity, inspire innovation, and encourage collaboration among researchers, practitioners, and students. The diagnostics right from conventional to advanced methodologies including high throughput sequencing has been comprehensively covered for detection of pathogens in this book. I would also like to express my appreciation to the authors who have compiled all the aspects of viruses infecting temperate fruit crops. I am confident that this bulletin will be very useful in implementation of the clean plant programme of apple in India recently launched by Government of India besides farmers, nursery growers, and policymakers.

Prof. V.K. Baranwal
(National Professor)
Advanced Center of Plant Virology
ICAR-IARI, New Delhi

Preface

In the ever-evolving landscape of agriculture and plant health, the quest for accurate, rapid, and efficient methods of disease diagnosis and pathogen detection remains paramount. The book *Conventional to Emerging Advances in Plant Disease Diagnosis and Pathogen Detection* delves into the rich tapestry of techniques that have shaped the past, it explores the current state of the art, and propels us into the future of plant pathology diagnostics.

The chapters within this volume offer a comprehensive journey through the historical perspectives and groundbreaking innovations that have shaped disease diagnosis. We traverse the conventional methods that have laid the foundation for our understanding, and then embark on a voyage through emerging technologies that promise to revolutionize the field. From the early days of microscopy as a fundamental tool for detecting plant pathogens to the current advancements in immunodiagnostics and molecular techniques, each chapter is a testament to the relentless pursuit of precision in identifying and combating plant diseases. The reader is guided through the intricacies of Polymerase Chain Reaction (PCR), Droplet Digital PCR, DNA Microarrays, and isothermal amplification assays, witnessing the transformation of diagnostic methodologies.

The advent of cutting-edge technologies such as next-generation sequencing, Lab-on-Chip platforms, biosensors, and CRISPR technology unfolds as we explore their potential applications in plant disease detection. The integration of high-throughput phenomics and artificial intelligence, alongside machine learning techniques, marks a paradigm shift in how we approach diagnosis, utilizing the power of data analytics to discern patterns.

Each chapter represents an unwavering commitment to precision in identifying and combatting plant diseases. This compilation aims to connect traditional knowledge with the latest technological advancements, offering a comprehensive view of the trajectory of plant disease diagnosis. Created through collaborative efforts of field experts sharing insights and experiences, this resource is not only informative but also forward-thinking. Join us on this journey through the history, present, and future of plant disease diagnosis, driven by innovation and a collective dedication to safeguarding plant health and global agriculture.

Sincerely
Sajad Un-Nabi. ARS, ***PhD***
Sheikh Mansoor, ***PhD***

About the Editors

Dr Sajad Un Nabi, (ARS, Ph.D.)
Scientist Plant Pathology,
ICAR-Central Institute of Temperate Horticulture
Srinagar J & K, India, 191132
Email: sajadnabi21@gmail.com

Dr Sajad Un Nabi (ARS, Ph.D.) is working as a Scientist-Senior scale in plant pathology at ICAR-Central Institute of Temperate Horticulture Srinagar J & K, India. He completed his Ph.D. in 2020 on apple viruses from ICAR-Indian Agricultural Research Institute, New Delhi India. He is a Life member of Indian Phytopathological Society, Indian virological society, world society of virology and honorary life member of Asian PGPR Society. He has published more than 100 research publications including research papers, review papers, popular articles, books, book chapters, bulletins, and pamphlets in nationally and internationally reputed journals and books. He has submitted more than 100 nucleotide sequences of plant pathogens to Genbank. He is the Recipient of IARI Gold Medal for M.Sc degree, M.J.Narshiman medal from Indian Phytopathological Society for best paper, Young Achiever Award from Agricultural Technology Development Society, Excellence in Plant Pathology Award by International Association of Research and Development Organization (IARDO). He identified a new potential vector *Hishimonus phycitis* (Distant) for sesame phyllody phytoplasma in India. He is presently working on apple cankers and temperate fruit viruses. He has identified a new species, *Diplodia bulgarica,* in India associated with canker disease of apple. In temperate fruit viruses, he identified new virus ApNMV and viroid AHVd associated with apple mosaic disease in India. He also confirmed the etiology of apple mosaic disease in India using RNA-Sequencing (RNA-Seq). He has also optimized tissue and season for early and rapid detection of apple latent viruses viz., ASGV and ASPV. He is presently involved in understanding the mechanism of mosaic and necrosis development in apple using transcriptome analysis and virome of stone fruits for developing easy diagnostics. He is also involved in developing scab-resistant apple varieties for which two scab-resistant apple hybrids have already been registered in NBPGR for national release. He is also involved in extension programmes on management of temperate fruit diseases by conducting trainings, TV talks, and consultancies.

Dr. Sheikh Mansoor. Ph.D.
Postdoctoral Researcher
Phenomics Laboratory
Department of Plant Resources and Environment,
Jeju National University South Korea
mansoorshafi21@gmail.com; mansoorshafi@jejunu.ac.kr
https://orcid.org/0000-0003-1902-4611
www.researchgate.net/profile/Sheikh-Mansoor

Dr. Sheikh Mansoor is a postdoctoral researcher at the Phenomics Laboratory, Jeju National University, South Korea. His research focuses on unraveling the complex mechanisms underlying plant responses to environmental stresses through multiomics approaches, including advanced phenotyping, transcriptomics, metabolomics, and genomics. With expertise in wet lab methodologies, he is dedicated to advancing innovative research and development initiatives in plant science. Before joining Jeju National University, Dr. Mansoor held positions as a Postdoctoral Research Associate at SKIMS Srinagar and as a Project Associate-I at CSIR-Indian Institute of Integrative Medicine (IIIM), Jammu. He earned a BSc degree from SP College of Science (University of Kashmir), an MSc degree from HNBGU Uttarakhand, and a PhD from Sher-e-Kashmir University of Agricultural Sciences and Technology of Jammu. Dr. Mansoor has made significant contributions to his field, with an impressive portfolio of over 70 research and review articles, five authored books, more than ten book chapters, and two patents. His achievements have been recognized through numerous national and international awards, including honors for Best Thesis, Young Scientist, and excellence in oral and poster presentations. He is also a recipient of a substantial grant under the Biotechnology Industry Research Assistance Council (BIRAC) awarded by the Department of Biotechnology (DBT), India. As an Associate Editor for Plant Cell Reports and a Youth Editorial Board Member for Phenomics, Dr. Mansoor actively contributes to the scientific community. He also serves as a reviewer for several leading journals in plant science, further demonstrating his commitment to advancing the field.

Contributors

Aniket Angira
Amity Institute of Virology & Immunology,
Amity University Uttar Pradesh, Sector 125 Noida 201313
Uttar Pradesh, India

Aadil Ayaz
ICAR-Central Institute of Temperate Horticulture,
Srinagar, 191132, India

Supreetha B G
Division of Fruit Science,
ICAR-CITH, RS, Dirang,
Arunachal pradesh-790101. India

Pooja Bhardwaj
Division of Plant Pathology
ICAR-Indian Agricultural Research Institute
New Delhi, India-110012

Farooq Ahmad Bhat
Division of Plant Pathology
FOA-SKUAST-K, Wadura, 193201
India

Muhammad Ashraf Bhat
Division of Plant Biotechnology
FOH-SKUAST-K,
Shalimar, 190025, India

V. K. Braranwal
ICAR-Central Institute of Temperate Horticulture,
Srinagar, 191132, India

Virendra Kumar Baranwal
Division of Plant Pathology
ICAR-Indian Agricultural Research Institute
New Delhi, India-110012

Vanita Chandel
Amity Institute of Virology & Immunology,
Amity University Uttar Pradesh, Sector 125 Noida 201313
Uttar Pradesh, India

Utkarsh Singh Chauhan
Center for Basic Sciences,
Pt. Ravishankar Shukla University, Raipur,
Chhattisgarh, India- 492010

Devendra Kumar Choudhary
Indira Gandhi Krishi Vishwavidyalaya,
Raipur Chhattisgarh 492012 India

Sunny Dhir
Department of Biotechnology, Markandeshwar (Deemed to be University),
Mullana, Ambala 133207
India

Damini Diksha
Advanced Centre for Plant Virology,
Division of Plant Pathology,
ICAR-Indian Agricultural Research Institute,
New Delhi-110012, India

Marimuthu Elangovan
Advanced Centre for Plant Virology
ICAR-Indian Agricultural Research Institute,
New Delhi- 110 012 India

Rashmi ER
Division of Plant Pathology
ICAR-CITH, RS, Mukteshwar
Uttarakhand-263138 India

Falak Fayaz
Department of Plant Pathology
Indian Agriculture Research Institute
New Delhi, India

Nitika Gupta
Division of Plant Pathology
ICAR-Indian Agricultural Research Institute
New Delhi, India-110012

Mohammad Waris Haider
Department of Plant Pathology
Indian Agriculture Research Institute
New Delhi, India

Mohammad Irfan
ICAR-Central Institute of Temperate Horticulture,
Srinagar-191132,
India

Ronit Jaiswal
ICAR-Central Institute of Temperate Horticulture
Srinagar 191132, India

Girish Kumar Jha
ICAR-Indian Agricultural Statistics Research Institute
New Delhi, India

Deeksha Joshi
Division of Plant Pathology
ICAR-Indian Agricultural Research Institute
New Delhi, India-110012

Ravpreet Kaur
ICAR-Indian Agricultural Research Institute
Pusa Campus, New Delhi 110012, India

Abnish Kumar
University School of Biotechnology,
GGSIPU, Sector 16C, Dwarka,
New Delhi, India-110078

Rajeev Ranjan Kumar
ICAR-Indian Agricultural Statistics Research Institute
New Delhi, India

Rakesh Kumar
Division of Plant Pathology
ICAR-Indian Agricultural Research Institute
New Delhi, India-110012

Yazing S. Mailem
Advanced Centre for Plant Virology,
Division of Plant Pathology,
ICAR-Indian Agricultural Research Institute,
New Delhi-110012, India

Naorem Loya Mangang
Advanced Centre for Plant Virology,
Division of Plant Pathology,
ICAR-Indian Agricultural Research Institute,
New Delhi-110012, India

Sheikh Mansoor
Department of Plant Resources and Environment
Jeju National University,
Jeju 63243,
Republic of Korea

Subaya Manzoor
Division of Plant Pathology
FOA-SKUAST-K, Wadura
J & K 193201, India

Mir Mehreen
Division of Plant Pathology,
SKUAST-K, Wadura, J & K 193201,
India

Zahoor Ahmad Mir
Department of Plant Science and Agriculture
University of Manitoba, R2M0TB
Canada

Javid Iqbal Mir
ICAR-Central Institute of Temperate Horticulture,
Srinagar-191132,
India

Diaa Abd El Moneim
Department of Plant Production (Genetic Branch)
Faculty of Environmental Agricultural Sciences,
Arish University, El-Arish 45511, Egypt

Fayaz. A. Mohiddin
Mountain Research Centre for Field Crops
Khudwani, Anantnag-SKUAST-Kashmir
J & K India-192102

Sajad Un Nabi
ICAR-Central Institute of Temperate Horticulture,
Srinagar, 191132, India

Zarka Nabi
Division of Plant Pathology
FOA-SKUAST-K,
Wadura, 193201, India

Nishant
Division of Vegetable Science
ICAR-IARI
New Delhi, India-110012

Shugufta Parveen
ICAR-Central Institute of Temperate Horticulture
Srinagar, 191132, India

Jyoti priya
Division of Plant Physiology
ICAR-CITH, Srinagar, J & K-191132, India

Reena Prusty
Division of Fruit Science, ICAR-CITH,
Srinagar, J & K-191132,
India

Kavitha R
Scientist (Fruit Science),
ICAR-Central Institute of Temperate Horticulture,
Srinagar-191132,
India

Wasim H Raja
ICAR-Central Institute of Temperate Horticulture,
Srinagar, 191132, India

Uzma Rashid
ICAR-Central Institute of Temperate Horticulture,
Srinagar-191132,
India

Tariq Rasool Rather
Division of Plant Pathology
FOH-SKUAST-K, Shalimar Srinagar-190025
India

Zohra Shabir
Division of Plant Pathology,
SKUAST-K, Wadura, J & K 193201,
India

Susheel Kumar Sharma
Division of Plant Pathology
ICAR-Indian Agricultural Research Institute
New Delhi, India-110012

Mailem Yazing Shimray
Division of Plant Pathology
ICAR-Indian Agricultural Research Institute
New Delhi, India-110012

Manoj Kumar Singh
Advance Centre for Plant Viruses
Indian Agriculture Research Institute
New Delhi, India

Pankhuri Singhal
Advanced Centre for Plant Virology
ICAR-Indian Agricultural Research Institute,
New Delhi- 110 012 India

Tripti Singhal
Amity Institute of Virology & Immunology,
Amity University Uttar Pradesh, Sector 125 Noida 201313
Uttar Pradesh, India

Nishant Srivastava
Division of Plant Pathology
ICAR-Indian Agricultural Research Institute
New Delhi, India-110012

Sunil Kumar Sunani
ICAR-IIPR Regional Station Bhubaneswar
Khordha -752055, India

Pooja Thapa
Advanced Centre for Plant Virology,
Division of Plant Pathology,
ICAR-Indian Agricultural Research Institute,
New Delhi-110012, India

Mahendra K Verma
ICAR-Central Institute of Temperate Horticulture,
Srinagar, 191132, India

1 Plant Disease Diagnosis

Past, Present, and Future

Aadil Ayaz, Sajad Un Nabi, Sheikh Mansoor, Mahendra K Verma, Wasim H Raja, Susheel Kumar Sharma, and V. K. Braranwal

1.1 INTRODUCTION

Plant diseases pose a significant challenge to agriculture, affecting crop yield and quality. The situation needs to be dealt swiftly and appropriate measures should be taken to overcome economic loss. Failing to address the issue promptly could have catastrophic consequences (Khakimov et al., 2020). The influence of pests and diseases has led to a significant decline in global food crop productivity, currently progressing at a rate of 20% to 40%. The FAO's 2012 report highlights the detrimental effects of these factors on agricultural output. Diseases alone contribute to 14% of the damage, resulting in an annual trade loss of over 220 billion dollars in the agricultural sector, as estimated by Agrios in 2005. Current statistics reveal that fungi, viruses, phytoplasmas, and bacteria are the culprits behind more than 80% of known plant diseases (Pavlovskaya et al., 2018). So early plant disease diagnosis and their effectual management is obligatory to avoid sharp declines in food crop production and productivity. Also, the contaminated products made from the diseased crops pose threats to farm animals and humans as well. While crop damage assessment is done under separate specifications for different diseases, it is stressed that prompt action and measures are the need of the hour, regardless of the level of specification in assessment of diseases to avoid huge losses in crops. The efficacy of plant disease detection systems very often relies on the accurate detection of causative agents and disease symptoms (Khakimov et al., 2022; Narayanasamy, P., & Narayanasamy, P. (2011). Plant disease diagnostic methods are improving day by day. Novel and emerging methods and technologies are being designed for quick and reliable detection of plant diseases. Despite the emergence of novel methods and technologies, current and traditional diagnostic methods retain their significance. They continue to support efficient and prompt diagnosis, complementing the advancements in the field. These diagnostic tools are essential for making well-suited and timely decisions concerning the management of food crop diseases.

DOI: 10.1201/9781003474036-1

1.2 COMMON PLANT DISEASES

Disease symptoms in plants, such as withering, rotting, staining, galls and powder development, mummification, warts, leaf curving, twisting, wrinkling, mucous leakage, etc., arise due to continuous pathological processes in plants caused by the influence of different pathogens. Plant decay, a pathological process, is initiated by the activity of bacteria such as *Pectobacteriumcarotovorum and Xanthomonoscompestris* and fungi from various genera such as Fusarium, Sclerotinia, Rhizopus, Pythium, and Rhizoctonia (Shkalikov, V.A., 2007: Peresipkin VF 1991). Wilting, often known as withering, is a prevalent disease caused due to damage of the conductive system in plants. The pathogen enters the stem's root system, clogging the veins and releasing toxins that lead to necrosis in the vein walls, which causes a disruption in the plant's water supply and eventually the plant withers. Both bacteria and fungi can induce wilt, which when caused by the bacteria is known as trexeobacteriosis, and that caused by fungi is known as tracheomycoses. Adverse environmental conditions like that of drought and root injury can also cause plant withering (Khakimov et al., 2022; Bagirova et al., 2012). The proliferation of damaged tissue under the pathogen's influence leads to tumours or galls. They can be found in a variety of plant organs, including the roots (stalks of cabbage), tubers (cancer of potatoes), root crops (cancer of beets), and others. The manifestation of their presence is a consequence of either an escalation in the quantity (hyperplasia) or the size of infected cells (hypertrophy). Both processes can occasionally take place simultaneously. Galls, warts, and tumours are common indicators of disease brought on by plant parasitic nematodes, bacteria, fungi, and viruses (Khakimov, et al., 2022). Pustules are circular or elliptical formations consisting of fungal spores and mycelium of varying sizes and colours. These pustules are responsible for causing rust disease in plants such as yellow rust, wheat brown, and white birch rust. Plant deformation or distortion occurs due to the diseases caused by fungi bacteria and phytoplasma. The condition arises due to the leakage of nutrients, disruption in assimilation process, and subsequently the failure of uniform growth rate of plant tissue. Plant deformation includes fruit ugliness, thicker or stiff leaves (doubling), creases on the leaves, curling leaves, and thread-like leaves. Because of disruptions to the conduction system and assimilate leakage, the leaves get overloaded with starch, causing leaf curl (Khakimov et al., 2022). The diseases listed and described above have varied symptoms, but overall, they are far less in number than typical plant diseases. That is, several agents (fungi, bacteria, or non-infectious diseases) might cause the same disease sign. It is challenging to correctly identify, detect, and diagnose diseases due to this condition. As such, relying on just one symptom to make the diagnosis is inappropriate. Thus, developing a robust and efficient plant disease detection system is the need of the hour.

1.3 DISEASE DIAGNOSIS IN PLANTS

Plants are vulnerable to different diseases that hamper the growth of the plant, so early diagnosis of disease may reduce the chance of additional damage to plants. Various detection systems are employed for the early detection of plant diseases and are thoroughly described in Figure 1.1.

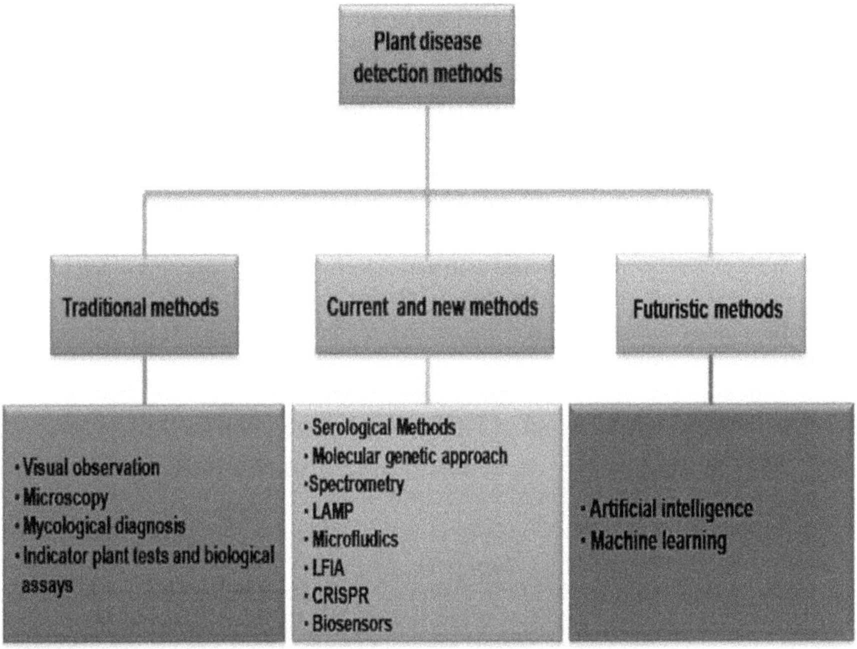

FIGURE 1.1 Methods used for plant disease detection.

1.3.1 Traditional Systems for Plant Disease Diagnosis

Traditional detection systems for plant diseases are continuously advancing, enabling the early identification of disease symptoms caused by various pathogens. The accuracy and speed of plant disease diagnosis are guaranteed by the advancement of new methods and technologies as well as in improvement of existing ones. However, traditional methods are still useful and can yield positive results when used alongside new approaches. Visual observation, microscopy, mycological diagnoses, testing using indicator plants, serological diagnosis, diagnosis by molecular genetic identification, and spectrophotometry are among the conventional methods still used to diagnose plant diseases (Khakimov et al., 2022; Narayanasamy, 2011; Dyakov and Elansky,(2019); Lukhmenev, (2012); Reich et al., 2013; Fang and Ramasamy, (2015); Mamiev et al., 2020; Kamilov et al., 2020; Zuparov et al., 2020).

1.3.1.1 Plant Disease Detection via Visual Observations

Initially, a visual inspection takes place at the cultivation site, where a healthy plant is contrasted with a diseased plant. This assessment involves noting any changes in the root system, plant height, colour, leaf shape, and branch density. Additionally, a magnifying lens is utilised to inspect different parameters, such as the presence of mycelium, sporangiophora, or sclerotia. The presence of necrosis or spots on the stems, leaves, etc., suggests the presence of a pathogenic fungus. Depending on the symptoms found in the plant sample, it is frequently easy to identify the major pathogenesis of fungus, bacteria, or virus through visual observation. A preliminary diagnosis may be made if the sample exhibits any distinctive symptoms. For instance, the *Agrobacterium tumefaciens* promotes host plant tissue growth and results in the easily observable formation of tumors. While plant parasitic nematodes (e.g., *Meloidogyne incognita*) produce "root knots" in the plant's thin roots and powdery mildew disease causes light powder in the tissues of plant organs (Putnam, 1995). When visualising, it is crucial to consider the plant's overall health (standing or turgidity); if it is withering, there may be an issue with the roots or conducting tissues. Both bacteria and phytopathogenic fungi, such as tracheomycoses and tracheobacteriosis, can cause these diseases. In many circumstances, visual observation is insufficient to make a conclusive diagnosis as the diagnosis's accuracy is decreased when inferences are made solely on the basis of a single morphological trait.

1.3.1.2 Microscopy as a Tool of Plant Disease Detection

High precision microscopes are important and frequently used instruments for accurate disease detection. The samples from infected plants are observed visually in fields and transported to laboratories for microscopic examination. With the aid of a stereo-zoom microscope (trinocular), the infected plant sample is subjected toprimary observation for disease symptoms. Upon observation under microscope, it is very often possible to reach a productive diagnostic conclusion by spotting hyphae, microslerosis, conidiophores, conidia, and bacterial cell clusters of pathogenic fungi. However, in some cases like inappropriate weather conditions visual observation and preliminary microscopic examination is not adequate enough for determining the occurrence of pathogens because under these conditions fungus may not be able to produce spores and also fungal mycelium may be absent on the affected surface (Dyakov and Elansky, 2019). Additionally, dead tissues may be devoid of pathogens but colonies of secondary parasites may be found (Putnam, 1995). Such cases demand additional mycological examination (moist chamber technique and pure culture isolation).

1.3.1.3 Mycological Diagnosis

Infected plant parts such as leaves, fruits, shoots, and roots are incubated in a high-humidity chamber at temperature ranging from 17 to 28°C to allow the fungus in the infected tissues to grow and show different symptoms. Through microscopic examination of the hyphae, macro, and microconidia a pathogenic fungal organism can be identified up to genus and species level (Campbell and Johnson, 2013).Visual observations, microscopy, and moist chamber techniques, however, might not always be sufficient to accurately identify all phytopathogens. Numerous phytopathogens, such as Ustilago spp., are dormant or latent and do not show abrupt symptoms. However, many infections have symptoms that are comparable to one another. According to Dyakov and Elansky (2019), a further issue is that certain dangerous fungi do not produce spores, even in damp environments. Any researcher conducting mycological study will commonly use the crucial, generally simple, and common (classical) approach of cultivating phytopathogenic bacteria in artificial nutritional media used for diagnosing plant diseases. It is not always possible to identify a pathogenic fungus species by its mycelium or spores. The form, colour, and other morphological traits of the colonies in the nutritional media are used as markers in several detection procedures to help in identifying the species. As a result, it is frequently required to isolate the fungi pure culture (Dyakov and Elansky, 2019; Bilay, 1982). Plant roots, soil, and infected surface portions of the plant can all be separated from pure culture. Following the pure culture isolation of a pathogen, its traits in the nutrient medium are ascertained based on the identification of micromorphological features using molecular-genetic identification or microscopy. Koch's postulates must be followed in order to identify the type of pathogen (Bos, 1981). Additionally, it must be shown that the pathogen's symptoms match those of the pathogen isolated from an infected plant under physiological conditions (Khakimov, et al., 2022).

1.3.1.4 Disease Diagnosis Involving Indicator Plant Tests or Biological Assays

The use of indicator plants for plant disease diagnosis involves the intentional infection of a known plant species with a particular pathogen in order to observe and identify the symptoms of the disease. This method is particularly useful when specific diagnostic tools or equipment for identifying a disease are not readily available. To carry out an indicator plant test, one selects a plant species that is known to be susceptible to the disease in question. This plant acts as an indicator or "sentinel" for the presence of the pathogen. Then, the selected indicator plant is exposed to the suspected pathogen, usually through methods such as inoculation with infected plant material, exposure to soil or water containing the pathogen, or by exposing the plant to vectors like insects that carry the disease (Bunsevich et al., 2010; Vlasov et al., 2016). Once the indicator plant has been exposed to the pathogen, it is monitored regularly for the development of any symptoms indicative of the disease. These symptoms can include characteristic patterns of leaf spotting, wilting, discolouration, stunted growth, or other visible signs of infection. More than 600 indicator plants from 43 families and 173 genera exist today. For instance, *Gomphrena globosa* and *Datura stramonium* are plant indicators of the potato X-virus; *Nicotiana glutinosa* and *Datura stramonium* are plant indicators of the tobacco mosaic virus; and *Chenopodium foetidum and Nicandra physaloides*species are plant indicators of the plum orca virus (Kartasheva, 2007). Indicator plants are inspected two to three days following infestation; for herbaceous indicator plants, the monitoring period is four weeks. During this time, the symptoms of viral disease develop on the plant under the testing. By observing and documenting these symptoms, one can make an initial diagnosis of the presence of the suspected disease. It is important to note that while indicator plants can provide valuable preliminary information about the presence of a particular pathogen, they are not without their challenges, and additional diagnostic methods such as laboratory testing, molecular analysis, or professional consultation with plant pathology experts may be needed to confirm the diagnosis. Furthermore, care should be taken to prevent the inadvertent spread of the disease to other plants in the surrounding environment during the indicator plant test. Also, the lengthy waiting period and the requirement for specialised, isolated cultivation chambers are the main drawbacks. Herbaceous indicator plants may require awaiting period of around two months, whereas trees may require up to two years (Upadyshev, 2018).

1.4 CURRENTLY EMPLOYED APPROACHES FOR DIAGNOSIS OF PLANT DISEASES

Innovative methods and tools are being created and implemented to guarantee the accuracy and speed of diagnosis. Nevertheless, the value of classic diagnostic techniques has not diminished because they enable early, fast, and highly accurate diagnosis of plant diseases using newly developed instruments and technology. This facilitates the making of appropriate and prompt decisions about the management of crop diseases. Phytopathological research relies on a range of traditional diagnostic methods. These methods include visual observation, microscopy, mycological diagnosis, biological diagnosis or testing in indicator plants, and various other techniques. It should be mentioned that diagnosing a disease based only on the host plant's outward symptoms is not always accurate. Many diseases share symptoms with physiological abnormalities brought on by external factors and certain phytopathogens can induce disease with minimally noticeable or non-existent symptoms when it first starts. The use of indicator plants in conjunction with pesticide is a long diagnostic process and can lead to a reduction in the reliability of the diagnosis. Furthermore, several sources (fungi, bacteria, or non-infectious disorders) might cause the same symptoms. It is challenging to correctly identify the microorganisms that cause infections in this setting. Therefore, it may be said that relying on just one symptom to make the diagnosis would not be accurate. For this reason, every technique is required in phytopathological research. Radically novel techniques for identifying and detecting pathogen presence in plant pathology have been introduced in response to the need to overcome issues with traditional plant disease diagnosis. Numerous cutting-edge techniques are employed in phytopathology to detect the pathogens causing plant diseases. These techniques include mass spectrophotometry, molecular genetic identification, and immunological diagnostics, among others (Khasanov et al., 2009; Ahmad et al., 2012); Reich et al., 2013; Fang and Ramasamy, (2015); Narayanasamy, 2011; Dyakov and Elansky, (2019). The most common prospective diagnostic techniques utilised in phytopathological research will be reviewed here.

1.4.1 Serological Methods for Diagnosis Plant Diseases

The basis of this technique is a unique antigen-antibody response. It finds the existence of viruses, tiny molecules, and the quantitative markers of those things. Enzyme-linked immunosorbent assay (ELISA) (Figure 1.2), immunoblotting, immunofluorescence, serological special electron microscopy, and many other techniques are used in immunodiagnostics or serological procedures (Dyakov, 2019). Previously serological methods were primarily used to identify viruses (Vlasov et al., 2016; Pandey et al., 2015). However, other plant pathogens, such as fungi, oomycetes, bacteria, and phytoplasmas, have also been identified using these methods (Pandey et al., 2015; Alarcon et al., 1990; Caruso et al., 2002; Van Vuurde et al.,

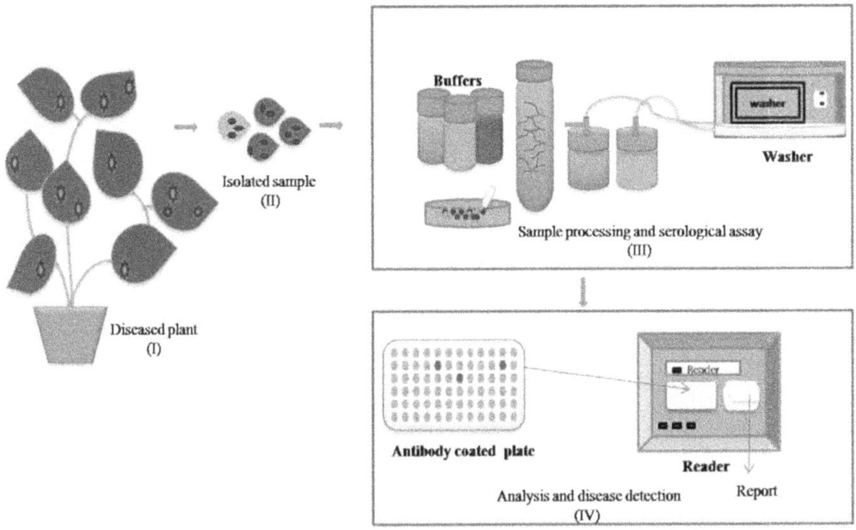

FIGURE 1.2 Serological method for plant disease diagnosis.

1987; Hampton Hampton et al., 1990). While there are other immunodiagnostics techniques, ELISA is one of the more sensitive and specific technique that provides a trustworthy diagnosis. Using this technique, a complex known as a "conjugate" is produced when an enzyme is linked to particular antibodies of a particular virus. The conjugation is enhanced with the sap of a virus-infected plant and antibodies labelled with enzymes interact with the viral antigen. The antibody-attached enzyme becomes inactive as a result of the antibodies' adsorption of the antigen. When an antibody does not react with the virus, the enzyme in the antibody is still active. By detecting the product's colour shift, a process known as colourimetry is used to quantify the amount of enzyme participating in the reaction. By comparing the decline in enzyme activity in the experimental and control groups, the presence and concentration of the viruses are ascertained (Khakimov et al., 2022). Enzymatic and immunochemical reactions serve as the foundation for the enzyme-linked immunosorbent assay. In an enzymatic reaction, an enzyme catalyses the conversion of a substance, leading to an immunological response and a subsequent change in the colour of the sample. In an immunochemical reaction, an antibody binds to a pathogen antigen. The material that the enzyme affects is referred to as the substrate, and the material that the enzyme produces is referred to as the enzymatic reaction's product. Equine perioxidase, beta-galactosidase, and alkaline phosphatases that are chemically attached (conjugated) are utilised as antibodies (Bagirova et al., 2015; Dyakov, 2019). Solid-phase ELISA can be performed in a number of ways, such as sandwich, indirect, and direct. The foundation of immunochromatographic analysis, specifically LFIA (Lateral Flow Immunoassay), lies in thin-layer chromatography. This immunochemical diagnostic technique operates by the interaction between the antigen and its corresponding antibody. Specialised test strips or test cassettes are employed to carry out this analysis (O'Farrell, 2008; Koczula and Gallotta, 2016).

1.4.2 Utilising Molecular-genetic Approaches to Identifying Diseases

Establishing the systematic location of plant diseases is crucial for diagnosis, as developing a control action against the pathogen is unfeasible without knowing its species composition. The conventional method of diagnosing plant diseases primarily relies on the identification of disease indicators and the pathogen's physical characteristics. While these techniques are valuable and essential, there are several requirements that must be met for their excellent application, most notably that the disease symptoms must be evident. Furthermore, an expert with a great deal of identifying experience is required. Traditional identification is further complicated when two or more disease symptoms are similar to one another. A conclusive diagnosis is frequently hampered by the appearance of novel pathogen forms for identification, including newly non-specific symptoms. The shortcomings of the traditional methods for phytopathogenic detection have led to more specific, accurate, and time-saving modern molecular approaches that offer more flexibility and excellence in pathogen detection. These molecular techniques include DNA microchips, conventional PCR, RT-PCR, and qRT-PCR (Figure 1.3), and nucleic acid hybridisation that are frequently used for identification of bacteria, fungi, and different classes of viruses in various plant systems (Narayanasamy, 2011; Dyakov and Elansky, 2019).

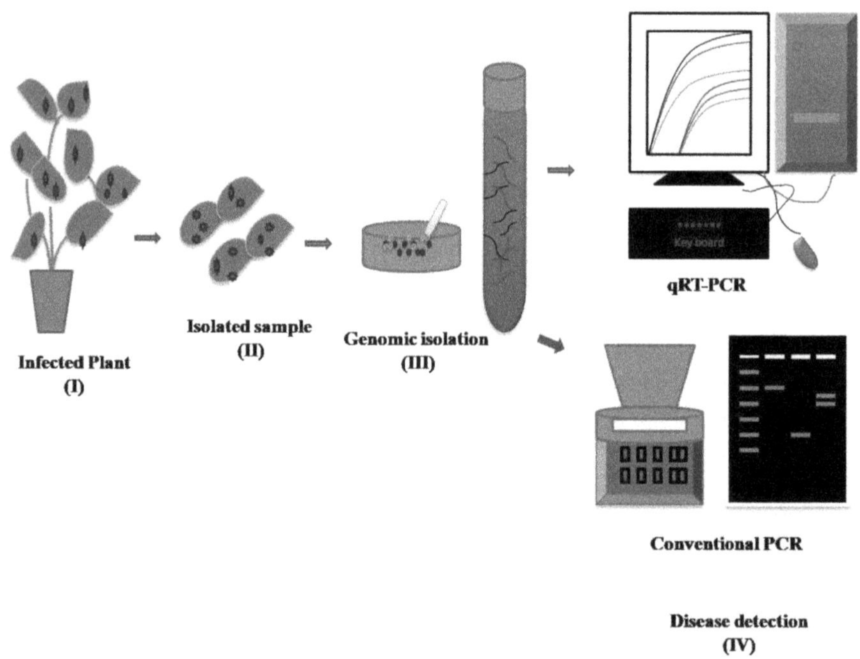

FIGURE 1.3 Diagrammatic representation of conventional and qRT-PCR.

Since its introduction in 1991, PCR has been frequently utilised in plant disease diagnostics due to its advantages over conventional methods. Rasmussen and Wolf (1991) utilised it to detect the bacterium *Pseudomonas syringae*. Specificity, sensitivity, efficiency, speed, and universality are the main characteristics of PCR-based diagnostics (Henson and French, 1993). Using a database as a guide, the genera and species of phytopathogenic organisms can be identified through the sequencing of a particular region of their genomes (Dyakov and Elansky, 2019). DNA barcoding is a molecular identification technique that uses short genetic markers found in DNA to identify organisms and establish if they are members of a specific taxon (Hebert et al., 2003). Molecular markers like TEF-1α (translation elongation factor 1-α), BT-3 (beta-tubulin), COI (cytochrome oxidase I), ITS (internal transcribed spacer), IGS (inter-genic spacer), RAPD (random amplification of polymorphic DNA), and TEF-1α (translation elongation factor 1-α) are used for the molecular identification of phytopathogenic fungi (Dyakov and Elansky, 2019; Fell and Blatt, 1999). Egamberdiev et al., 2014; Stielow et al., 2015; Egamberdiev et al., 2017; Wurzbacher et al., 2019; Khasanov and Sherimbetov, 2023). The primary genetic marker for fungi is ITS, that, is an intergeneral interval in a tandem repeating gene cluster of ribosomal DNA. It is an intermediate sequence consisting of approximately 600 nucleotide pairs. The subunits 18S, which is a small subunit of rDNA (SSU), and 28S, which is a large subunit of rDNA (LSU), are located next to each other (Begerow et al., 2010; Schoch et al., 2012). TEF-1α, a secondary genetic marker for fungus in DNA-based taxonomy or DNA barcoding, has a length of 66–3000 bp in sequences at GenBank (Meyer et al., 2019; Stielow et al. 2015). The primary method for identifying bacteria using nucleotide sequences is the 16S rRNA gene (Dyakov and Elansky, 2019; Lane et al., 1985; Woese, 1987; Cai et al., 2003; Zeigler, 2003). Other markers that have been proposed include RNA polymerase, DNA gyrases, and some heat shock protein dependent genes. The detection of pathogens in plants through nucleotide sequencing involves different steps:DNA isolation, electrophoresis, and determination of nucleotide sequences. Molecular marker selection for PCR, primer formation (based on the selected genetic marker), polymerase chain reaction, isolation of DNA fragment, and DNA purification are some of the steps involved in the identification of phytopathogens by nucleotide sequences (in traditional PCR) (Dyakov and Elansky, 2019; Egamberdiev et al., 2017). The DNA sequences are edited and corrected when the nucleotide sequence is ascertained using specialised bioinformatic software such as Sequencing Analysis (Applied Biosystems), BioEdit (Ibis Biosciences, USA), UGENE (UniPro, Russia), and ClustalX. Using the BlastN algorithms in the international resource NCBI (National Centre for Biotechnology Information, USA) (http://www.ncbi.nlm.nih.gov), the nucleotide sequences of the studied samples are compared with the nucleotide sequences interpreted from genetic databases (Egamberdiev et al., 2017; Larkin et al., 2007; Kumar et al., 2016). Apart from the NCBI resource, there are several international databases available for molecular biology, genetics, and biomedicine. These databases include the European Molecular Biology Laboratory (EMBL), the European Nucleotide Archive (ENA), the DNA Data Bank of Japan (DDBJ), FungiDB (Fungal and Oomycetes Genomics

Resources), and MycoBank (Centraal bureau voor Schimmel cultures fungal biodiversity centre).

1.4.3 Plant Disease Diagnosis using Spectrometry

A quick, precise, and economical technique for routinely identifying a variety of microorganisms is matrix-assisted laser desorption ionisation (also known as time of flight mass spectrometry, or MALDI-TOF MS) (Kumar et al., 2016, Dingle and Butler-Wu, 2013). Using a laser pulse absorbing matrix, ionising a tiny piece of the analyte is achieved by desorption in MALDI-TOF mass spectrometry (Patel, 2015). The matrix is made up of crystallised molecules that ionise the analyte and lessen the damaging qualities of laser radiations. Synaptic acid, 2,5-dihydroxybenzoic acid (DHB), and α-cyano-4-hydroxycinnamic acid (α-CHCA) are utilised as a matrix (Hillenkamp et al., 1991). For the investigation of non-volatile high-molecular substances (peptides, proteins, polysaccharides, oligonucleotides, synthetic polymers, organic complex molecules, etc.), MALDI-TOF mass spectrometry is extensively used (Fukuyama et al., 2008; Seng et al., 2009; Korfmacher, 2009). The process of identifying microbes using MALDI-TOF mass spectrometry involves several sequential steps. Initially, a pure culture of bacteria, fungi, or fungal organisms is obtained by either directly inoculating them in a solid agar medium or cultivating them in a liquid medium. If the colonies are grown in a fluid medium, they undergo a series of treatments including mixing with buffer solutions, centrifugation, dipping, and removal of the initial supernatant. The resulting purified culture is then applied onto a metal tablet cell, specifically a 96-cell metal tablet, and allowed to dry. Following this, a matrixliquid is applied onto the dried culture in the cell of the metal tablet, thoroughly mixed, and dried once again. Finally, the metal tablet is placed in the equipment mold for further analysis (Ahmad et al., 2012; Sandrin et al., 2013; Lévesque et al., 2015).When the device is turned on, the laser light's effect on the metal tablet's cells starts. This causes a set of spectra unique to each species to emerge in the computer programme, serving as a unique "fingerprint" for that species. By comparing the sample's spectrum with those stored in the database, the programme indicates the kind to which it belongs. The top producers of MALDI-TOF mass spectrometry microorganism identification equipment today are bioMérieux (France), Shimadzu (Japan), and Bruker Daltonics (USA). About 7000 mass spectra of different species and strains of microorganisms are available in the BrukerMALDIBiotyper Microorganism Identification System database maintained by Bruker Daltonics (www.bruker.com). In contrast, over 50,000 spectra of biological objects are available in the SARAMIS (Spectral Archive And Microbial Identification System) database maintained by the AXIMA Microorganism Identification System (www.shimadzu.com). It is clear that clinically significant bacteria and fungi are the primary focus of the producers of all of these instruments for MALDI-TOF mass spectrometry microorganism spectrum bases (Reich et al., 2013; Dingle and Butler-Wu, 2013; Sandrin et al., 2013; Lévesque et al., 2015). Over time, distinct spectra of phytopathogenic bacteria have been continuously added to these databases, and the quantity of these microorganisms increases yearly (Ahmad et al., 2012: Hou et al., 2019; Siricord and O'Brien, 2008: Ziegler et al., 2012).

For example, 62 isolates of the genus Fusarium, representing nine species, were identified by MALDI-TOF and genetic analysis (TEF-1 gene sequence) in a study carried out in France (Santos et al., 2016). Out of the numerous species that were documented, 57 isolates (*Fusarium solani, F. oxysporum, F. verticilloides, F. proliferatum,* and *F. dimerum*) that were verified using molecular identification (www.ncbi.nlm.nih.gov/BLAST/) had 92% accurate identifications. Four species that were absent from the MALDI-TOF database *(F. chlamydosporum, F. equiseti, F. polyphialidicum,* and *F. succhari*) could not be identified. Of the six isolates whose morphological and molecular identities did not match, five had consistent identifications using molecular-genetic and MALDI-TOF techniques. This quick technique, which is based on mass spectral diagnostic, can be very helpful in identifying members of the Fusarium family at the species level. It is only appropriate to add spectra of uncommon species to the database. Furthermore, accurate identification was obtained for 95.5% of gram-positive bacteria (Faron et al., 2015), 92.5% of fungus (Rychert et al., 2013), 99.8% of clinically relevant gram-negative bacteria, and 98.2% of species (Marinach-Patrice et al., 2009).

1.5 NOVEL AND EMERGING METHODS OF PLANT DISEASE DETECTION

1.5.1 Loop-mediated Isothermal Amplification (LAMP) Assay

The LAMP assay is a single-tube reaction technique that amplifies DNA at a specific temperature using an isothermal polymerase. In a LAMP test, four to six distinct primers are utilised, binding to six to eight different primers on the target gene. The primer set includes the two outer primers, F3 and B3, along with the forward inner primer (FIP), the backward inner primer (BIP), and loop forward and loop backward primers. This reaction occurs in an isothermal environment and employs the Bst DNA polymerase enzyme. (Wong et al., 2018). Plant pathogen detection has made significant advancements through the utilisation of LAMP tests. A LAMP test has been developed to identify the pathogen responsible for late blight on potatoes and tomatoes (Ristaino et al., 2020). This developed assay was compared to ddPCR, real-time LAMP, and conventional PCR. The SYBR Green real-time LAMP assays were found to be specific for P. infestans in potatoes and tomatoes, and they exhibited ten times more sensitivity than conventional PCR. Another group developed a LAMP method to identify Pseudomonas syringae PV. The hrpZ gene (B-Loop) was designed using two backward loop primers, two internal primers, and two exterior primers (F3/B3). The LAMP detection of Pst was optimised for a 60-minute completion time at 63°C. The limit of detection (LOD) for genomic DNA was determined to be 1.61x10 fg/L, while for bacterial suspension without DNA extraction, it was 1.05x103 CFU/mL. In the case of a Penicillium oxalicum LAMP test used to detect fungus on purposely infected grapes, the LOD for genomic DNA was 100 pg per response. Recently, a LAMP test was developed for the detection of Phytophthora agathidicida, a root disease causing oomycete for kauri dieback (Winkworth et al., 2020). The total DNA LOD of P. agathidicida was found to be 1 fg. Similarly, a LAMP-based system designed for the detection of Bull's eye rot (BER) can detect DNA amounts as low as 0.001 ng/L and provides a 10-minute response time (Enicks et al., 2020). A rapid, sensitive, and focused LAMP-based assay has been reported for the detection of the widespread plant disease Sclerotinia sclerotiorum, with a LOD of 10 pg of DNA (Grabicoski et al., 2020). Highly specific LAMP primers for Fusarium proliferatum in the TEF-1 area have been identified (Wang et al., 2020). The same group also described a LAMP assay for the rapid identification of *F. proliferatum* in maize ear and kernel rot. The identification of *Colletotrichum nymphaeae* strain CCTUCch32 infection in asymptomatic strawberry plants was made easy by using a LOD of 50 fg/L DNA recovered from infected leaves (Karimi et al., 2019). LAMP assays have also been employed to detect RNA by utilising the reverse transcriptase enzyme, which converts RNA into complementary DNA and amplifies it with a polymerase. In a separate study, the reverse transcriptase LAMP test successfully identified the Apple chlorotic leaf spot virus (ACLSV) by employing a set of four primers based on conserved sequences of the coat protein. With an LOD of 0.02 g/L at 64 °C, this assay proved to be effective for on-site diagnostics (Peng et al., 2017). Kokane et al. (2021) described a highly specific, single-step RT-LAMP technique for rapid detection of 35 isolates of Indian citrus ringspot virus (ICRSV). They utilised four different primers to target the coat protein gene of the virus. Tseng et al. (2021) developed a test to identify six different viroids using one-step RT-PCR and RT-LAMP with ubiquitous primers. These viroids included columnea latent viroid (CLVd), pepper chat fruit viroid (PCFVd), potato spindle tuber viroid (PSTVd), tomato apical stunt viroid (TASVd), tomato chlorotic dwarf viroid (TCDVd), and tomato plant macho viroid (TPMVd). The detection of viroids was based on visualising colour changes after the addition of SYBR Green I during the process. The study demonstrated that a one-step RT-PCR using two degenerate primers could successfully amplify up to six pospiviroids from seeds and plants in the Solanaceae family. The amount of viroid RNA required for successful identification ranged from 1 fg to 10 ng, depending on the specific virus of interest. Furthermore, the LAMP assays were designed specifically to detect needle blights, a specific type of fungal disease that poses a threat to native pine trees. These assays targeted three primary infections: Lecanostictaacicola, Dothistromapini, and Dothistromaseptosporum. To ensure accuracy, species-specific fluorescent probes were developed for each infection, exclusively for this assay. As stated by Agietti et al. (2021), the limit of detection (LOD) values for L. acicola, D. pini, and D. septosporum in the assay were 0.128 pg/μL, 0.64 pg/μL, and 3.2 pg/μL, respectively. In a recent study by Ren et al. (2021), a rapid LAMP assay was introduced to identify Marssonina coronaria, the primary cause of apple blotch in a majority of apple crops. What sets this assay apart is the use of a ribosomal DNA internal transcribed spacer primer, specifically designed to differentiate M. coronaria from other fungi responsible for apple diseases. The LAMP test demonstrated a detection limit of 100 fg/L of M. coronaria genomic DNA, with a detection time of 70 minutes. Moreover, a team of researchers developed a swift virus diagnostic technique based on the LAMP assay, targeting the Sri Lankan cassava mosaic virus (SLCMV). This newly created assay utilised six primers to enhance sensitivity and improve detection capabilities. In fact, its sensitivity was found to be up to 10,000 times higher than that of the conventional PCR method, potentially aiding in early plant disease diagnosis. Notably, a novel method was also devised to enable one-go analysis without the need for a thermal cycler. This method utilised chemicals from a commercial kit that were designed to remain stable when dried, as highlighted by Uke et al. (2022). The Wisteria vein mosaic virus, which affects wisteria, a decorative flowering plant, was detected using a reverse transcription LAMP test. In comparison to traditional PCR, the sensitivity was amplified by a factor of 100. The RT-LAMP primers were designed based on the coat protein gene sequence of the wisteria vein mosaic virus (Yao et al., 2022). A single-phase colourimetric RT-LAMP assay was conducted, targeting the coat protein region of the prune dwarf virus genome. The study successfully identified positive LAMP reactions with the naked eye, where the change in colour from pink to yellowish indicated a positive outcome of the virus infection in prunes. The sensitivity of this approach was ten times higher than that of conventional PCR (Çelik, 2022). Another group of researchers developed a LAMP assay to detect the pea enation mosaic virus (PEMV), which infects plants in the Leguminosae family, using specific primer sets as inner and outer primers. Additionally, the application of the immunological capture RT LAMP test (IC-RT-LAMP) to identify a highly pathogenic strain of a plant virus in field samples was disclosed (Mehetre et al., 2021). To reduce diagnostic costs and increase efficiency, an IC step was incorporated during the development of the IC-RT-LAMP test. The assay successfully identified the two most prevalent viruses that infect lily plants worldwide: cucumber mosaic virus (CMV) and lily symptomless virus (LSV). The viral antigens were captured using antibodies against recombinant LSV or CMV coat proteins, eliminating the need for RNA isolation. RT-LAMP was then employed to identify the viral antigens. The sensitivity of the assay was 100 times higher than that of conventional PCR.

1.5.2 Lateral-flow Immune-chromatographic Analysis Test

The lateral flow assay (LFA), also referred to as the lateral-flow immune-chromatographic analysis test, employs a paper-based platform to analyse the concentration and quality of a substance in a complex mixture. Within a timeframe of 5to 10minutes, the test strip reveals the result after the sample is added. Test equipment's low production cost, convenience of use, and capacity for quick findings have led to an increase in its utilisation (Boisen, 2015). Saliva, serum, blood, and other fluids can be analysed in a variety of biological samples using the lateral flow immune-chromatographic analysis test (Boisen, 2015; Carrio et al., 2015; Magambo et al., 2014; Schramm et al., 2015). This analysis test is now frequently used in plant disease detection tests (Danks and Barker, 2000; Boonham et al., 2008; López-Soriano et al., 2017). In order to validate immunochromatographic analysis, a test strip is utilised. This strip provides a straightforward means of examining a liquid sample or extract. The sample is placed into a trough and then proceeds to flow along polymer lines. This flow enables the molecules within the sample to interact with the solution without any interference from external sources. The test strip consists of the sample, conjugate pad, and membranes, all of which are securely affixed to the rear card.

The conjugate pad is loaded with a specific antibody that is highly responsive to the sample being tested. This antibody is designed to target the pathogen of interest and is attached to fluorescent or coloured fragments, colloidal gold, and latex microspheres. The sample fluid binds to the antibody and travels through the detection zone along the line. Nitrocellulose-based porous membrane is used to immobilise biological components, primarily antibodies or antigens. Their job is to react with a conjugated antibody that is attached to an analyte, which is a biological entity that is being studied. The test strip's result shows that the samples are moving as intended, and the control line's colouring indicates a positive test sample result, alerting the user to the possibility of pathogens in the material (Xu et al., 2014; Yen et al., 2015). The "ladder bars" analysis works on the basis of a stepwise colourometric conjugate-antigen complex that follows the immobilised antibody. The number of streaks that appear on the strip indicates the concentration of the substance (analyte) under investigation (Fung et al., 2009; Fang et al., 2011).

1.5.3 Use of Microfluidics for Diagnosing Plant Diseases

A multidisciplinary method for manipulating fluids in small volumes is called microfluidics. Microfluidic chips can be used as a platform for automation and diagnosis in conjunction with a variety of methodologies, including electrochemistry, PCR, and LAMP. Using a microfluidic system, the DNA of *Ganoderma boninense*, a pathogenic fungus that infects oil palm trees in Malaysia, was retrieved. The UV–Visspectroscopy study was used to confirm the excitation peaks that were seen between 260 and 280 nm. Peak at 260 nm showed that the synthetic ssDNA sample was pure, but a shift in peak to 280 nm suggested that the sample had contaminants from other molecules. The accuracy of microfluidic devices was demonstrated by the current-voltage measurement results for both synthetic and actual *G. boninense* samples. Effective DNA extraction was demonstrated by UV-Vis tests and current-voltage measurements (Ayoib et al., 2020). A microfluidic chip was developed for the real-time monitoring of crop fungal disease. It comprises an inertial impactor, a low-pressure collection chamber, and a pre-treatment channel to enrich the airborne fungal spores. Numerical analysis and experimentation were employed to determine the optimal microfluidic device design. The pre-treatment channel was filled with radially stacked collected bio-aerosols. The main impactor's function was to divide the gathered bio-aerosols into different sizes. Last but not least, the aerosols are unable to bounce thanks to the low-pressure collection chamber (Wang et al., 2020). A reusable microfluidic bioassay was developed using two DNA targets for fungal infections, *Botrytis cinerea* and *Didymellabryoniae*. The oligonucleotide sequence was then recreated for reuse by thermal denaturing the DNA. Using the laser-induced fluorescence detection technique, a rapid evaluation of multiple target detection was completed in less than 10 minutes (Qu et al., 2017). Using azelaic acid as the target moiety, an enzyme colourimetric test was created to identify disease in grapes. For optical detection, thin-film silicon photosensors were incorporated into the microfluidic device. Tyrosinase was rendered immobile on the sensor, which prevented AzA from acting. It was observed that the assay time of 10 minutes yielded

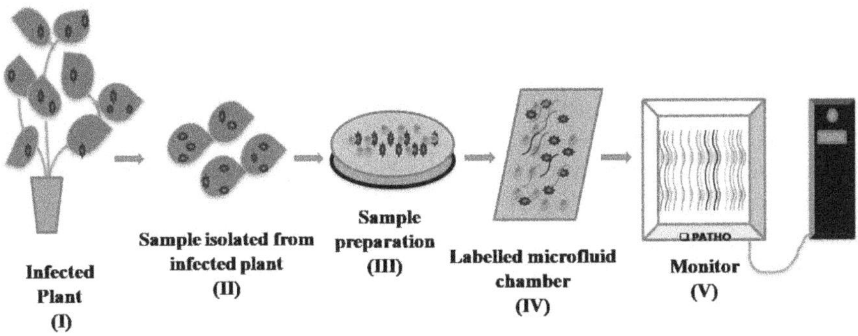

FIGURE 1.4 Pictorial representation of Microfludic disease detection system.

a detection limit of 5–10 nM (Bras et al., 2019). A microfluidic disposable device was created using magnetic beads embellished with antibodies and horseradish peroxidase. Using the immunosensor and immunomagnetic separator, *Citrus tristeza* virus was detected with a LOD of 0.3 fg/mL in samples from both healthy and infected plant samples (Freitas et al., 2019). Another breakthrough was the detection of virus-infected *Phalaenopsis amabilis* leaves and flowers using a microfluidic device in conjunction with subterranean optical fibres. A microfluidic stirring device was made of polydimethyl siloxane (PDMS), and optical fibres were added to it to enhance the optical signal for detection. The apparatus was able to separate and purify RNA. A test of the concept was conducted using capsicum infection with the *Chlorosis Tospovirus* (CaCV) (Lin et al., 2015). By varying the resistors in the circuit, a thin-film amorphous silicon photodiode is used in a point-of-care portable microfluidic system prototype to measure transmission due to optical transduction. This arrangement also enables the sensor to read out in both high and low light circumstances. With its compact design, the device could measure currents as low as 1013 A. The system may also be able to operate an LED light source and a peristaltic pump that moves fluids inside the channel. The prototype was used in a bioassay to demonstrate its utility in the portable detection of fungal infections in plants (Bras et al., 2020). A simple concept of a microfluidic system for phytopathogen detection is shown in Figure 1.4.

1.5.4 CRISPR-Based Assays for Identifying Plant Diseases

CRISPR-Cas technology, also referred to as cluster regularly interspersed short palandromicrepeate, is a a collection of CRISPR-associated proteins like Cas 9, Cas 12, Cas 13, etc. (Mahmud et al., 2015). It has emerged as a powerful tool in the field of plant disease diagnosis (Wheatley and Yang, 2021).This revolutionary gene editing technology has the potential to dramatically transform the ways we detect and manage plant diseases. The use of CRISPR-Cas in plant disease diagnosis involves the precise targeting and editing of specific DNA sequences associated with plant pathogens. By harnessing the natural defence mechanism of bacteria scientists have been able to adapt CRISPR-Cas for use in identifying and diagnosis of plant disease with a high degree of accuracy and specificity (Zhang et al.,

2022). The unparalleled precision in detecting pathogens is the key advantage of CRISPR-Cas in plant disease diagnosis. The conventional methods of pathogen and disease diagnosis often rely on laborious and time-consuming techniques such as culturing, serological tests, and PCR-based assays (Lau et al., 2017). These methods can be costly, take days to yield results, and may lack the necessary sensitivity and specificity required for early and accurate disease detection. CRISPR-Cas-based diagnostic tools offer a promising alternative by providing a rapid, specific, and cost-effective approach to detecting plant pathogens. By programming CRISPR-Cas systems to recognise and cleave target DNA sequences unique to particular plant pathogens, researchers have developed innovative tools capable of identifying diseases in real time. Furthermore, the use of CRISPR-Cas in plant disease diagnosis has the potential to revolutionise disease management strategies. Early and precise detection of plant pathogens is crucial for implementing effective control measures and minimising the spread of diseases. By enabling rapid and reliable diagnosis, CRISPR-Cas technologies can help farmers and agricultural professionals respond swiftly to emerging threats, thereby reducing the economic and environmental impact of plant diseases. In addition to its diagnostic potential, CRISPR-Cas technology holds promise for the development of disease-resistant crops. By leveraging the precision and versatility of CRISPR-Cas, researchers are exploring the possibility of engineering plants with enhanced resistance to a broad spectrum of pathogens. This approach could offer a sustainable and environmentally friendly means of disease control, reducing the reliance on chemical pesticides while ensuring food security for a growing global population. The use of CRISPR-Cas in plant disease diagnosis is not without challenges. One of the key considerations is the need to ensure the specificity and accuracy of CRISPR-Cas-based diagnostic tools. While CRISPR-Cas systems are highly precise, off-target effects and the potential for unintended genetic modifications remain important areas of concern. Addressing these challenges requires ongoing research and development to refine and optimise CRISPR-based diagnostic platforms for reliable and safe use in agricultural settings. Ethical and regulatory considerations also play a critical role in the application of CRISPR-Cas technology in agriculture. As with any gene-editing technology, the responsible and transparent use of CRISPR-Cas in plant disease diagnosis requires careful consideration of potential environmental impacts, consumer safety, and regulatory frameworks governing the release of genetically modified organisms (GMOs) into the environment. Despite these challenges, the potential benefits of using CRISPR-Cas in plant disease diagnosis are substantial. From early and accurate disease detection to the development of disease-resistant crops, CRISPR-Cas technology offers a promising avenue for enhancing global food security, sustainable agriculture, and environmental stewardship. Overall the use of CRISPR-Cas in plant disease diagnosis represents a significant leap forward in the field of agricultural biotechnology. By harnessing the precision and versatility of CRISPR-Cas systems, researchers and agricultural professionals are poised to revolutionise the way we detect, manage, and mitigate the impact of plant diseases. While facing challenges related to safety, ethics, and regulation, the potential benefits of CRISPR-Cas technology in plant disease diagnosis are vast, promising a future where agricultural sustainability and food security are enhanced through innovative and precise disease management strategies.

1.5.5 Digital Droplet PCR (ddPCR): An Emerging Method for Plant Disease Diagnosis

A sample partitioning method called ddPCR can detect a single copy of a gene. When it comes to measuring plant pathogens including fungus, bacteria, viruses, and viroid, it is more user-friendly, reproducible, and specific than traditional PCR approaches (del Pilar Martínez-Diz et al., 2020). A highly rapid and precise PCR-based technique has been developed to identify up to two genomic copies of *Botrytis cinerea* DNA (Suarez et al., 2005). Another example is the creation of an isothermal amplification method for the sensitive, dependable, quick, and affordable detection of *Candidatus Liberibacter asiaticus* that is based on Recombinase Polymerase Amplification and LFA (Ghosh et al., 2018). Gold nanoparticles (AuNPs) have also been used to create colourimetric nano-biosensing devices for pathogen detection using localised surface plasmon resonance (Ghosh et al., 2018). ddPCR is a contemporary technique that replicates a single molecule in a highly diluted droplet. A probe that has been fluorescently tagged is used to locate the target molecule. The low sensitivity of ddPCR to inhibitors accounts for its effectiveness as a reliable method for detecting even low target concentrations. Using this method, the target sample is divided into 20,000 droplets, and the template molecules are amplified by PCR in each droplet. The ddPCR method was reported to be able to detect *Tilletiacontroversa teliospores* in soil for the first time with sensitivity LOD of 2.1 copies/μL, in contrast to PCR and real-time PCR (Liu et al., 2020). Diz et al. (2020) used the ddPCR approach to discover *Ilyonectria liriodendra*, the causal agent of black foot disease impacting grapevine nurseries. Researchers occasionally assert that the method's accuracy, efficiency, and specificity are on par with real-time PCR, and that the degree of correlation with ddPCR was determined to be 0.95. A study team researched a ddPCR format for *Acidovoraxcitrulli*, a serious cucurbit disease causing agent. When compared to qPCR, that used 102 CFU/mL bacterial cells and 2 fg/DNA, the LOD was significantly more sensitive (Yu et al., 2020). The benefits and the limitations of the various detection methods mentioned so far are summarised in Table 1.1.

1.5.6 Biosensors: A Novel Plant Disease and Pathogen Detection System

Biosensors are the analytical devices that convert a biological response into an electrical signal. The quantifiable signal is generated by biological transducer when the targets analyte binds with biorecognition element (Hameed et al., 2018; Bridle and Desmulliez, 2021). Biosensors offer a promising solution for point-of-care applications due to their affordability, user-friendly nature, and ability to provide rapid results (Bridle and Desmulliez, 2021). Transducers have widespread applications across various domains. In the case of electrochemical transducers, the detection of binding events is accomplished by monitoring voltage, impedance, or conductance changes. Mass-based transducers, on the other hand, detect alterations in resonance

TABLE 1.1
Advantages and disadvantages of plant disease detection systems

Detection system/ Method	Advantages	Disadvantages
Visual Inspection and Microscopy	Easy method, real time and rapid detection, detection based on morphological features.	Restrict definite diagnosis; detection is possible only after major damage has been done to crop, hence limiting the treatment. Error in pathogen type detection.
Enzyme Linked Immunosorbent Assay (ELISA-DAC-DAS)	Good specificity and sensitivity; high throughput; diverse sample type.	Temporary readouts; limited antigen information; instrument and chemical cost is higher. Requires good laboratory setup and trained professionals.
Polymerase chain reaction (PCR) and its variants rt-PCR, qRT-PCR	Crucial to many strategies wielded in genetic testing; Pathogen identification; Accurate quantification; Requires low starting material; Fast sensitive and specific.	Good lab setup; Genomic extractions; Thermal cycling; Skilled support staff; Prior knowledge of primers; Cost of chemicals.
Loop mediated Isothermal Amplification(LAMP) Assay	Robust, specific, easy and fast detection system requires simple equipments; Short time duration and constant temperature.	Requires trained specialists for designing primers; primer designing is complicated for LAMP.
Lateral flow Immunoassay (LFIA)	Rapid and cost-effective detection system requires less sample volume; Point-of-care diagnostics; Longer shelf-life, no requirement of pre-treatment.	Raising a good antibody is a concern; sample type defines the analysis time.
Microfluidics	Requires low sample volume; High specificity, rapid response and portability.	External factors like PH, temperature, and pressure hampers the analysis.
Digital Droplet Polymerase Chain Reaction (ddPCR)	High sensitivity; Absolute quantification	Instrument cost is high; Requirement of rigorous QC plans.
CRISPR-basedassays	Easy to design; High efficiency	Unexpected mutations arise due to non-specific site cleavage.

frequency resulting from the binding of the target analyte to the biorecognition element. Optical transducers detect binding events by analysing discrepancies in the reflection of incoming light (Fang and Ramasamy, 2015; Hameed et al., 2018; Bridle and Desmulliez, 2021). Various biorecognition elements are employed to identify a target analyte, with aptamers, enzymes, nucleic acid probes, and antibodies being the most commonly used. Antibodies, which can be easily labeled, undergo a conformational change upon binding to the target analyte. Aptamers, on the other hand, are highly suitable for use in biosensors (Toh et al., 2015; Khater et al., 2017; Shahdordizadeh et al., 2017; Bridle and Desmulliez, 2021). Enzymes are utilised based on their ability to convert a particular target analyte with high specificity and affinity. However, enzyme-based methods are more suitable for identifying specific substrates rather than detecting pathogens. The choice of the most appropriate biorecognition element depends on the type of transducer and the targeted molecule. Nucleic acid probes are employed to target specific DNA sequences, while aptamers and antibodies are used to target specific antigens. The simplicity of labeling and immobilization, as well as the type of sample being analysed, are crucial factors in selecting the best biorecognition element, as they impact the biosensor's final cost, robustness, and sensitivity. The immobilisation of biorecognition components on the sensing surface significantly affects the effectiveness of biosensors (Cardoso et al., 2022). Fixing the biorecognition element to the electrode and ensuring the recognition components are oriented and packed at the ideal density are the two main objectives of immobilisation(Cesewski and Johnson, 2020). There are several ways to immobilise materials, including covalent attachment, avid in and biotin systems, adsorption-based approaches, and self-assembled monolayers. Yüksel et al. (2015) provided an example of a DNA hybridisation-based biosensor that may be used to detect Phytophthora ramorum in rhododendron leaves. Nucleotide capture probes unique to P. ramorum's ypt1 gene were mounted on AuNPs. Through the use of surface enhanced Raman spectroscopy, hybridisation between the probes and the target amplicon was found. There was no mention of the technique's sensitivity, although there was no evidence of cross-reactivity with the closely related P. lateralis species. In a separate investigation, the detection of Pseudomonas syringae was achieved using an electrochemical biosensor that relied on RPA amplification (Lau et al., 2017). Once the DNA from the sample was extracted, an RPA was performed using the forward primer, which possessed a 5' addition that was complementary to the oligonucleotide probes attached to gold particles. Additionally, the reverse primer was tagged with biotin. The resulting amplification products were then subjected to treatment with AuNPs that were coupled with oligonucleotides, enabling them to hybridise with the forward primer. Moreover, magnetic beads coupled with streptavidin were utilised to bind the biotin-labeled reverse primer. By applying a magnetic field, the magnetic beads (which were either bound with the amplified product or the reverse primer alone) could be separated. The AuNPs were then released from the items by heat treatment, and they are used as a label to coat the detection probe. Differential pulse voltammetry is used to detect the AuNPs, and the resulting signal is proportional to the quantity of AuNPs, which serves as a proxy for the pathogen concentration in the sample. In comparison to a traditional PCR assay, the assay's sensitivity was almost 10,000 times higher. There are two factors that are believed to be responsible for the significant increase in sensitivity. Firstly, the use of RPA allows for the amplification of even small amounts of DNA. Secondly, the electrochemical biosensor used in this study was found to be 100 times more sensitive than gel visualisation. Additionally, the assay itself can be completed within a relatively short time frame of 60 minutes.

In another example, an immunoassay biosensor was utilised to identify the presence of Citrus tristeza virus in infected citrus samples (Freitas et al., 2019). This particular assay involved the use of a magnetic bead coated with a secondary antibody, which was coupled to an HRP enzyme. The probe surface was coated with capture antibodies. By utilising the magnetic beads attached to the secondary antibody, the primary antibody-antigen-secondary antibody complex was successfully separated. The biosensor, which relie on amperometry, detected the presence of the target antigen by sensing the redox reaction induced by the HRP enzyme. The limit of detection for this biosensor was determined to be 0.3 fg/ml, and the entire assay could be completed in just 50 minutes. Lastly, a non-invasive volatile organic compound (VOC) biosensor, which was incorporated into a smartphone, was employed to identify tomato leaf late blight (Li et al., 2019). This innovative biosensor allowed for the detection of the specific VOCs associated with the disease. By utilising a smartphone, the identification process became more accessible and convenient. A plant afflicted with P. infestans will release certain volatile organic compounds (VOCs). Cysteine-functionalised AuNPs are capable of detecting these markers because they aggregate when exposed to specific VOCs, which results in the creation of colour. The device based on a smartphone can be equipped with a disposable paper strip that has been functionalised with cysteine. When certain VOCs characteristic of a particular disease are present, the device's tiny battery-operated pump can pull air from the sample environment over the paper strip, causing the nanoparticles to agglomerate. The colourimetric shift that results can then be examined by the smartphone's camera to accurately and promptly identify various plant VOCs linked to illness, including late blight in tomato leaves. Blind testing has been used to confirm the device utilising infected leaves obtained from the field as well as artificially contaminated tomato leaves (Paul et al., 2023).

1.6 FUTURISTIC APPROACH FOR PLANT DISEASE DETECTION

With the advancement in technology, there has been a growing interest and need for futuristic approaches to address agriculture-related challenges (Araújo et al., 2021). One such challenge is the early detection of plant diseases, which can have detrimental effects on crop yield and quality (Mahlein, 2016). The use of futuristic technologies for plant disease detection holds great promise in revolutionising the way farmers and agricultural scientists manage crop health. One of the most promising futuristic approaches for plant disease detection is the use of remote sensing and imaging technologies (Kouadio et al., 2023). Hyperspectral imaging, for example, allows for the collection of high-resolution spectral information from plants, which can be used to identify subtle changes in their physiological condition associated with disease presence (Lowe et al., 2017). This technology can be deployed on drones or satellites to scan large agricultural areas, providing early detection of diseases before visible symptoms appear. Additionally, thermal imaging can detect variations in plant temperature associated with disease infection, offering a non-invasive and rapid means of identification (Pineda et al., 2020). Artificial intelligence (AI) and machine learning (ML) have the potential to revolutionise plant disease detection (Das and Sharma, 2023). By training AI algorithms on large datasets of plant images, spectral data, and environmental parameters, it is possible to develop intelligent systems capable of recognising disease patterns and making accurate diagnoses in real time (Joseph et al., 2023). These AI-powered systems can continually learn and improve their accuracy as they are exposed to more data, making them invaluable tools for farmers and agricultural experts (Tetala, 2023). Nanotechnology and biosensors represent another futuristic approach for plant disease detection. Nanoscale sensors can be designed to detect specific disease-related biomarkers in plants, allowing for ultra-sensitive and rapid diagnosis of infections. These sensors can be integrated into handheld devices or attached to unmanned aerial vehicles, providing farmers with on-site and real-time information about the health status of their crops (Gan et al., 2023). The integration of Internet of Things (IoT) devices in agriculture, coupled with big data analytics, offers futuristic solutions for plant disease detection (Fan et al., 2021). IoT sensors can collect data on various environmental parameters such as temperature, humidity, and soil moisture, which is then analysed using big data techniques to identify disease risk factors and early indicators. This real-time monitoring and analysis can enable proactive disease management strategies, minimising the impact of infections on crop yield and quality (Kishorebabu and Sravanthi, 2020).While these futuristic approaches hold great promise, several challenges need to be addressed for their effective implementation. Ensuring affordability and accessibility of these technologies to small-scale farmers, addressing privacy and data security concerns, and integrating these advanced solutions into existing agricultural practices are important considerations to enable widespread adoption.

CONCLUSION

In the context of agriculture and technology, plant disease detection has a huge opportunity to transform crop monitoring and crop protection. With continued development, it is projected that these technologies will be essential to contemporary farming methods, strengthening and enhancing world food systems. The management of plant diseases and yielding high-quality crops are only with accurate and timely detection of pathogens. A vital component of contemporary agriculture is the identification of plant diseases, which is essential for maintaining crop productivity and food security. A vast array of techniques firmly anchored in practical knowledge, cultural customs, and empirical data make up traditional ways for diagnosing plant diseases. These techniques, which still hold true in modern agricultural contexts include visual examination, field testing kits, chemical reagent testing, experience knowledge, disease indicators, herbal and cultural practices, environmental observations, information transmission, and oral traditions. Traditional approaches provide a comprehensive approach to diagnosing plant diseases and help to maintain plant health in a sustainable way when combined with contemporary diagnostic tools. Accurate and efficient diagnosis and detection of plant diseases have been greatly improved by the integration of cutting-edge technologies, including machine learning, image processing, and sensor-based systems. Early identification, prompt response, and efficient treatment of plant diseases have been demonstrated by the invention and application of automated disease detection systems. These technologies help agriculturalists minimise production losses and lessen the overall environmental effect of agricultural activities, in addition to supporting them in making well-informed decisions about crop protection. It is not necessary to disregard conventional procedures in light of the widespread introduction and use of new, contemporary, fast, and dependable approaches but to cross integrate them. Modern prospective techniques for diagnosing plant diseases, such as molecular-genetic identification, lateral flow immunoassay, ELISA, biosensor and CRISPR-based and MALDI-TOF mass spectrometry, ensure accurate and timely diagnosis and facilitate prompt decision-making. Identifying the type and isolates of the pathogen quickly, accurately, and reliably is essential for the creation of disease control systems, regardless of the technique of detection. Notwithstanding, it is imperative to recognise the persistent obstacles in this domain, such as the requirement for resilient datasets, continuous enhancement of algorithm precision, and adjustment to varied environmental circumstances. Refinement of current detection techniques and investigation of new avenues will necessitate sustained research in the future. The ultimate objective is to develop easily attainable, affordable, and practical solutions that farmers throughout the globe may readily use, therefore encouraging sustainable agricultural methods and guaranteeing global food security. To overcome these obstacles and refine the area of plant disease detection even further, cooperation between researchers, agricultural specialists, and technology developers is the need of the hour.

REFERENCES

Aglietti, C., Meinecke, C. D., Ghelardini, L., Barnes, I., Van der Nest, A., and Villari, C. (2021). Rapid detection of pine pathogens Lecanostictaacicola, Dothistromapini and D. septosporum on needles by probe-based LAMP assays. *Forests*, 12(4), 479.

Agrios, G. N. (2005). Plant Pathology. San Diego: Academic Press, 952.

Ahmad, F., Babalola, O. O., and Tak, H.I. (2012). Potential of MOLDI-TOF mass spectrometry as a rapid detection technique in plant pathology: identification of plant-assoiated microorganisms. *Analytical and Bioanalytical Chemistry*, 404(4), 1247–1255.

Alarcon, B., López, M. M., Cambra, M., Gorris, M. T., and Guerri, J. (1990). Differentiation of Erwinia carotovora subsp. carotovora and Erwinia carotovora subsp. atroseptica isolated from potato by Western blot and subsequent indirect ELISA. *Journal of Applied Bacteriology*, 69(1), 17–24. –122.

Aman, R., Mahas, A., and Mahfouz, M. (2020). Nucleic acid detection using CRISPR/Cas biosensing technologies. *ACS Synthetic Biology*, 9(6), 1226–1233.

An Introduction to ELISA–Basics Guide I Bio-Rad [Electronic resource] URL: www.bio-rad-antibodies.com/download-request-copy-elisa-guide.html

Araújo, S. O., Peres, R. S., Barata, J., Lidon, F., and Ramalho, J. C. (2021). Characterising the agriculture 4.0 landscape—emerging trends, challenges and opportunities. *Agronomy*, 11(4), 667.

Ayoib, A., et al. (2020). Design and fabrication of PDMS microfluidics device for rapid and label-free DNA detection. *Applied Physics A*, 126, 1–8.

Bagirova, S. F., Javakhia, V. G., Dyakov Yu, T., Ozeretskovskaya, O. L., Provorov, N. A., Tikhonovich, I. A., and Shcherbakova, L. A. (2012). *Fundamental Phytopathology (Fundamental nayafitopatologiya)*. Moscow: URSS: Krasand.

Balodi, R., Sunaina, B., Ghatak, A., and Rao, R, H. (2017). Plant disease diagnosis: technological advancements and challenges. *Indian Phytopathology,* 70(3), 275–281.

Bartlett, J. M. S., and Stirling D. A. (2003). In PCR Protocols, 3–6.

Begerow, D., Nilsson, H., Unterseher, M., and Maier, W. (2010). Current state and perspectives of fungal DNA barcoding and rapid identification procedures. *Applied Microbiology and Biotechnology,* 87(1), 99–108.

Bilay, V. I. (1982). Experimental Mycology Methods. Kiev: Naukova dumka.

Boisen, M. L., Viral Hemorrhagic Fever Consortium, Oottamasathien, D., Viral Hemorrhagic Fever Consortium, Jones, A. B., Viral Hemorrhagic Fever Consortium, ... and Viral Hemorrhagic Fever Consortium. (2015). Development of prototype filovirus recombinant antigen immunoassays. *Journal of Infectious Diseases*, 212, S359–S367.

Boonham, N., Glover, R., Tomlinson, J., and Mumford, R. (2008). Exploiting generic platform technologies for the detection and identification of plant pathogens. *European Journal of Plant Pathology*, 121(3), 355–363.

Bos, L. (1981). Hundred years of Koch's Postulates and the history of etiology in plant virus research. *Netherlands Journal of Plant Pathology*, 87, 91–110.

Brás, E. J., Fortes, A. M., Chu, V., Fernandes, P., and Conde, J. P. (2019). Microfluidic device for the point of need detection of a pathogen infection biomarker in grapes. *Analyst*, 144(16), 4871–4879.

Brás, E. J., Pinto, R. M., Chu, V., Fernandes, P., & Conde, J. P. (2020). A versatile and fully integrated hand-held device for microfluidic-based biosensing: A case study of plant health biomarkers. *IEEE Sensors Journal*, *20*(23), 14007–14015.

Bridle, H., and Desmulliez, M. (2021). Biosensors for the detection of waterborne pathogens. In Waterborne pathogens (pp. 189–235). Academic Press.

Buja, I., Sabella, E., Monteduro, A. G., Chiriacò, M. S., De Bellis, L., Luvisi, A., & Maruccio, G. (2021). Advances in plant disease detection and monitoring: From traditional assays to in-field diagnostics. *Sensors*, 21(6), 2129.

Bunsevich, L. L., Kostyuk, M.A. and Danilyuk, Y. P. (2010). Fruit growing and viticulture of the South of Russia [Electronic resource] Krasnodar: NCZRIHV6(5), Access mode: www.journal.kubansad.ru/pdf/10/05/12.pdf

Cai, H., Archambault, M., and Prescott, J. F. (2003). 16S ribosomal RNA sequence—based identification of veterinary clinical bacteria. *Journal of Veterinary Diagnostic Investigation*, 15(5), 465–469.

Campbell, C. K., and Johnson, E. M. (2013). Identification of pathogenic fungi. John Wiley & Sons.

Cardoso, R. M., Pereira, T. S., Facure, M. H., dos Santos, D. M., Mercante, L. A., Mattoso, L. H., and Correa, D. S. (2022). Current progress in plant pathogen detection enabled by nanomaterials-based (bio) sensors. *Sensors and Actuators Reports*, 4, 100068.

Carrio, A., Sampedro, C., Sanchez-Lopez, J. L., Pimienta, M., and Campoy, P. (2015). Automated low-cost smartphone-based lateral flow saliva test reader for drugs-of-abuse detection. *Sensors,* 15(11), 29569–29593.

Caruso, P., Gorris, M. T., Cambra, M., Palomo, J. L., Collar, J., and López, M. M. (2002). Enrichment double-antibody sandwich indirect enzyme-linked immunosorbent assay that uses a specific monoclonal antibody for sensitive detection of Ralstonia solanacearum in asymptomatic potato tubers. *Applied and Environmental Microbiology*, 68(7), 3634–3638.

Çelik, A. (2022). A novel technology for the one-step detection of prune dwarf virus: Colorimetric reverse transcription loop-mediated isothermal amplification assay. *Crop Protection*, 155, 105910.

Cesewski, E., and Johnson, B. N. (2020). Electrochemical biosensors for pathogen detection. *Biosensors and Bioelectronics*, 159, 112214.

Chang, W., Liu, W., Liu, Y., Zhan, F., Chen, H., Lei, H., and Liu, Y. (2019). Colorimetric detection of nucleic acid sequences in plant pathogens based on CRISPR/Cas9 triggered signal amplification. *Microchimica Acta*, 186, 1–8.

Chao, Q. T. Lee, T. F., Teng, S. H., Peng, L. Y., Chen, P. H., Teng, L. J., and Hsueh, P. R. (2014). Comparison of the accuracy of two conventional phenotypic methods and two MALDI-TOF MS systems with that of DNA sequencing analysis for correctly identifying clinically encountered yeasts. (2014)*PLoSONE,* 9(10),e109376.

Chen, Z. D., Kang, H. J., Chai, A. L., Shi, Y. X., Xie, X. W., Li, L., and Li, B. J. (2020). Development of a loop-mediated isothermal amplification (LAMP) assay for rapid detection of Pseudomonas syringaepv. tomato in planta. *European Journal of Plant Pathology*, 156, 739–750.

Danks, C., and Barker, I. (2000). On-site detection of plant pathogens using lateral-flow devices. *EPPO Bulletin,* 30(3–4), 421–426.

Das, S., and Sharma, A. (2023, June). Apple Leaves Disease Detection Using Multilayer Convolutional Neural Network. In 2023 3rd International Conference on Intelligent Technologies (CONIT) (pp. 1–7). IEEE.

del Pilar Martínez-Diz, M., Andrés-Sodupe, M., Berbegal, M., Bujanda, R., Díaz-Losada, E., and Gramaje, D. (2020). Droplet digital PCR technology for detection of Ilyonectrialiriodendri from grapevine environmental samples. *Plant Disease*, 104(4), 1144–1150.

Dingle, T. C., and Butler-Wu, S. M. (2013). MALDI-TOF mass spectrometry for microorganism identification. *Clinics in Laboratory Medicine*, 33(3), 589–609.

Dyakov, Y. T., and Elansky, S. N. (2019). General Phytopathology. Moscow: Yurayt Publishing House.

Egamberdiev, S. S., Salakhutdinov, I. B., Radjabov, F. S., Kurbanov, A. Y., and Abdurakhmonov, I. Y. (2017). Methodical instruction on the detection of pathogen of Fusarium genera, identifying more aggressive forms and their control measures. Center of Genomics and Bioinformatics of SA of RUz "Fan vatalimpoligraf. LLC Tashkent.

Egamberdiev, S. S., Salakhutdinov, I., Abdullaev, A. A., Ulloa, M., Saha, S., Radjapov, F., Mullaohunov, B., Mansurov, D., Jenkins, J. N., and Abdurakhmonov, I. Y. (2014). Detection of Fusarium oxysporum f. sp. v asinfectum race 3 by single-base extension method and allele-specific polymerase chain reaction. *Canadian Journal of Plant Pathology,* 36(2), 216–223.

Enicks, D. A., Bomberger, R. A., and Amiri, A. (2020). Development of a portable LAMP assay for detection of Neofabraeaperennans in commercial apple fruit. *Plant Disease*, 104(9), 2346–2353.

Fan, J., Zhang, Y., Wen, W., Gu, S., Lu, X., and Guo, X. (2021). The future of Internet of Things in agriculture: Plant high-throughput phenotypic platform. *Journal of Cleaner Production*, 280, 123651.

Fang, C., Chen, Z., Li, L., and Xia, J. (2011). Barcode lateral flow immunochromatographic strip for prostate acid phosphatase determination. *Journal of Pharmaceutical and Biomedical Analysis*, 56(5), 1035–1040.

Fang, Y., and Ramasamy, R. (2015). Current and prospective methods for plant disease detection. *Biosensors*, 3(5), 537–561.

Fang, Y., and Ramasamy, R. P. (2015). Current and prospective methods for plant disease detection. *Biosensors*, 5(3), 537–561.

Faron, M. L., Buchan, B. W., Hyke, J., Madisen, N., Lillie, J. L., Granato, P. A., et al. (2015). Multicenter evaluation of the Bruker MALDI Biotyper CA system for the identification of clinical aerobic gram-negative bacterial isolates. *PLOS ONE,* 10(11), e0141.350.

Fell, J. W. and Blatt, G. M. (1999). Separation of strains of the yeasts Xanthophyllomyces dendrorhous and Phaffia rhodozyma based on rDNA IGS and ITS sequence analysis. *Journal of Industrial Microbiology & Biotechnology,* 23(1), 677–681.

Freitas, T. A., Proença, C. A., Baldo, T. A., Materon, E. M., Wong, A., Magnani, R. F., and Faria, R. C. (2019). Ultrasensitive immunoassay for detection of Citrus tristeza virus in citrus sample using disposable microfluidic electrochemical device. *Talanta*, 205, 120110.

Fukuyama, Y., Nakaya, S., Yamazaki, Y., and Tanaka, K. (2008). Ionic liquid matrixes optimized for MALDI-MS of sulfated/sialylated/neutral oligosaccharides and glycopeptides. *Analytical Chemistry*, 80(6), 2171–2179.

Fung, K K, Chan, C P., and Renneberg, R. (2009). Development of enzyme-based bar code style lateral-flow assay for hydrogen peroxide determination. *Analytica Chimica Acta*, 634(1), 89–95.

Gan, Z., Zheng, C., and Wang, J. (2023). Challenges and applications of volatile organic compounds monitoring technology in plant disease diagnosis. *Biosensors and Bioelectronics*, 27(6), 115540.

Ghosh, D. K., Kokane, S. B., Kokane, A. D., Warghane, A. J., Motghare, M. R., Bhose, S., ... and Reddy, M. K. (2018). Development of a recombinase polymerase based isothermal amplification combined with lateral flow assay (HLB-RPA-LFA) for rapid detection of "Candidatus Liberibacter asiaticus". *PloSOne*, 13(12), e0208530.

Global pact against plant pests marks 60 years in action. FAO celebrates anniversary of creation of the International Plant Protection. www.fao.org/news/story/en/item/131114/icode/ (3 April 2012).

Grabicoski, E. M. G., Jaccoud-Filho, D. D. S., Lee, D., Henneberg, L., and Pileggi, M. (2020). Real-time quantitative and ion-metal indicator LAMP-based assays for rapid detection of Sclerotinia sclerotiorum. *Plant Disease*, 104(5), 1514–1526.

Hameed, S., Xie, L., and Ying, Y. (2018). Conventional and emerging detection techniques for pathogenic bacteria in food science: A review. *Trends in Food Science & Technology*, 81, 61–73.

Hampton, R., Ball, E., and De Boer, S. (1990). Serological methods for detection and identification of viral and bacterial plant pathogens: A laboratory manual. APSPress St Paul Minn.

Hebert, P. D., Cywinska, A., Ball, S. L., and deWaard, J. R. (2003). Proceedings biological sciences. *The Royal Society*, 270 (1512), 313–321.

Henson, J. M., and French, R. (1993). The polymerase chain reaction and plant disease diagnosis. *Annual Review of Phytopathology*, 1(31), 81–109.

Hillenkamp, F., Karas, M., Beavis, R. C., and Chait, B. T. (1991). Matrix-assisted laser desorption/ionization mass spectrometry of biopolymers. *Analytical Chemistry*, 63(24), 1193A–1203A.

Hou, T. Y., Chiang, N. C., and Teng, S. H. (2019). Current status of MALDI-TOF mass spectrometry in clinical microbiology. *Journal of Food and Drug Analysis*, 27(2), 404–414.

Joseph, D. S., Pawar, P. M., and Pramanik, R. (2023). Intelligent plant disease diagnosis using convolutional neural network: a review. *Multimedia Tools and Applications*, 82(14), 21415–21481.

Kamilov, S. G., Nuraliev, K. K., Sattarova, R. K., and Khakimov, A. A. (2020). The history of mycology and phytopathology development in Uzbekistan. *Mikologiya I Fitopatologiya,* 54(5), 313–319.

Karimi, K., Arzanlou, M., and Pertot, I. (2020). Development of novel species-specific primers for the specific identification of Colletotrichum nymphaeae based on conventional PCR and LAMP techniques. *European Journal of Plant Pathology*, 156, 463–475.

Kartasheva, I. A. (2007). *Agricultural Phytovirusology Study Guide*. Moscow: Kolos Stavropol: AGRUS.

Khakimov, A. A., Omonlikov, A. U., and Utaganov, S. B. (2020). Current status and prospects of the use of biofungicides against plant diseases. *GSC Biological and Pharmaceutical Sciences,* 13(03), 119–126.

Khakimov, A., Salakhutdinov, I., Omolikov, A., and Utaganov, S. (2022). Traditional and current-prospective methods of agricultural plant diseases detection: A review. In *IOP Conference Series: Earth and Environmental Science* (Vol. 951, No. 1, p. 012002). IOP Publishing.

Khasanov, B. A., and Sherimbetov, A. G. (2020). Traditional and current-prospective methods of agricultural plant diseases detection: A review. *Uzbekistan Biology Journal,* 2, 22–32.

Khasanov, B. O., Ochilov, R. O., and Gulmurodov, R. A. (2009). Diseases of Vegetables, Potatoes and Melon Crops and Their Control Measures Tashkent.

Khater, M., De La Escosura-Muñiz, A., and Merkoçi, A. (2017). Biosensors for plant pathogen detection. *Biosensors and Bioelectronics*, 93, 72–86.

Kim, J. H., Lee, S., Park, E. R., and Jang, W. C. (2022). Development of a highly sensitive and rapid detection method for Pea enation mosaic virus using loop-mediated isothermal amplification assay. *Journal of Virological Methods*, 300, 114427.

Kishorebabu, V., and Sravanthi, R. (2020). Real time monitoring of environmental parameters using IOT. *Wireless Personal Communications*, 112(2), 785–808.

Koczula, K. M., and Gallotta, A. (2016). Lateral flow assays. *Essays in Biochemistry*, 60(1), 111–120.

Kokane, A. D., Kokane, S. B., Warghane, A. J., Gubyad, M. G., Sharma, A. K., Reddy, M. K., and Ghosh, D. K. (2021). A rapid and sensitive reverse transcription–loop-mediated isothermal amplification (RT-LAMP) assay for the detection of Indian Citrus Ringspot Virus. *Plant Disease*, 105(05), 1346–1355.

Korfmacher, W. A. (2009). *Using Mass Spectrometry for Drug Metabolism Studies*. CRC Press.

Kouadio, L., El Jarroudi, M., Belabess, Z., Laasli, S. E., Roni, M. Z. K., Amine, I. D. I., ...and Lahlali, R. (2023). A review on UAV-based applications for plant disease detection and monitoring. *Remote Sensing*, 15(17), 4273.

Kumar, S., Stecher, G., and Tamura, K. (2016). MEGA7: molecular evolutionary genetics analysis version 7.0 for bigger datasets. *.Molecular Biology and Evolution,* 33, 1870–1874.

Lane, D. J., Pace, B., Olsen, G. J. et al. (1985). *Proceedings of the National Academy of Sciences of the USA*, 82, 6955–6959.

Larkin, M. A., Blackshields, G., Brown, N. P., Chenna, R., McGettigan, P. A., McWilliam, H., ... and Higgins, D. G. (2007). Clustal W and Clustal X version 2.0. *Bioinformatics*, 23(21), 2947–2948.

Lau, H. Y., and Botella, J. R. (2017). Advanced DNA-based point-of-care diagnostic methods for plant diseases detection. *Frontiers in Plant Science*, 8, 2016.

Lecomte, P., Manceau, C., Paulin, J. P., and Keck, M. (1997). Identification by PCR analysis on plasmid pEA29 of isolates of Erwinia amylovora responsible of an outbreak in Central Europe. *European Journal Plant Pathology*, 103, 91–98.

Levesque, S., Dufresne, P. J., Soualhine, H., Domingo, M.-C., Bekal, S., Lefebvre, B., et al. (2015). A side by side comparison of Bruker Biotyper and VITEK MS: utility of MALDI-TOF MS technology for microorganism identification in a public health reference laboratory. *PLoS ONE,* 10(12), e0144878.

Li, W., Lee, S. Y., Back, C. G., Ten, L. N., and Jung, H. Y. (2019). Loop-mediated isothermal amplification for the detection of Xanthomonas arboricolapv. pruni in peaches. *Plant Pathology Journal*, 35(6), 635.

Lin, C. L., Chang, W. H., Wang, C. H., Lee, C. H., Chen, T. Y., Jan, F. J., and Lee, G. B. (2015). A microfluidic system integrated with buried optical fibers for detection of Phalaenopsis orchid pathogens. *Biosensors and Bioelectronics*, 63, 572–579.

Liu, J., Li, C., Muhae-Ud-Din, G., Liu, T., Chen, W., Zhang, J., and Gao, L. (2020). Development of the droplet digital PCR to detect the teliospores of Tilletiacontroversa Kühn in the soil with greatly enhanced sensitivity. *Frontiers in Microbiology*, 11, 4.

Lopez-Soriano, P., Noguera, P., Gorris, M. T., Puchades, R, Maquieira, A, Marco-Noales, E., and Lopez, M. M. (2017). Lateral flow immunoassay for on-site detection of Xanthomonas arboricola pv. pruni in symptomatic field samples. *PLOS ONE*, 12(4), e0176201.

Lowe, A., Harrison, N., and French, A. P. (2017). Hyperspectral image analysis techniques for the detection and classification of the early onset of plant disease and stress. *Plant Methods*, 13(1), 80.

Lukhmenev, V. (2012). Phytopathology: A tutorial. Orenburg: OSAU Publishing Center.

Magambo, K. A., Kalluvya, S. E., Kapoor, S. W., Seni, J., Chofle, A. A., Fitzgerald, D. W., & Downs, J. A. (2014). Utility of urine and serum lateral flow assays to determine the prevalence and predictors of cryptococcal antigenemia in HIV-positive outpatients beginning antiretroviral therapy in Mwanza, Tanzania. *African Journal of Reproduction and Gynaecological Endoscopy*, 17(1).

Mahas, A., Hassan, N., Aman, R., Marsic, T., Wang, Q., Ali, Z., and Mahfouz, M. M. (2021). LAMP-coupled CRISPR–Cas12a module for rapid and sensitive detection of plant DNA viruses. *Viruses*, 13(3), 466.

Mahlein, A. K. (2016). Plant disease detection by imaging sensors–parallels and specific demands for precision agriculture and plant phenotyping. *Plant Disease*, 100(2), 241–251.

Mahmud, S., Ahmed, J., Aziz, M. A., Hasan, M. R., Shaon, S. M., Bhuiyan, M. N. A. H., ...and Islam, M. S. (2015). Clustered regularly interspaced short palindromic repeats cas systems: A comprehensive review. *International Journal of Basic & Clinical Pharmacology,* 4(4), 613.

Mamiev, M. S., Khakimov, A. A., Zuparov, M. A., and Rakhmonov, U. N. (2020). IOP Conference Series: Earth and Environmental Science 614 012112.

Marinach-Patrice, C., Lethuillier, A., Marly, A., Brossas, J. Y., Gene, J., Symoens, F., Datry, A., Guarro, J., Mazier, D. and Hennequin, C. (2009). *Clinical Microbiology and Infection*, 7(15), 634–642.

Mehetre, G. T., Leo, V. V., Singh, G., Sorokan, A., Maksimov, I., Yadav, M. K., ... and Singh, B. P. (2021). Current developments and challenges in plant viral diagnostics: A systematic review. *Viruses*, 13(3), 412.

Meyer, W., Irinyi, L., Hoang, M. T., Robert, V., Garcia-Hermoso, D., Desnos-Ollivier, M., et al. (2019). *Genome*, 62(3), 160–169.

Mu, K., Ren, X., Yang, H., Zhang, T., Yan, W., Yuan, F., ... and Zeng, Q. (2022). CRISPR-Cas12a-based diagnostics of wheat fungal diseases. *Journal of Agricultural and Food Chemistry*, 70(23), 7240–7247.

Narayanasamy, P. (2011). Detection of virus and viroid pathogens in plants. Microbial Plant Pathogens-Detection and Disease Diagnosis. *Viral and Viroid Pathogens,* 3, 7–220.

O'Farrell, B. (2008). Evolution in lateral flow–based immunoassay systems. *Lateral Flow Immunoassay*, 1–33.

Paul, R. (2023, April). Design and Performance Investigation of SiGe Heterojunction Based Vertical Dual Pocket Biosensor Design Based on Charge Plasma TFET. In *2023 IEEE 8th International Conference for Convergence in Technology (I2CT)* (pp. 1–7). IEEE.

Pandey, P., Pandey, N. S., Shamim, Md., Srivastava, D., Dwivedi, D. K., Awasthi, L. P., Singh, K. N., and Pandey P. (2015). Molecular tools and techniques for detection and diagnosis of plant pathogens. *Recent Advances in the Diagnosis and Management of Plant Diseases*, 253–271.

Parkinson, N., Cowie, C., Heeney, J., and Stead, D. (2009). Phylogenetic structure of Xanthomonas determined by comparison of gyrB sequences. *International Journal of Systematic and Evolutionary Microbiology,* 59(Pt 2), 264–274.

Patel, R. (2015). MALDI-TOF MS for the diagnosis of infectious diseases. *Clinical Chemistry*, 61(1), 100–111.

Pavlovskaya, N. E., Gagarina, I. N., Borodin, D. B., Gneusheva, I. A., Gorkova, I. V., Solokhina, I. Yu., Kostromicheva, E. V., Lushnikov, A. V., Yakovleva, I. V., and Ageyeva, N. Y. (2018). *Agrobiological Substantiation of the Technology of Growing Vegetable Products with the Use of Biological Means of Protection: Monograph.* Orel: Publishing house of the Federal State Budgetary Educational Institution of Higher Education OSAU.

Peng, D., Xie, J., Qiang, W., Ling, K. S., Guo, L., Fan, Z., and Zhou, T. (2017). One-step reverse transcription loop-mediated isothermal amplification assay for detection of Apple chlorotic leaf spot virus. *Journal of Virological Methods*, 248, 154–158.

Peresipkin, V. F. (1991). Diseases of Agricultural Crops 3 Diseases of vegetable and fruit crops. Kiev: Urojay.

Pililshikova, N. (2013). Diagnostics of plant diseases and modern technologies. https://biomolecula.ru/articles/diagnostika-boleznei-rastenii-i-sovremennye-tekhnologii

Pineda, M., Barón, M., and Pérez-Bueno, M. L. (2020). Thermal imaging for plant stress detection and phenotyping. *Remote Sensing*, 13(1), 68.

Putnam, M. L. (1995). Evaluation of selected methods of plant disease diagnosis. *Crop Protection,* 14(6), 517–525.

Qu, X., Li, M., Zhang, H., Lin, C., Wang, F., Xiao, M., ... and Li, L. (2017). Real-time continuous identification of greenhouse plant pathogens based on recyclable microfluidic bioassay system. *ACS Applied Materials & Interfaces*, 9(37), 31568–31575.

Rasmussen, O. F., and Wulff, B. S. (1991). Proceedings of 4th International Working Group Kluwer Academic Publishers Dordrecht and Nederland, 367–376.

Reich, M., Bosshard, P.P., Stark, M., Beyser, K., and Borgmann, S. (2013). Species identification of bacteria and fungi from solid and liquid culture media by MALDI-TOF mass spectrometry. *Journal of Bacteriology and Parasitology,* S5–002.

Ren, W., Liu, N., and Li, B. (2021). Development and application of a LAMP method for rapid detection of apple blotch caused by Marssonina coronaria. *Crop Protection*, 141, 105452.

Ristaino, J. B., Saville, A. C., Paul, R., Cooper, D. C., and Wei, Q. (2020). Detection of Phytophthora infestans by loop-mediated isothermal amplification, real-time LAMP, and droplet digital PCR. *Plant Disease*, 104(3), 708–716.

Rocafort, M., Arshed, S., Hudson, D., Sidhu, J. S., Bowen, J. K., Plummer, K. M., ... and Mesarich, C. H. (2022). CRISPR-Cas9 gene editing and rapid detection of gene-edited mutants using high-resolution melting in the apple scab fungus, Venturia inaequalis. *Fungal Biology*, 126(1), 35–46.

Rychert, J., Burnham, C. A., Bythrow, M., Garner, O.B., Ginocchio, C. C., Jennemann, R., et al. (2013). Multicenter evaluation of the Vitek MS matrix-assisted laser desorption ionization–time of flight mass spectrometry system for identification of Gram-positive aerobic bacteria. *Journal of Clinical Microbiology*, 51(7), 2225–2231.

Sandrin, T. R., Goldstein, J. E. and Schumaker, S. (2013). MALDI TOF MS profiling of bacteria at the strain level: a review. *Mass Spectrometry Reviews,* 32(3), 188–217.

Santos, C., Ventura, J. A., and Lima, N. (2016). New insights for diagnosis of pineapple fusariosis by MALDI-TOF MS technique. *Current Microbiology,* 73(2), 206–213.

Schoch, C. L., Seifert, K. A., Huhndorf, S., Robert, V., Spouge, J. L., Levesque, C. A., ... & White, M. M. (2012). Nuclear ribosomal internal transcribed spacer (ITS) region as a universal DNA barcode marker for Fungi. *Proceedings of the National Academy of Sciences of the United States of America*, 109(16), 6241–6246.

Schramm, E. C., Staten, N. R., Zhang, Z., Bruce, S. S., Kellner, C., and Atkinson, J. P. (2015). A quantitative lateral flow assay to detect complement activation in blood. *Analytical Biochemistry*, 477, 78–85.

Seng, P., Drancourt, M., Gouriet, F., La Scola, B., Fournier, P. E., Rolain, J. M. and Raoult, D. (2009). MALDI-TOF-mass spectrometry applications in clinical microbiology. .*Clinical Infectious Diseases*, 49(4), 552–553.

Shahdordizadeh, M., Taghdisi, S. M., Ansari, N., Langroodi, F. A., Abnous, K., andRamezani, M. (2017). Aptamer based biosensors for detection of Staphylococcus aureus. *Sensors and Actuators B: Chemical*, 241, 619–635.

Shkalikov, V. A. (2007). *Rots Russian Big Encyclopedia*, 7, Moscow.

Siricord, C. and O'Brien, P. A. (2008). MALDI-TOF mass spectrometry can be used for detection of pathogenic microorganisms in soil. *Australasian Plant Pathology,* 37(6), 543–545.

Stielow, J. B., Lévesque, C. A., Seifert, K. A., Meyer, W., Iriny, L., Smits, D., et al. (2015). One fungus, which genes? Development and assessment of universal primers for potential secondary fungal DNA barcodes. *Persoonia-Molecular Phylogeny and Evolution of Fungi*, 35, 242–263.

Suarez, M. B., Walsh, K., Boonham, N., O'Neill, T., Pearson, S., and Barker, I. (2005). Development of real-time PCR (TaqMan®) assays for the detection and quantification of Botrytis cinerea in planta. *Plant Physiology and Biochemistry*, 43(9), 890–899.

Tetala, S. S. D. R. (2023). Artificial Intelligence Powered Personalized Agriculture (Doctoral dissertation, Colorado State University).

Tian, Q., Zhao, Z., Lu, S., Zhu, S. and Li, S. (2016). DNA barcoding for efficient species- and pathovar-level identification of the quarantine plant pathogen Xanthomonas. *PLOS ONE,* 11(11), e0165995.

Toh, S. Y., Citartan, M., Gopinath, S. C., and Tang, T. H. (2015). Aptamers as a replacement for antibodies in enzyme-linked immunosorbent assay. *Biosensors and Bioelectronics,* 64, 392–403.

Tseng, Y. W., Wu, C. F., Lee, C. H., Chang, C. J., Chen, Y. K., and Jan, F. J. (2021). Universal primers for rapid detection of six pospiviroids in solanaceae plants using one-step reverse-transcription PCR and reverse-transcription loop-mediated isothermal amplification. *Plant Disease*, 105(10), 2867–2872.

Uke, A., Khin, S., Kobayashi, K., Satou, T., Kim, O. K., Hoat, T. X., ... and Ugaki, M. (2022). Detection of Sri Lankan cassava mosaic virus by loop-mediated isothermal amplification using dried reagents. *Journal of Virological Methods*, 299, 114336.

Upadyshev, M. T. (2018). Diagnosis of viruses of fruit and small fruit crops is an important stage in the production of certified planting material. *Horticulture and Viticulture*, 2, 43–48.

Van Vuurde, J. W., Ruissen, M. A., and Vruggink, H. (1987). Principles and Prospects of New Serological Techniques Including Immunosorbent Immunofluorescence, Immunoaffinity Isolation and Immunosorbent Enrichment for Sensitive Detection of Phytopathogenic Bacteria Plant Pathogenic Bacteria. *Current Plant Science and Biotechnology in Agriculture Springer Dordrecht,* 4, 835–842.

Venter, J. C. (2001). The sequence of human genome. *Science*, 291(5507), 1304–1351.

Vlasov, Y. I., Larina, E. I., and Truskinov, E. V. (2016). Agricultural Phytovirusology St. Petersburg–Pushkin: ARRIPP. *Appendix of Journal "Plant Protection Bulletin"*, 17.

Wang, P., Yuan, S., Yang, N., Wang, A., Fordjour, A., and Chen, S. (2020). The collection method for crop fungal spores based on an efficient microfluidic device. *Aerosol and Air Quality Research*, 20(1), 72–79.

Wang, Y., Wang, C., Wang, L., Zhang, X., Yan, J., Wang, J., and Wang, M. (2020). Development of loop-mediated isothermal amplification (LAMP) assay for rapid detection of Fusarium proliferatum causing ear and kernel rot on maize. *Crop Protection*, 132, 105142.

Wheatley, M. S., and Yang, Y. (2021). Versatile applications of the CRISPR/Cas toolkit in plant pathology and disease management. *Phytopathology®*, 111(7), 1080–1090.

Winkworth, R. C., Nelson, B. C., Bellgard, S. E., Probst, C. M., McLenachan, P. A., and Lockhart, P. J. (2020). A LAMP at the end of the tunnel: A rapid, field deployable assay for the kauri dieback pathogen, Phytophthora agathidicida. *PLoS One*, 15(1), e0224007.

Woese, C. R. (1987). *Microbiological Reviews,* 51, 221–271.

Wong, Y. P., Othman, S., Lau, Y. L., Radu, S., and Chee, H. Y. (2018). Loop-mediated isothermal amplification (LAMP): a versatile technique for detection of micro-organisms. *Journal of Applied Microbiology*, 124(3), 626–643.

Wurzbacher, C., Larsson, E., Bengtsson-Palme, J., Van den Wyngaert, S., Svantesson, S., and Kristiansson, E. (2019). Introducing ribosomal tandem repeat barcoding for fungi. *Molecular Ecology Resources*, 19(1), 118–127.

Xu, Y., Liu, Y., Wu, Y., Xia, X., Liao, Y., and Li, Q. (2014). Fluorescent probe-based lateral flow assay for multiplex nucleic acid detection. *Analytical Chemistry,* 86(12), 5611–5614.

Yao, M. C., Zhu, P. X., Zhang, Q. Q., Zhang, Q. P., Ji, Z. L., and Zhu, F. (2022). Development of a reverse transcription-loop mediated isothermal amplification assay for detection of wisteria vein mosaic virus. *European Journal of Plant Pathology*, 163(1), 113–123.

Yen, C. W., de Puig, H., Tam, J. O., Gomez-Marquez, J., Bosch, I., and Hamad-Schifferli, K. (2015). Multicolored silver nanoparticles for multiplexed disease diagnostics: distinguishing dengue, yellow fever, and Ebola viruses. *Lab on a Chip*, 15(7), 1638–1641.

Yu, L. U., Zhang, H. J., Zhao, Z. J., Wen, C. L., Ping, W. U., Song, S. H., ... and Xu, X. L. (2020). Application of droplet digital PCR in detection of seed-transmitted pathogen Acidovorax citrulli. *Journal of Integrative Agriculture,* 19(2), 561–569. *Journal of Integrative Agriculture*, 20(8), VIII–VIII.

Yüksel, S., Schwenkbier, L., Pollok, S., Weber, K., Cialla-May, D., and Popp, J. (2015). Label-free detection of Phytophthora ramorum using surface-enhanced Raman spectroscopy. *Analyst*, 140(21), 7254–7262.

Zeigler, D. R. (2003). Gene sequences useful for predicting relatedness of whole genomes in bacteria. *International Journal of Systematic and Evolutionary Microbiology,* 53(6), 1893–1900.

Zhang, L., Jiang, H., Zhu, Z., Liu, J., and Li, B. (2022). Integrating CRISPR/Cas within isothermal amplification for point-of-care assay of nucleic acid. *Talanta*, 243, 123388.

Zhang, Y., Xie, Z., Wang, Y., Wang, R., Guo, Z., Qiu, Y., and He, Y. (2022). Development of immunocapture reverse transcription loop-mediated isothermal amplification assays to detect both lily symptomless virus and cucumber mosaic virus. *Annals of Applied Biology*, 180(1), 132–139.

Ziegler, D., Mariotti, A., Pflüger, V., Saad, M., Vogel, G., Tonolla, M., et al. (2012). In situ identification of plant-invasive bacteria with MALDI-TOF mass spectrometry. *PLoS ONE,* 7(5), e37189.

Zuparov, M. A., Khakimov, A. A., Mamiev, M. S., and Allayarov, A. N. (2020). In vitro efficacy testing of fungicides on Botrytis cinerea causing gray mold of tomato. *International Journal on Emerging Technologies*, 11(5), 50–55.

2 Microscopy as a Tool for the Detection of Plant Pathogens

Zohra Shabir, Sajad Un Nabi, Mir Mehreen, and Sheikh Mansoor

2.1 INTRODUCTION

Imagine a world where invisible adversaries threaten the very foundation of our food supply. In the sphere of plant pathology, this world is not a fantasy; it is our reality, it resonates with real-world challenges faced by agriculture, where various threats to crops, such as pests, diseases, and environmental factors, jeopardize food security. The situation can be attributed to a variety of reasons; damage caused by pets and pathogens play a significant role in crop losses throughout the world. The losses in the crop yield due to the pathogen infections range from 20% to 40% (FAO, 2021). In such a world, the importance of advanced technologies for detection, monitoring, and prevention should be heightened. Plant pathology is the science dedicated to studying and managing plant diseases that can devastate crops and disrupt ecosystems. At the heart of the battle against plant diseases stands a silent hero: The Microscope. This humble instrument has unveiled the hidden world of pathogens, enabling scientists to identify, characterize, and develop strategies to manage these silent invaders. Microscopy is a fundamental tool in plant pathology, enabling researchers to examine the morphology, structure, and behavior of plant pathogens, as well as the plants response to infection. In the ever-evolving landscape of agriculture, the complex interaction between plants and pathogens has long been a focal point of scientific inquiry. Plant diseases are a significant constraint to crop production worldwide and exert particularly severe impacts in developing countries, where agricultural systems may be less resilient than in developed ones (FAO, 2021). Our drive to safeguard and preserve our agricultural resources depends heavily on our capacity to identify and understand the minute interactions that occur between plants and the microscopic invaders that target them. The lens of microscopy has become an indispensable tool in this context, providing a glimpse into a world that would otherwise remain unseen. In this chapter, we delve into the profound significance of microscopes in understanding, diagnosing, and managing plant diseases. At the heart of plant pathology lays the challenge of identifying, characterizing, and combating the pathogens that endanger the well-being and yield of crops. Microscopy is the lighthouse that illuminates the complexities of these often-invisible adversaries, whether they are, bacteria, viruses, fungi, or nematodes. It provides the means to unravel their forms, structures, and behaviors, unveiling a level of detail that is crucial in deciphering their impact on

DOI: 10.1201/9781003474036-2

plant health. The sphere of plant pathology is inexorably tied to our ability to comprehend the nuanced intricacies of disease development, spread, and containment. Microscopy grants us a front-row seat to witness the interplay between pathogens and their plant hosts, enabling a comprehensive understanding of infection mechanisms and the subsequent cellular and molecular alterations. Plant pathogen detection by the help of microscopes aims to identify the presence of plant pathogens as early as possible so control measures can be implemented to alleviate their impacts on crop production (Harakannanavar *et al.*, 2022). Armed with the knowledge and insights gleaned from microscopic observations, scientists, researchers, and agricultural practitioners can devise targeted disease management strategies. By distinguishing between different pathogen strains, understanding their vulnerabilities, and assessing disease severity, microscopes become instrumental in the development and implementation of effective control measures. They are associated with applying multiple techniques and approaches to identify and understand the presence and cause of plant diseases (Gullino and Bonants, 2014). Detection of plant pathogens involves identifying the existence of these harmful agents in a particular place or region. This procedure includes the examination of apparent disease symptoms in plants, gathering plant samples for subsequent analysis in a laboratory, or employing remote sensing methods to identify the existence of pathogens (Liu and wang, 2021).

Ever since the initial examination of microscopic entities through light microscopy by Leeuwenhoek in 1673, and the subsequent identification of the first virus (tobacco mosaic virus) in plant sap using transmission electron microscopy (TEM) by Kausche *et al.* in 1939, the use of microscopy to confirm microbes as causal agents in plant diseases has been pivotal. This application serves as a crucial diagnostic tool and contributes significantly to our comprehension of interactions between plants and microbes. In the 1980s, the integration of digital imaging and confocal microscopy utilizing vital stains was followed by the application of diverse genetic and molecular tools, including genomic sequencing, mutant analysis, and the introduction of fluorescent proteins for tracking microbes and their proteins (Hepler and Gunning 1998; Hickey *et al.* 2004). These advancements resulted in rapid progress in comprehending both beneficial and pathogenic cellular processes. The evolution of imaging methods and technologies, encompassing light, electron, X-ray, and probe-based techniques, especially within the multiscale and three-dimensional (3D) domain, is continuously accelerating. This progress aligns with the evolving possibilities of these innovative tools to enhance our insights into the cellular intricacies of plant-microbe interactions. This chapter aims to provide a comprehensive review of the role of microscopy in plant pathology, emphasizing its applications, strengths, and limitations.

2.1.1 Unveiling the Intricacies: How Microscopy Takes Center Stage in Detection of Plant Pathogens

In the intricate realm of plant pathology, microscopy emerges as the undisputed protagonist, casting its spotlight on the subtle nuances that govern the health and afflictions of plants. This powerful investigative tool transcends mere observation, immersing itself in the microscopic landscapes where the narrative of pathogenic interactions unfolds.

Microscopy, with its diverse array of techniques, serves as the key to unlocking the mysteries at the cellular and subcellular levels. From light microscopy, providing a broader perspective, to electron microscopy (EM), enabling a closer scrutiny of minute structures, and the innovative world of probe-based techniques, offering unprecedented insights, the toolkit of microscopic methods facilitates a holistic comprehension of both advantageous and harmful processes occurring within plant cells. In the realm of disease detection, a range of imaging techniques are employed, and one notable method is photo-acoustic imaging. This approach harnesses the absorption of light by tissues, utilizing the property of light absorption and its conversion into heat, resulting in the generation of photo-acoustic signals. The pressure distribution emitted by tissues is then mapped and utilized for imaging purposes. In the process of detection of diseases, fluorescence spectroscopy has also emerged as a recent method, relying on the measurement of fluorescence released after light excitation in the area of interest. Laser-induced fluorescence, particularly prevalent in vegetative studies and plant disease detection, is extensively employed. This method proves vital for identifying different physiological states in various plants, particularly assessing chlorophyll fluorescence or blue-green fluorescence in green leaves. This facilitates the early detection of impairments resulting from significant nutrient deficiencies (Sankaran *et al.*, 2010). Another effective imaging methodology in recent times is hyper-spectral imaging. This technique utilizes the electromagnetic spectrum of an images pixel to detect the presence of plant diseases. Employing a broad spectrum of light for the analysis of each images pixel allows for a more comprehensive and reasonable detection of various diseases across different plants. When combined with microscopy, hyper-spectral imaging extends its capabilities, providing higher-resolution, clearer images for microscopic studies at the genotypic level of diverse plant leaves (Rumf *et al.*, 2010).

The multidimensional capabilities of modern microscopy, extending into the 3D domain, propel research into new frontiers. The detection and identification of diseases in crops can be achieved through both direct and indirect methods. Direct detection methods involve molecular and serological approaches, particularly useful for high-throughput analysis when dealing with a large number of samples. In these methods, the identification of the disease-causing pathogens, such as bacteria, fungi, and viruses, occurs directly, ensuring precise identification of the disease or pathogen. In contrast, indirect methods ascertain plant diseases by examining various parameters, including morphological changes, alterations in temperature, shifts in transpiration rates, and the release of volatile organic compounds by infected plants (Fang *et al.*, 2015). As technology continues to evolve, the pace of discovery in plant pathology accelerates, unveiling hidden intricacies and paving the way for innovative approaches to disease detection and management. In essence, microscopy stands as the gateway to a deeper comprehension of the dynamic interplay between plants and pathogens. By assuming a pivotal role in the field of plant pathology, it not only unveils the microscopic narrative but also steers the formulation of strategies to protect the resilience of plant life amidst continually evolving challenges. Microscopes play a crucial role in the management of plant diseases through various essential avenues (Figure 2.1).

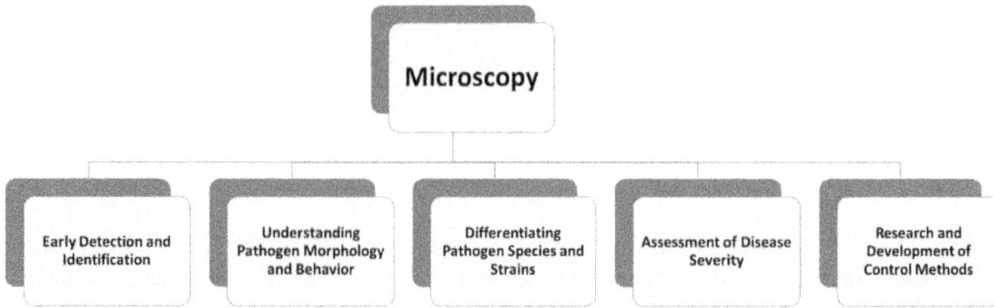

FIGURE 2.1 Microscopy enables early pathogen detection, identification, and management in plant pathology. By revealing pathogen morphology, infection mechanisms, and colonization at cellular levels, it guides precise control strategies, assesses disease severity, and aids in developing resistant varieties. Microscopy also monitors treatment efficacy in real-time for effective disease management.

Essentially, microscopes are essential instruments in the management of plant diseases, enabling not only the early identification of pathogens but also the creation of customized strategies to control and minimize their effects on crops and plant well-being.

2.2 A DIVERSE ARRAY OF MICROSCOPES REVOLUTIONIZING PLANT PATHOGEN DETECTION

Microscopy assumes a pivotal role in the detection and identification of plant pathogens. Various microscopy techniques allow for the visualization and characterization of pathogens, offering insights into their morphology, structure, and behavior. Table 2.1 gives an overview of the various microscopes used in plant pathology, emphasizing their unique features and applications in studying plant diseases.

2.2.1 Light Microscopy (LM)

Light microscopy is used to make small structures and samples visible by providing a magnified image of how they interact with visible light (e.g., their absorption, reflection, and scattering). The light microscope is an instrument for visualizing fine detail of an object. It does this by creating a magnified image through the use of a series of glass lenses, which first focus a beam of light onto or through an object, and convex objective lenses to enlarge the image formed. In the majority of light microscopes, the image is viewed directly through a binocular eye piece that acts as a secondary lens in the form of a magnifying glass to observe the projected image (Lavanya *et al.*, 2017). Such instruments are termed "compound microscopes," and the total magnification is the sum of the objective magnification and the eye piece magnification. The magnification range extends from 10to 1000, with a resolving power of the order of 0.2mm, depending on the type and numerical aperture (area available for passage of light) of the objective lenses (Holgate and Webb, 2003). This is useful to understand what the sample looks like and what it is made of, but also allows us to see processes of

TABLE 2.1
Overview of various microscopes used in plant pathology, emphasizing their unique features and applications in studying plant diseases

S, no	Microscope Type	Description	Application in Plant Pathology
1	Light microscope	Uses visible light to observe specimens	Studying general morphology of plant tissues and pathogens and helps identifying symptoms caused by fungi, bacteria, and viruses.
2	Fluoresecnce microscope	Uses fluorescense to highlight specific structures	Detecting plant pathogen-specific molecules and proteins and helps in visualizing host response such as induced fluorescence.
3	Confocal microscope	Uses lasers and pinholes to create 3D images	Examining fluorescence labeled structures in thick specimens and studying spatial distribution of pathogens within plant tissues.
4	Phase contrast microscope	Enhances contrast in transparent speciemens	Viewing live, unstained plant cells and microorganisms and monitoring dynamic processes, such as fungal hyphal growth.
5	Transmission electron microscope	Uses electrons to pass through thin specimens	Visualizing ultrastructures of pathogens and host cell organelles and studying virus particles and internal structures of pathogen.
6	Scanning electron microscope	Scans specimen surface with a focussed beam of electrons	Observing detailed surface structure of pathogens and tissues and examining fungal spores, bacterial cells and plant surfaces.

the microscopic world, such as how substances diffuse across a cell membrane. It is commonly used for initial observations. Stained samples are examined under visible light to observe the morphology of pathogens, such as fungi, bacteria, or nematodes. This helps in identifying the type of pathogen causing damage to the plants. LM is a fundamental and widely used tool in the field of biology, including plant pathology. It utilizes visible light to magnify and examine specimens, providing insights into the structure, morphology, and behavior of various biological entities. In the context of plant pathology, LM serves as an essential technique for the detection and identification of pathogens that affect plant health (Gudkov *et al.*, 2023). It involves the use of visible light and lenses to magnify the image of a specimen, allowing researchers to observe cellular and subcellular structures. Stained samples, such as plant tissues or pathogen cultures, can be observed under different magnifications, enabling scientists

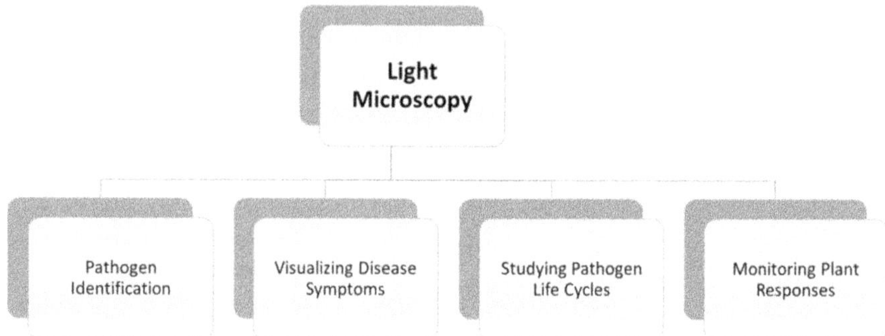

FIGURE 2.2 Light microscopy (LM) in plant pathology enables pathogen identification through morphological examination, visualizes disease symptoms within plant tissues, and aids in studying pathogen life cycles. LM also monitors plant cellular responses to infections, providing insights into defense mechanisms and informing targeted disease management strategies.

to study the characteristics of various pathogens like fungi, bacteria, viruses, and nematodes. In plant pathology, LM helps in (Figure 2.2).

Yellow Leaf Disease (YLD), one of the most drastic diseases affecting the areca nut palms of Kerala and Karnataka, causing extensive crop loss. The disease was associated with the presence of phytoplasma. A classic diagnostic technique was the observation of symptoms on the leaves; however, microscopic studies could be used for pathogen visualization along with molecular methods. Although the direct observation of phytoplasmas required the magnification achieved only by EM, light and FM could also be used to detect the presence of phytoplasmas in diseased palm tissues (Sumi *et al.*, 2023). Dienes' stain and the 4′,6-diamidino-2-phenylindole (DAPI) fluorescent microscopic method developed for confirming the presence of phytoplasma in hand-cut or freezing microtome sections of infected tissues were employed. YLD-affected plants gave positive results for phytoplasma-specific staining. Fluorescent particles of the DAPI-stained phytoplasma were limited to sieve tube elements. The phloem of healthy root sections remained unstained with Dienes' stain, but the phloem of the infected root sections contained many dark blue DAPI-stained areas. The results from the study indicated that these methods could be used as preliminary steps in phytoplasma presence detection and were found to be very effective in highly infected plants. The microscopic observation of phytoplasma cells could be used as a confirmative step for subsequent application of molecular diagnostic techniques (Gopinath *et al.*, 2016).

A study was done by Guilherme *et al.*, 2014 to investigate whether commercial fruit seeds of variety Red Globe could be a potential bacterial colonization and settlement niche, using light and scanning EM associated with immunolocalization techniques. Samples were examined with a light microscope. Based on LM analysis of transversal and longitudinal sections stained with toluidine blue, it was possible to distinguish different tissue systems in the anatomy of the grape seed var. Red Globe. Moreover, by using such a technique, it was possible to localize the bacteria inside the seed tissue, colonizing the inner space between the endosperm and the tegument. Finally, confirmation of the etiological agent for grapevine bacterial canker associated with asymptomatic fruit seeds was possible. In an another study done by Zelalem *et al.*, 2023 the light and electron microscopes were used to study the mechanisms of tomato leaf infection (Black Leaf Mould). The study confirmed that there was prevalence of primary and secondary infection hyphae caused by a fungal pathogen *Pseudocercospora fuligena.*

While LM provides valuable insights, it does have limitations regarding resolution and magnification compared to more advanced techniques such as EM. Nonetheless, its accessibility, ease of use, and cost-effectiveness make it an indispensable tool for routine analysis and initial observations in the field of plant pathology. Although light microscopy offers valuable insights it does have limitations in terms of resolution and magnification compared to more advanced techniques such as EM (Peddie *et al.*, 2022). Nevertheless, its accessibility, ease of use, and cost-effectiveness make it an indispensable tool for routine analysis and initial observations in the field of plant pathology. While light microscopes are valuable tools in plant pathology, they have certain limitations that can impact their utility in specific contexts.

Despite these limitations, light microscopes remain indispensable tools in plant pathology for routine observations, disease diagnosis, and preliminary investigations. Researchers often use a combination of different microscopy techniques to overcome these limitations and obtain a comprehensive understanding of plant diseases.

2.2.2 Fluorescence Microscopy

This technique utilizes specific fluorescent dyes that bind to certain structures within pathogens. It helps in visualizing specific parts of the pathogen (e.g., cell walls, organelles) and is used for more precise identification. Fluorescence microscopy (FM) is a powerful imaging technique that leverages the fluorescent properties of certain molecules to visualize specific structures within biological samples, including those in plant pathology. Unlike conventional light microscopy, FM uses specific fluorescent dyes or fluorophores that emit light of a specific wavelength when excited by a particular wavelength of light. This method allows for more targeted and precise visualization of structures or components within plant tissues and pathogens (Hickey *et al.*, 2021). FM offers several significant advantages and applications in plant pathology. It allows for the specific labeling and visualization of targeted components within plant cells, such as DNA, proteins, or organelles, through the use of specialized fluorescent dyes. This selective approach helps researchers study particular structures within both plant pathogens and host tissues. The technique is particularly valuable for examining the interactions between pathogens and plants, enabling the observation of pathogen invasion into plant cells or tissues and providing insights into infection mechanisms.

FM also supports subcellular imaging, revealing detailed views of organelles and cellular structures within both plant and pathogen cells. This capability allows for the study of specific components' behavior and movement, shedding light on pathogen biology and host-pathogen interactions. Additionally, by using different fluorescent labels, researchers can differentiate between various pathogen strains or species, facilitating precise identification and classification, which is essential for effective disease management (Singh and Kenney 2024). The technique's live cell imaging capabilities enable real-time observation of pathogen-host interactions, offering valuable insights into the progression of infections. Furthermore, FM can be paired with analytical tools for quantitative analysis, allowing researchers to measure pathogen spread, gene expression, and other biological processes.

Fluorescence method has been used to detect bacteria in seeds (Franken 1992; Van Vuurde and Bovenkamp 1995). FM can also be used to study the viability of bacterial cells. Fluorescent dyes such as propidium iodide and SYTO9 can be used for this purpose. SYTO9 can permeate the membranes and thus can identify all the bacterial population, while propidium iodide is membrane-impermeable. It has been used to detect onion crop infection by a fungus *Botrytis cinerea* (Dewey, 1996). It has also been combined with other techniques such as FISH *for Solanum dulcamara* detection, which causes crown rot in potatoes (Wullings, 1998). While FM offers exceptional benefits in visualizing specific structures and interactions, it is important to note that sample preparation and the selection of appropriate fluorescent markers are critical for successful imaging. Also, like other optical microscopy techniques, it is limited by the resolution of light, although it offers significant improvement in specificity and detailed visualization compared to conventional LM in plant pathology research. The disadvantages of this technique are: 1) it does not discriminate between live and dead cells; 2) it depends on the quality of the antiserum; 3) the examination of immunofluorescence (IF) slides is time consuming and should be performed by experienced technicians.

2.2.3 Confocal Microscopy

Confocal microscopy (CM) provides 3D images of pathogens within plant tissues. It allows for a detailed look at the interaction between the pathogen and the plant, and it is particularly useful for studying infection mechanisms. CM is an advanced imaging technique used extensively in plant pathology and biological research. It overcomes the limitations of traditional LM by utilizing a focused laser light source and a pinhole system to capture high-resolution, 3D images of biological specimens. In plant pathology, CM provides several important advantages and applications, particularly for detailed, high-resolution imaging. It enables researchers to visualize the 3D structure of plant tissues and pathogens, allowing for a comprehensive examination of cells, tissue organization, and pathogen interactions within the plant. The technique's optical sectioning capability, achieved by capturing images at precise focal planes, is essential for studying specific cell layers and tissue sections without interference from surrounding areas, providing a clear view of intricate plant structures and infections. CM is especially valuable for visualizing pathogen infections in detail, allowing researchers to track the spread, localization, and impact of pathogens within plant tissues, as well as observe plant cellular responses to invasion (Elliott, A. D. 2020).

Additionally, CM supports live cell imaging, making it possible to observe dynamic processes in real time, such as cellular changes, infection progression, and host-pathogen interactions. With fluorescence labeling, researchers can selectively highlight and analyze specific structures or components within plant tissues and pathogens, enhancing the precision of their observations. The technique also facilitates quantitative analysis, enabling measurements of fluorescent signal intensity or structure volume, and provides image stacks for 3D reconstruction, which aids in understanding complex biological structures. Beyond pathology, CM is widely used in plant physiology and developmental research, supporting studies on processes such as root growth, leaf anatomy, and other aspects of plant biology.

Nelson *et al.,* 2023 set out a study to develop a straightforward confocal staining and imaging technique that yields consistent results between samples, with improved visualization for both the plant and fungal pathogen using contrasting fluorescent stains that eliminate the need for gene transformation. As proof of concept, the non-fluorescently transfected foliar pathogens *Parastagonospora nodorum* (septoria nodorum blotch), *Pyrenophora teres f. teres* (net form net blotch), and *Cercospora beticola* (Cercospora leaf spot) of wheat, barley, and sugar beet, respectively, were used to show the efficacy of this technique. The staining technique was subsequently coupled with machine learning to analyze fungal volume, an indicator of pathogen fitness in plant, and nuclear breakdown, a classical early indicator of programmed cell death (PCD). Machine learning volume analysis propels fungal volume detection methods forward, as other methods, like qPCR, are laborious and only serve as a proxy to esti mate the fungal biomass, while this method shows the actual estimate of fungal volume (Mapuranga *et al.*, 2022). This technique is simple, robust, durable, and consistent, and can be used to characterize model and non-model plant-fungal interactions without transformation, making it useful to anyone using CM to characterize plant-fungal interactions.

Recently (Brain 2022) utilized advanced imaging techniques (CM), it revealed asymmetries in the spores of *Colletotrichum graminicola*, a significant corn pathogen. The crucial observation that these spores could only attach to their new host on one face had signified a pivotal discovery. This attachment becomes a prerequisite for initiating the disease cycle in corn, making it an essential target for disrupting fungal diseases in this agricultural context.

While CM is a powerful tool, it requires specialized expertise for operation and data analysis. Additionally, the equipment is costly and may not be readily available in all research settings. However, its ability to provide detailed 3D imaging and live observations makes it an indispensable technique in plant pathology and broader biological research.

2.2.4 Phase Contrast Microscopy

It is beneficial for observing live and unstained samples. This method enhances the contrast of transparent or nearly transparent specimens, making it easier to visualize certain pathogens. Phase contrast microscopy is an optical microscopy technique that enhances the contrast of transparent or nearly transparent specimens without the need for staining. It is particularly useful in visualizing live, unstained biological samples, including those relevant to plant pathology, where observation of structures without damaging or altering the specimen's natural state is desired.

In plant pathology, phase contrast microscopy provides numerous advantages for examining live, unstained specimens, allowing researchers to observe living cells, plant tissues, and pathogens while preserving their natural structures. This technique is particularly valuable for enhancing contrast in transparent or nearly transparent samples, commonly found in plant tissues and certain pathogens, enabling clear visualization of different structures. It facilitates the observation of cellular components such as cell walls, vacuoles, chloroplasts, and nuclei, as well as the structures of diverse plant pathogens, aiding in the study of their morphology and organization. Additionally, phase contrast microscopy allows for real-time observation of dynamic biological processes, such as cell division, organelle movement, and changes in pathogen morphology, offering insights into the behavior and interactions of these entities. Widely utilized in education and routine laboratory observation, this method introduces students to microscopy and serves as a practical tool for initial sample analysis in plant pathology research. Moreover, it is a fast and non-destructive imaging technique that avoids time-intensive sample preparation and is relatively cost-effective compared to advanced microscopy methods, making it an accessible and efficient option for many laboratories (Franco-Duarte *et al.*, 2019).

While phase contrast microscopy offers benefits in observing live and unstained samples, it is important to note that it has limitations in terms of resolution compared to more advanced techniques like confocal or EM. However, its ability to observe live and natural samples without staining or fixing makes it a valuable tool for initial observations and studying biological samples in their near-natural state within the field of plant pathology.

2.2.5 Electron Microscopy (EM)

Transmission electron microscopy (TEM) and scanning electron microscopy (SEM) provide extremely high-resolution images. These can reveal fine details of pathogen structures, down to the level of subcellular components, aiding in accurate identification. (Gooday *et al.*, 1988). These techniques have been instrumental in elucidating the morphology and characteristics of various plant pathogens.

In an electron microscope a beam of electrons makes the surface or content of a miniscule object visible. The use of electrons instead of photons, as used in LM, results in a higher resolution and magnification. EM is, for example, used to study micro-organisms, such as viruses, for morphological characteristics.

EM, with a resolution at least 2000 times higher than light microscopy, enables the visualization of extremely small objects and is applicable to both biological and inorganic specimens. In plant pathology, it is particularly useful for analyzing virus particle structures, allowing rapid detection of viruses in plants even before other diagnostic methods yield definitive results. This high-resolution technique plays a significant role across medical, animal, and plant sciences for the discovery, characterization, and description of unidentified viruses. Additionally, the shape of virus particles observed under EM can provide insights into the viral group to which they belong, supporting more precise identification and study (Zeltins, 2013).

The investigation done by Su jung in 2005 focused on hydroponically cultivated ivyleaf morning glory (*Ipomoea hederacea*) seedlings that were subjected to inoculation with deleterious rhizobacteria (DRB), with specific attention given to the root colonization process using scanning and transmission EM. Two DRB strains, *Bradyrhizobium japonicum* isolate GD3 and *Pseudomonas putida,* isolate GD4, renowned for their high production of indole-3-acetic acid (IAA), were compared with a plant growth-promoting rhizobacterium (PGPR), Bacillus megaterium isolate GP4. SEM unveiled a consistent distribution of isolates GP4 and GD4 on the root surface, whereas isolate GD3 exhibited a deeper localization within the surface furrows of the roots. TEM revealed notable alterations in root cells, including vesiculation, partial cell wall degradation, and cytoplasm disorganization. The population density of isolate GD4 on the root surface proved to be approximately 10 and 100 times greater than GP4 and GD3, respectively. Following seven days of growth, seedlings inoculated with GD3 and GD4 exhibited a significant inhibition of root elongation by approximately 26% and 90%, respectively, in comparison to the control. The findings suggested that the inhibition of morning glory root growth induced by isolate GD3 could be linked to the elevated production of IAA, although other phytotoxins likely contributed to the suppression of root elongation in morning glory inoculated with isolate GD4. The study proposes that rhizobacteria with the capacity to impede morning glory growth may serve as effective biological control agents, supplementing traditional herbicide weed management in crops where morning glory proves challenging to control.

While electron microscopes present exceptional capabilities in terms of resolution and magnification, they are associated with certain drawbacks. It is important to note that despite these challenges, electron microscopes remain indispensable tools for researchers across various disciplines, providing invaluable insights into the nanoscale structure of materials and biological specimens. Ongoing technological advancements aim to address some of these limitations, ensuring the continued evolution and utility of EM in scientific research.

CONCLUSION

In conclusion, microscopy plays a pivotal role in the detection and identification of plant pathogens, offering a spectrum of techniques that contribute to our understanding of pathogen biology and the development of effective disease management strategies. In short, the microscopy is indispensable for plant disease detection, offering a wealth of information that ranges from early identification to in-depth insights into the dynamics of host-pathogen interactions. Its role extends beyond diagnosis to encompass research, education, and the development of effective strategies for disease management in agriculture. As technology advances, microscopy continues to be a cornerstone in the ongoing efforts to safeguard global food security by mitigating the impact of plant diseases.

REFERENCES

Agrios, G. N. (2005). *Plant Pathology* (5th ed.). Academic Press.

Brain, S. (2022). Understanding the environmental microbiome using confocal microscopy. *States News Service*, 29 Nov., p. NA. *Gale Academic OneFile Select*, link.gale.com/apps/doc/A728257568/EAIM?u=anon~89310858 & sid=sitemap & xid=c8526c49.

de Oliveira Tostes, G., de Paula Araujo, J. S., Gomes Farias, A. R., Rossinol Frade, D. Â., & Olivares. F. L. (2014). Detection and cellular localization of *Xanthomonas campestris pv. viticola* in seeds of commercial 'Red Globe' grapes. *Tropical Plant Pathology*, **39**(2): 134–140.

Elliott, A. D. (2020). Confocal microscopy: principles and modern practices. *Current Protocols in Cytometry*, *92*(1), e68.

Fang, Y., & Ramaraja P. Ramasamy. (2015). Current and prospective methods for plant disease detection. *Biosensors,* **5**(3): 537–561

FAO. (2021). *The State of Food and Agriculture 2021: Making Agrifood Systems More Resilient to Shocks and Stresses*. The State of Food and Agriculture (SOFA); FAO: Rome, Italy; ISBN 978-92-5-134329-6.

Franken, A. A. J. M. (1992). Comparison of immunofluorescence microscopy and dilution-plating for the detection of *Xanthomonas campestris pv. campestris* in crucifer seeds. *Netherlands Journal of Plant Pathology*, **98**: 169–178.

Gooday, G. W., Mims, C. W., & Dickman, M. (1988). *Ultrastructural Pathology of Fungal Infections*. Cambridge University Press.

Gopinath, R., Purath Deesma, K., Purayil Soumya, V., Mohanan, C., & Manimakeli, R. (2016). Application of light microscopic staining techniques for the detection of phytoplasmas in yellow leaf disease affected arecanut palms in India. *Phytopathogenic Mollicutes,* **6**(2): 77.

Gudkov, S. V., Matveeva, T. A., Sarimov, R. M., Simakin, A. V., Stepanova, E. V., Moskovskiy, M. N., Dorokhov, A.S., & Izmailov, A. Y. (2023). Optical methods for the detection of plant pathogens and diseases. *Agriengineering*, **5**(4), 1789–1812.

Gullino, M. L., & Bonants, P. J. M. (2014). (Eds.). *Detection and Diagnostics of Plant Pathogens*. Springer: Dordrecht, The Netherlands; ISBN 978-94-017-9019-2.

Holgate, H., & Webb, J. (2003). MICROSCOPY | Light Microscopy and Histochemical Methods, Editor(s): Benjamin Caballero, *Encyclopedia of Food Sciences and Nutrition* (Second Edition), Academic Press, pp. 3917–3922. ISBN 9780122270550,

Harakannanavar, S. S., Rudagi, J. M., Puranikmath, V. I., Siddiqua, A., & Pramodhini, R. (2022). Plant leaf disease detection using computer vision and machine learning algorithms. *Global Transitions Proceedings*, **3**: 305–310.

Hepler, P. K., & Gunning, B. E. S. (1998). Confocal fluorescence microscopy of plant cells. *Protoplasma,* **201**: 121–157.

Hickey, P. C., Swift, S. R., Roca, M. G., & Read, N. D. (2004). Live-cell imaging of filamentous fungi using vital fluorescent dyes and confocal microscopy. *Methods of Microbiology*, **34**: 63–87.

Hickey, S. M., Ung, B., Bader, C., Brooks, R., Lazniewska, J., Johnson, I. R., Sorvina, A., Logan, J., Martini, C., Moore, C.R. and Karageorgos, L. (2021). Fluorescence microscopy—an outline of hardware, biological handling, and fluorophore considerations. *Cells*, *11*(1), 35.

Kausche, G. A., Pfankuch, E., & Ruska, H. (1939). Die Sichtbarmachung von pflanzlichem virus im Übermikroskop. *Naturwissenschaften*, **27**: 292–299.

Kim, S.-J., & Kremer, R. J. (2005). Scanning and transmission electron microscopy of root colonization of morningglory (*Ipomoea spp.*) seedlings by rhizobacteria. S*ymbiosis*, **39**: 117–124.

Lavanya, A., Sowmya, S. V., Rao, R. S., Augustine, D., Haragannavar, V. C., & Nambiar, S. (2017). Troubleshooters in light microscopy. *World Journal of Dentistry*, *8*(6): 511–518.

Leeuwenhoek, A. V. (1673). A specimen of some observations made by a microscope, contrived by M. Leewenhoeck in Holland, lately communicated by Dr. Regnerus de Graaf. *Philosophical Transactions of the Royal Society,* **8**: 6037–6038.

Liu, J., & Wang, X. (2021). Plant diseases and pests detection based on deep learning: A Review. *Plant Methods*, 17–22.

Mapuranga, J., Chang, J., Zhang, L., Zhang, N., & Yang, W. (2022). Fungal secondary metabolites and small RNAs enhance pathogenicity during plant-fungal pathogen interactions. *Journal of Fungi*, *9*(1), 4.

Mersha, Z., Birru, G., & Hau, B. (2023). Light and electron microscope studies elucidating mechanism of tomato leaf infection by Pseudocercospora fuligena. *Plant Pathology Journal*, **39**(2): 181–190.

Peddie, C.J., Genoud, C., Kreshuk, A., Meechan, K., Micheva, K.D., Narayan, K., Pape, C., Parton, R.G., Schieber, N.L., Schwab, Y., and Titze, B., (2022). Volume electron microscopy. *Nature Reviews Methods Primers*, *2*(1), 51.

Rumpf, T., Mahlein, A.-K., Steiner, U., Oerke, E.-C., Dehne, H.-W., & L. Plümer. (2010). Early detection and classification of plant disease with support vector machine based on hyperspectral reflectance. *Computer Electronics in Agriculture*, **74**: 91–99.

Sankaran, S., Mishra, A., Ehsani, R., & Davis, C., (2010). A review of advance techniques for detecting plant diseases. *Computer Electronics in Agriculture*, **72**(1): 1–13.

Savary S., Ficke A., Aubertot J., & Hollier C. (2021). Crop losses due to diseases and their implications for global food production losses and food security. *Food Security,* **4**: 519–537.

Schena, L., Hughes, K. T., & Kahl, G. (2008). *Advances in Plant Pathology: Vol. 6. Plant-Microbe Interactions* (1st ed.). Springer.

Singh, M. K., & Kenney, L. J. (2024). Visualizing the invisible: novel approaches to visualizing bacterial proteins and host-pathogen interactions. *Frontiers in Bioengineering and Biotechnology*, *12*, 1334503.

Stelzer, E. H. K. (2015). Light sheet fluorescence microscopy for quantitative biology. *Nature Methods.* **12**(1): 23–26.

Sumi, K., Rao, G. P., Abeysinghe, S., Tiwari, A. K., Manimekalai, R., Hegde, V., & Babu, M. K. (2023). Up-to-date information of phytoplasma diseases associated with palm species in Asian countries. In *Phytoplasma Diseases of Major Crops, Trees, and Weeds* (pp. 141–166).

Talbot, N. J., McCafferty, H. R. K., & Ma, M. (1993). Molecular and cellular biology of pathogenicity in fungi. In H. D. Osiewacz (Ed.), *Molecular Biology of Fungal Development* (pp. 33–58). CRC Press.

Taylor, J. D. (1978). Detection of seed-borne bacteria. In: Report on the Sixteenth International workshop on Seed Pathology, pp. 28–29.

Van Vuurde, J. W. L., & Bovenkamp, G. W. (1995). Detection *of Pseudomonas syringe pv. phaseolicola* in bean. In Saettler, A. W., Schaad, N. W., and Roth, D. A. (eds). *Detection of Bacteria in Seed and Other Planting Material.* St. Paul, Minnesota: USA, pp. 30–40.

Zeltins, A. (2013). Construction and characterization of virus-like particles: A review. *Molecular Biotechnology*, *53*(1), 92–107.

3 Immunodiagnostics

From Traditional to Recent Advances in Plant Virus Detection

Aniket Angira, Tripti Singhal, Manoj Kumar Singh, Sunny Dhir, and Vanita Chandel

3.1 INTRODUCTION

Plant viruses are renowned for inflicting substantial economic losses on a wide range of agricultural and horticultural crops globally. The estimated annual financial impact of plant viruses amounts to several billion dollars, posing a significant threat to the overall productivity and food security of crops worldwide. Conventional serological techniques hinge on the interplay between antibodies and antigens, wherein each antibody demonstrates a distinct affinity for the particular antigen against which it is generated.

Amongst the different serological methods used to diagnose plant viruses are conventional methods under which few assays fall such as chloroplast agglutination, gel diffusion, tube precipitin, and microprecipitin and these assays involve visually assessing specific precipitates resulting from the interaction between viruses and antibodies, whether in liquid solutions or agar gels. These approaches usually demand substantial quantities of reagents and frequently face issues related to sensitivity, specificity, and time efficiency in the accurate detection of viruses.

In response, researchers have turned to modern serological techniques that utilise antibodies to detect specific viral proteins or nucleic acids. Enzyme-linked immunosorbent assays (ELISAs), lateral flow assays, and immunofluorescence assays (IFAs) have become essential tools in plant virus detection. The key advantage of modern serological techniques lies in their high sensitivity and specificity. They can detect low virus concentrations and differentiate closely related virus strains, crucial for early disease management and quarantine measures. Furthermore, these methods are amenable to high-throughput screening, making them suitable for large-scale virus monitoring and eradication programs.

Additionally, the development of monoclonal antibodies and recombinant antibodies (nanobodies) has enhanced the precision of serological assays. These antibodies can be tailored to target specific viral epitopes, ensuring minimal cross-reactivity and false-positive results. Molecular serological techniques, such as PCR-based serology and serological microarrays, offer even greater sensitivity and

 DOI: 10.1201/9781003474036-3

multiplexing capabilities. These methods enable the simultaneous detection of multiple viral species, facilitating comprehensive disease diagnosis and management.

Overall, modern serological techniques have transformed plant virus detection, providing agriculturists, researchers, and policymakers with powerful tools to mitigate the devastating effects of viral infections on crops. These techniques offer rapid and accurate results, enhance our understanding of virus epidemiology, and inform timely and effective disease control strategies. As plant viruses continue to threaten global food production, the adoption of these innovative serological techniques is vital for sustainable agriculture and food security.

3.2 TRADITIONAL SEROLOGICAL METHODS

3.2.1 Precipitin Tests

Plant viruses exhibit polyvalency, meaning that each viral particle can interact with multiple antibody protein molecules, resulting in the formation of a lattice-like precipitate. When virus particles interact with their corresponding antibodies, this leads to the creation of precipitates due to the reaction between multivalent virus antigens and antibodies. Numerous virus antigens interconnect to establish a lattice structure, becoming insoluble once the complex reaches a certain size. The ratio of reactant concentrations plays a crucial role in precipitate formation, with an extreme excess of one reactant preventing precipitation. In cases involving elongated viruses, the precipitate zones tend to be broader, while isometric viruses yield narrower zones. The results of these tests are affected by multiple factors, including electrolyte levels, pH, temperature, and the presence of foreign substances. Antigens are typically produced in a mild buffer or buffered saline, while antisera are typically diluted in saline. These assays use either pure viral preparations or cleared plant extracts. There are many ways to do precipitin tests to find plant viruses, but the most widely used ones include tube precipitin, micro-precipitin, chloroplast agglutination, and agar double diffusion tests.

3.2.1.1 Tube precipitin

The tube precipitin test is primarily employed for assessing the potency of an antiserum, and it is only occasionally used for making comparative assessments. This method utilises slender glass tubes containing antiserum and antigen with a two-fold dilution. The tubes are partially submerged in a water bath, which facilitates convection currents, expediting the precipitation process. Observations of precipitation are made at regular intervals. Isometric viruses yield a finely granular precipitate, while elongated viruses generate a more flocculent precipitate (Gibbs, A., Harrison, 1976; Hill SA, 1984; Walkey DGA, 1985). The titre is determined based on the highest dilution at which precipitation becomes evident. It is important to note that the titre value can vary depending on the method employed for its assessment, thus warranting the inclusion of the specific method used. Moreover, titres do not offer absolute measurements of antigen or antibody concentrations; a greater quantity of antibodies is required to precipitate small isometric viruses compared to elongated rod-shaped viruses.

3.2.1.2 Micro-precipitin

A refinement of the tube precipitin method, the micro-precipitin test offers cost-effective antiserum and antigen utilisation without sacrificing sensitivity. Antigen and antiserum are mixed in varying dilutions, and the precipitate's quality and rate of production are evaluated. Because only a few drops of reactants are mixed thoroughly, the amounts involved are kept to a minimum. It is possible to set up a grid titration on a Petri dish that is comparable to the one used in tube precipitation. After the combined reactants are placed in the grid as droplets, a layer of mineral oil is applied to keep them from drying out during the incubation time. The results are examined on a dark background with a stereo microscope or a dark-field microscope (Hill SA, 1984; Van Slogteren DHM, 1955; Walkey DGA, 1985).

3.2.2 Chloroplast Agglutination

The chloroplast agglutination test, utilised in the diagnosis and identification of plant viruses, is uncomplicated but lacks high sensitivity. It is most efficient in cases involving viruses with elongated particles, such as tobamo-, carla-, and potexviruses, commonly present in high concentrations within the plant (Van Slogteren DHM, 1955). In this procedure, small amounts of antiserum specific to the target virus are mixed with droplets of crude plant extract. With a sufficient quantity of antibodies for extensive chloroplast linkage, these organelles aggregate. This aggregation is visually discernible, even to the naked eye, as chloroplasts clump within the mixture (Dijkstra, J., de Jager, 1998). It is significant to remember that this type of precipitation is called agglutination because bigger particles, like cell organelles, are involved in the reaction. Reactions with molecular antigens, such viruses, are explicitly referred to as "precipitation" reactions. A further, more accurate variation of the micro-precipitin test is known as the "latex agglutination test." This method uses latex particles that have been sensitised with a particular antiserum to increase sensitivity. Visible agglutination happens when latex particles sensitised with a particular antiserum are combined with a sample containing a particular antigen (viral particles). The antigen-antibody combination is amplified by latex particles. This is a simple, low-cost solution that does not require any specialised equipment. For the purpose of identifying viruses that infect potatoes and other horticultural crops, kits based on latex serological agglutination tests are offered (Gella FJ, Serra J, 1991; Talley J, Warren FHJB, Torrance L, 1980).

3.2.3 Agar Double Diffusion (Ouchterlony) Test

This assay uses a gel media instead of a liquid to facilitate the interaction between antibodies and antigens. A precipitation line where the antigen and antibody converge in the ideal proportions appears when they diffuse over the gel. Agar gels are used for the test, which is done on glass slides or Petri plates (Gibbs, A., Harrison, 1976). Selecting the right pH and salt levels for each system is essential, as is adding a preservative such sodium azide. The antigen is positioned in wells around a central well holding antiserum, regardless of whether it is a crude sap, clarified extract, or purified virus. SDS is used to break down virions into antigenically active diffusible pieces for some virus species that include rod-shaped particles. It can also be added to the agar gel medium.

The initial precipitin line's location primarily depends on the diffusibility and concentration of the reactants. When the reactants diffuse to the reaction site at equal concentrations, the system is termed balanced, a prerequisite for sharp, stationary precipitin lines. Under these balanced conditions, characteristic patterns of precipitin lines form, enabling conclusions about various antigenic relationships. Conversely, under unbalanced conditions, the excess reactant moves beyond the initial precipitin line, leading to re-precipitation on the side with lower concentration. Under unbalanced conditions, observations such as band splitting and the formation of multiple bands may occur (Hill SA, 1984; Noordam D, 1975; Ouchterlony O, 1968; Walkey DGA, 1985).

3.3 MODERN SEROLOGICAL TEST

3.3.1 Enzyme-linked immunosorbent assay (ELISA)

Enzyme-Linked Immunosorbent Assay (ELISA) is a powerful, quantitative, and widely used immunological technique for the detection of plant viruses. It involves the assessment of antigen-antibody reactions using enzyme-based measurements. Enzymes like alkaline phosphatase or horseradish peroxidase facilitate the conversion of colourless substrates into coloured ones, signalling the presence of antigen-antibody binding. The antigen of interest is immobilised on a polystyrene plate containing 96 micro wells, and unwanted components are meticulously removed at each stage, leaving only the specific reactants. Ultimately, these reactants are treated with an enzyme and quantified using an ELISA reader or spectrophotometer. ELISA can be broadly categorised into two main types, direct and indirect, based on the labelling of the antibody (Voller et al., 1976). Direct ELISA, known for its high specificity, involves the primary antibody directly binding to the test antigen. In contrast, indirect ELISA employs a secondary antibody that binds to the primary antibody, which has already attached to the test antigen (shown in Figure 3.1). While direct ELISA is highly specific, indirect ELISA is preferred for its broader applicability in detecting a wide range of viruses.

There are several types of ELISA assays that play a pivotal role in plant virology and agriculture, enabling the rapid and accurate identification of viral infections in plants, aiding in the management of crop diseases, including direct, indirect, double-antibody sandwich, triple antibody sandwich, Protein A sandwich, Dip stick, tissue-printing and Immuno-precipitation ELISA (IP-ELISA). Here's an overview of these ELISA types.

3.3.1.1 Direct ELISA

This sensitive and specific technique involves several key steps. First, plant samples suspected of harbouring a virus are homogenised to create an extract. Microtiter plates are then coated with specific antibodies known to target the virus of interest. After blocking to prevent non-specific binding, the plant extract is added to the wells.

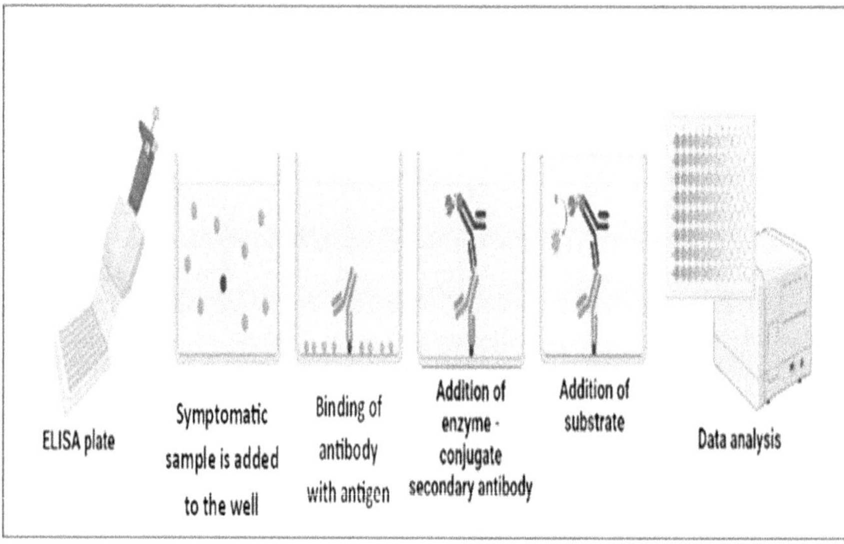

FIGURE 3.1 Diagrammatic representation of indirect ELISA for plant virus detection.

TABLE 3.1
Plant viruses detected using direct ELISA

S.No.	Viruses	References
1	Banana Bract Mosaic Virus	(Chetanchidambar N Mangalore, D Sri Hari, B V K. Bhagavan, SalomiSuneeta, K Uma Krishna, 2020)
2	Cucumber Mosaic Virus	
3	Triticum mosaic virus	(Tatineni et al., 2013)
4	Tobacco streak virus	(S. Rageshwari, P. Renukadevi, V.G. Malathi, 2017)
5	Potato viruses Y and X	(Kapoor et al., 2014)
6	Soybean mosaic virus	(A. Ahangaran, Gh. Mosahebi Mohammadi, M. Koohi Habibi, S. Khezri, 2009)

If the virus is present, it binds to the immobilised antibodies. Following an incubation period and a series of washes to remove unbound materials, a secondary antibody linked to an enzyme is introduced. This secondary antibody recognises the primary antibody bound to the virus and, when incubated, catalyses a colour-producing reaction. The intensity of the resulting colour change is directly proportional to the amount of the virus in the sample. By measuring the optical density, the concentration of the virus can be quantified (Table 3.1).

3.3.1.2 Indirect ELISA

In this method, a microtiter plate is used as a solid support matrix to immobilise the target antigen. After incubation and washing steps to remove unbound

TABLE 3.2
Plant viruses detected using indirect ELISA

S.No.	Viruses	References
1	Tomato mottle mosaic virus	(A. Ahangaran, Gh. Mosahebi Mohammadi, M. Koohi Habibi, S. Khezri, 2009)
2	Sorghum mosaic virus	(H. Chen et al., 2020)
3	Soybean mosaic virus	(Ren et al., 2022)
4	Cucumber mosaic virus	(Wagih et al., 2021)

TABLE 3.3
Plant viruses detected using DAS-ELISA

S.No.	Viruses	References
1	Potato virus Y, Potato virus A, Potato virus M, and Potato leafroll virus	(Halabi et al., 2021)
2	Bean common mosaic virus	(ÇULAL KILIÇ & ÇAT, 2022)
3	Pepino mosaic virus	(He et al., 2020)
4	Sugarcane bacilliform virus	(Krishna et al., 2023)
5	Prunus necrotic ring spot virus	(Chandel et al., 2011)

materials, a secondary antibody, labelled with an enzyme like horseradish peroxidase or alkaline phosphatase, is introduced. This secondary antibody specifically recognises the primary antibody and binds to it, effectively amplifying the signal. The enzyme-linked secondary antibody then catalyses a colour-producing reaction, with the intensity of the colour directly proportional to the concentration of the primary antibodies, which indirectly reflects the presence of the target antigen. The use of ELISA readers or spectrophotometers allows for precise quantification (Table 3.2).

3.3.1.3 Double-Antibody Sandwich ELISA (DAS-ELISA)

It is a sophisticated variant of the traditional ELISA technique. In DAS-ELISA, a microtiter plate is coated with a primary antibody that binds to the target antigen. After antigen binding, a second antibody, which is also specific to the antigen but different from the primary antibody, is introduced. This secondary antibody is conjugated to an enzyme, such as horseradish peroxidase or alkaline phosphatase. The use of two antibodies that recognise different epitopes on the antigen increases specificity. The enzyme-linked secondary antibody catalyses a color-producing reaction upon substrate addition, and the intensity of the color is proportional to the amount of antigen present. DAS-ELISA is particularly useful when high specificity is required, as it significantly reduces the chances of false-positive results and cross-reactivity, making it a preferred choice in various fields, including disease diagnostics and pathogen detection (Table 3.3).

3.3.1.4 Triple Antibody Sandwich ELISA (TAS-ELISA)

TAS-ELISA is an advanced modification of the traditional enzyme-linked immunosorbent assay (ELISA). In TAS-ELISA, three antibodies are employed, which enhances the specificity and sensitivity of the test. The first antibody binds to the target antigen, while a second labelled antibody recognises a different epitope on the antigen. Finally, a third antibody, also labelled, attaches to the second antibody. This triple antibody structure minimises the risk of false-positives and cross-reactivity, making it highly suitable for precise antigen detection. TAS-ELISA is particularly valuable when the utmost accuracy is required, and it finds applications in various fields, including disease diagnostics and pathogen detection (Table 3.4).

3.3.1.5 Protein A Sandwich ELISA (PAS-ELISA)

It is a specialised variation of the enzyme-linked immunosorbent assay used for the highly specific detection of target proteins or antigens. In PAS-ELISA, a microtiter plate is coated with a primary antibody specific to the target antigen. Subsequently, a sample containing the antigen is added, allowing for binding to the immobilised primary antibody. Then, a secondary antibody, conjugated with a specific protein A molecule, is introduced. Protein A is known for its strong affinity for the constant (Fc) region of antibodies, making it an effective linker in the assay. After washing to remove unbound materials, the enzyme-linked secondary antibody is used for signal amplification. The enzyme catalyses a colour-producing reaction when a substrate is added, and the intensity of the colour directly corresponds to the antigen's concentration. PAS-ELISA is particularly valuable for its high specificity and is often employed in biomedical research and diagnostics for detecting specific antibodies or antigens in a sample (Table 3.5).

3.3.1.6 Immuno-Precipitation ELISA (IP-ELISA)

IP-ELISA is known for its heightened sensitivity, making it particularly suitable for the detection of low concentrations of target antigens or proteins, even in complex biological samples, such as cell lysates or tissue extracts, to study interactions and detect specific antigens in their native environments. In this method the infected plant extract and antiserum are combined in equal proportions and then subjected to centrifugation. The resulting pellet is subsequently reconstituted in an ELISA extraction buffer, following the standard protocol for conventional indirect ELISA. The "precipitation" step in IP-ELISA, where immune complexes are physically removed

TABLE 3.4
Plant viruses detected using TAS-ELISA

S.No.	Viruses	References
1	Sri Lankan cassava mosaic virus	(Charoenvilaisiri et al., 2021)
2	Pepper mild mottle virus	(Phatsaman et al., 2020)
3	Melon necrotic spot virus	(Miras et al., 2020)
4	Tomato yellow leaf curl Thailand virus	(Seepiban et al., 2017)

TABLE 3.5
Plant viruses detected using PAS-ELISA

S.No.	Viruses	References
1	Sugarcane mosaic virus Maize chlorotic mottle virus	(Stewart et al., 2017)
2	Tomato Spotted Wilt Virus	(J. L. Sherwood, M. R. Sanborn, G. C. Keyser, 1988)
3	Cowpea aphid-borne mosaic virus	(Eni et al., 2013)
4	Yam mosaic virus	(A.O. Eni, J.D.A. Hughes, 2012)

from the sample, enhances specificity and minimises background noise. This method is applicable for the detection of viruses from various families, including Cowpea severe mosaic virus (CPSMV) and Squash mosaic virus (SQMV) (John, Alby, Deepa R. Chandran, n.d.).

3.3.1.7 Dip Stick ELISA

It is a rapid and simplified version of the traditional ELISA. In this method, a paper or membrane strip, resembling a dipstick, is used as the solid support instead of a microtiter plate. The dipstick is typically coated with specific antibodies or antigens that can capture target molecules, such as antigens or antibodies, from a test sample. When the dipstick is dipped into the sample, these target molecules bind to the immobilised antibodies or antigens, forming a visible line or signal on the dipstick. Dip Stick ELISA is known for its convenience, speed, and suitability for on-site testing in various fields, including medical diagnostics and environmental monitoring (Darshan, K., Amrurthalakshmi, M., 2021).

3.3.1.8 Biotin-Streptavidin ELISA

It is an enhanced variation of the traditional ELISA that leverages the strong and specific binding affinity between biotin and streptavidin. In this method, biotinylated antibodies or antigens are used, and they form a stable complex with streptavidin, an avidin-related protein. This biotin-streptavidin interaction significantly amplifies the sensitivity of the assay. When the biotinylated antibodies or antigens bind to streptavidin, the resulting complex enhances the detection signal, leading to a more sensitive and accurate quantification of target molecules, such as plant viral antigens. Biotin-Streptavidin ELISA is particularly valuable when ultra-sensitive detection is required in plant virus research and diagnostics, offering an improved and reliable assay platform (A. Konate & B. Neya, 1996; C. desbiez and H. lecoq, 1997).

3.4 IMMUNOBLOTTING:

This is a serological detection technique where the antigen-antibody complex is anchored onto a nitrocellulose membrane. By employing an enzyme-conjugate and a

TABLE 3.6
Plant viruses detected using DIBA

S.No.	Viruses	References
1	Maize chlorotic mottle virus	(C. Zhang et al., 2023)
2	Potato virus Y	(Zhao et al., 2023)
3	Fig mosaic emaravirus	(Shahmirzaie et al., 2019)
4	Southern rice black-streaked dwarf virus	(Z. Chen et al., 2012)

specific substrate, the presence of the complex is visualised through the emergence of coloured spots on the membrane. Various immunoblotting methods, such as DIBA, TIBA, and dipstick assays, fall under this category. These techniques are instrumental in the precise detection of specific antigens and antibodies, offering versatility in applications ranging from clinical diagnostics to research.

3.4.1 Dot Blots Immunobinding Assay (DIBA)

DIBA, an alternative to ELISA, substitutes the traditional polystyrene plate with a nitrocellulose membrane. To prepare the antigen, tissue samples are ground in Tris-buffered saline, often containing additional additives, and a small volume of the test sample is spotted onto the membrane. Bovine Serum Albumin (BSA) serves as the blocking agent. The membrane is then immersed in a diluted solution of antigen-specific crude antiserum at a temperature of 37°C for approximately 30 minutes. During this time, antibodies that bind with the antigen are captured. Subsequently, alkaline phosphatase-labelled antibodies are introduced, and the addition of the substrate NBT (nitro blue tetrazolium) leads to the conversion of the enzyme linked to IgG into a coloured substance (Table 3.6).

3.4.2 Tissue blot immunoassay (TBIA)

An expedited approach involves directly pressing cut tissues onto the membrane to transfer the antigen, and the subsequent steps follow a process akin to DIBA (Abd El-Aziz, 2019; Fegla, G. I., I. A. El-Samra, H. A. Younes, 2001; Mahmoud Hamdy Abd El-Aziz Hosny Aly Younes, 2019).

3.5 IMMUNOSORBENT ELECTRON MICROSCOPY (ISEM)

ISEM is a specialised technique that combines the principles of electron microscopy and immunology. In ISEM, specific antibodies are used to label and visualise target molecules or structures within a sample under the high-resolution scrutiny of an electron microscope. This method allows researchers to precisely locate and identify antigens or proteins within cells, tissues, or other biological samples. ISEM is particularly valuable for studying subcellular structures, viral particles, or specific molecules in biological specimens. It provides detailed visual information that can enhance our understanding of cellular processes, pathogen-host interactions, and various biological phenomena.

In 1973, Derrick pioneered the use of the ISEM technique to identify TMV and PVY. ISEM represents an amalgamation of electron microscopy and serology, merging the precision of serological assays with the imaging capabilities of an electron microscope. This method enables the selective capture of virus-coated particles on carbon-coated grids, followed by the addition of antibodies to the viral particles. Skaff and Carroll, in 1995, introduced a procedure for the purification of barley yellow streak mosaic virus in barley and wheat. They produced polyclonal antibodies from concentrated virus preparations, employing them in both ELISA and ISEM to detect viral infections in host plants, as demonstrated by (Darshan, K., Amrurthalakshmi, M., 2021).

3.6 FLOW CYTOMETRY

Like standard flow cytometry (FCM) staining techniques, viruses are normally extracted, stained, and then examined using a flow cytometer that can identify single viral particles as well as their characteristics within infected plant cells. The first plant virus to be identified using FCM was not discovered until 1996 (Iannelli et al., 1996), despite the fact that the initial report on the FCM analysis of plant material was among the very first FCM publications (Heller, 1973). Multiple target cells can be detected at once when different sizes or colours of antibody-coated beads are used (D'Hondt et al., 2011). For example, it is reported that three distinct plant viruses are detected using multiplex FCM with different diameters of latex beads (Iannelli et al., 1996, 1997). FCM can quantify virus-infected cells through antigen-antibodies interaction followed by specific probes labelling can be performed by fluorescein isothiocyanate (FITC), which is an organic fluorophore.

FCM enables the simultaneous detection of multiple viruses in a sample by using (i) specific viral antibodies tethered to different-sized latex particles or (ii) antibodies to separate viral antigens coupled to different fluorochromes. Because each antibody is specific to a different virus size particle, variations in the forward-scattered light signal indicate the presence of distinct viral antigens. As seen in Table 3.1, for instance, Iannelli et al. (1997) used 3-, 6-, and 10-μm-diameter latex particles to concurrently detect the potato, tomato, and cucumber mosaic viruses. Despite the fact that this technique was designed to identify plant viruses, it did not seem particularly useful or more sensitive than already exciting techniques. This method demonstrated the possibility that FCM can detect and applied to plant viruses that are more difficult to identify than CMV (Iannelli et al., 1996). Luminex created a microsphere-based flow cytometric technology that has been shown to be effective in concurrently detecting numerous viral genes in the tissues of lily plants (Lim et al., 2016). A lot of viruses can be found in a single test using this bead-based array detection technique, which is similar to the ELISA approaches (Boonham et al., 2014). Like multiplex real-time RT-PCR, this recently developed approach has been denoted to be more cost-effective, specific, and sensitive (Gong et al., 2023) (Table 3.7).

TABLE 3.7
Plant viruses analysed with flow cytometry (FCM) for detection and enumeration

S.No.	Viruses	Bead and label system	Matrix	Sensitivity detection limit (ml^{-1})	Reference
1	Cucumber mosaic virus Tomato mosaic virus Potato virus Y Plum pox potyvirus	Latex beads 1, 3 & 6 μm PE & FITC	Leaf sap	10pg	(Iannelli etal.,1996, 1997)
2	Potato leafroll virus Potato virus Y Potato virus X	Luminex polystyrene 5.3μm Alexa fluor 532 Luminex paramagnetic 6.5μm	Leaf sap	10x senstive than ELISA	(Bergervoet et al., 2008)
3	Cowpea chlorotic mottle virus	fluorescence-labelled	Leaf extract		(Xue et al., 2023)
	Lily symptomless virus Cucumber mosaic virus Lily mottle virus	Luminex-based liquid bead	Leaf extract		(Lim et al., 2016)

3.7 BIOSENSORS

Biosensors for the detection of viruses of plant play a vital role in agriculture, helping to monitor and control viral diseases that can have devastating effects on crop yields. When developing a biosensor for the identification of plant viruses, factors such as sensitivity, specificity, speed, cost-effectiveness, and ease of use should be considered. Furthermore, the biosensor design should be optimised to ensure reliable and accurate detection of a wide range of plant viruses (Dyussembayev et al., 2021). Multiple types of biosensors have been evolved for this purpose, utilizing different detection principles and technologies.

3.7.1 Immunological biosensors

These biosensors can enable label-free as well as sensitive options for fast qualitative and quantitative analysis of viruses. These biosensors use specific antibodies or antigens to detect the presence of viral particles. The binding of the virus to the corresponding antibody or antigen triggers a measurable signal, such as a colour change or an electrical signal, indicating the presence of the virus (J. Zhang & Zhao, 2019).

3.7.2 Resonant biosensors

Resonant biosensors are designed to function as the bioreceptor, here an antibody, is coupled with an acoustic wave transducer. The mass of the membrane changes when the analyte (or antigen) molecule binds. The transducer connected to the target molecule eventually encounters a shift in resonance frequency or wavelength, as a result of the ensuing mass change (Guo et al., 2018).

3.7.3 Electrochemical biosensors

Integrating enzyme immunoassay with electrochemical detection techniques has increased the sensitivity and selectivity of diagnosing plant infections. (Paternolli et al., 2004). Label-free technologies (impedimetric) and enzymatic label-based voltametric methods using carbon, gold, and mercury electrodes are the basis of plant virus detection, as shown in Table 3.2. The electrochemical impedance spectroscopy (EIS) used in impedimetric-based immunosensors examines impedance changes imposed on the biomolecular interaction and focuses on alterations in the electrode-solution interface such as protein immunocomplex formation and DNA hybridisation (Newman et al., 1986).

3.7.4 Optical biosensors

Optical biosensors measure a chemical's effect on a biological system using light. These biosensors are suited for detecting intracellular and intercellular physiological and biological factors in the microenvironment due to their small size. Once the biorecognition molecules are embedded on the sensor, they interact with the optics to generate a change in optical signal, which is then used to detect the immune response, as shown in Table 3.2.

3.7.5 Surface Plasmon Resonance Biosensors (SPR)

These sensors can be utilised for the identification of label-free and real-time plant viruses by monitoring the specific binding between antibodies and viral antigens. Refractive index variations in proximity to a sensor surface are detected, providing information about the binding kinetics and affinity between the antibody and the target viral antigen (Gutiérrez-Aguirre et al., 2014; Zeng et al., 2013). SPR can be employed to study the binding of viral proteins to plant host components, leading to a deeper knowledge of the molecular mechanisms associated with plant-virus interactions. SPR technology is instrumental in elucidating the intricacies of plant-virus relationships, ultimately contributing to improved strategies for the management of viral diseases in agriculture (Table 3.8) (Das et al., 2021; Razmi et al., 2019).

TABLE 3.8
Various antibody-based biosensing methods applied for plant virus detection

S.No.	Viruses	Detection	Assay format	Substrate	Sensitivity	References
1	Cucumber mosaic virus	Electrochemical/AuNPs tag Electrochemical/Enzyme label	Enzyme-based detection Voltametric	o-phenylenediamine (OPD) o-aminophenol (OAP)	0.5ng/ml	(Jiao et al., 2000)
	Plum pox virus Maize chlorotic mottle virus Prunus necrotic ringspot virus	Electrochemical/label-free	Electrochemical impedance spectroscopy (EIS)-based detection	Self-assembled monolayer (SAM) formation on gold electrodes	250 ng/ml 10 pg/ml	(Jarocka et al., 2011)
2	Potato virus X	Optical/AuNPs tag	Lateral flow immunoassays		2 ng/ml	(Drygin et al., 2012)
3	Chilli vein-banding mottle virus Melon yellow spot virus Watermelon silver mottle virus	Optical/fluorescent tag	Fluorescent approaches		35.3 ng/ml	(Charlermroj et al., 2013)
4	Odontoglossum ringspot virus Cymbidium mosaic virus	Optical/label free	Surface plasmon resonance (SPR) systems		42 pg/ml 48 pg/ml	(Lin et al., 2014)

3.8 IMMUNO-CAPTURE REVERSE TRANSCRIPTION POLYMERASE CHAIN REACTION (IC-RT-PCR)

This is a method that is applied in identifying viruses that do not require a nucleic acid extraction process. The targeted virus accumulates on a PCR tube that has been pre-coated with a particular antibody, assuring that the template carries only viral nucleic acid, as shown in Figure 3.2. Two primers (one for forward strand and one for reverse strand) were designed from high sequence diversity regions. First-generation IC-RT-PCR assays have been developed to detect a variety of fruit tree viruses that are economically significant (i.e., plum pox virus and apple chlorotic leaf spot virus) (Candresse et al., 1994; Wetzel et al., 1992) and also employed to confirm the presence of numerous other viruses, such as yam mosaic virus (YMV) and yam mild mosaic virus (YMMV) (Mumford & Seal, 1997), tomato (ToMV) and tobacco mosaic (TMV) (Jacobi et al., 1998), pepper mild mottle virus (PMMV), bean yellow mosaic virus (BYMV), cherry leafroll virus (CLRV),cucumber mosaic virus (CMV), citrus tristeza virus (CTV), grape fan leaf virus (GFLV), potato leaf roll virus (PLRV), and tomato spotted wilt virus (TSWV) along with CMV satellite RNA (Narayanasamy & Narayanasamy, 2011), grapevine fanleaf virus (GFLV) (Emmanuelle Vigne, Shahinez Garcia, Véronique Komar, 2018), apple stem pitting virus (ASPV) and apple mosaic virus (ApMV) (Dhir et al., 2021), plum pox virus-M (Teber et al., 2023), and sugarcane streak mosaic virus (Daugrois et al., 2023). IC-RT-PCR is more reliable assay for the simultaneous identification and differentiation of two viruses at once, which can not be done in ELISA. It is reported that IC-RT-PCR identified target virus at concentration of 10–100 fg/ml while, ELISA had detection sensitivity of 1 ng/ml (Jacobi et al., 1998). The technique is significant because

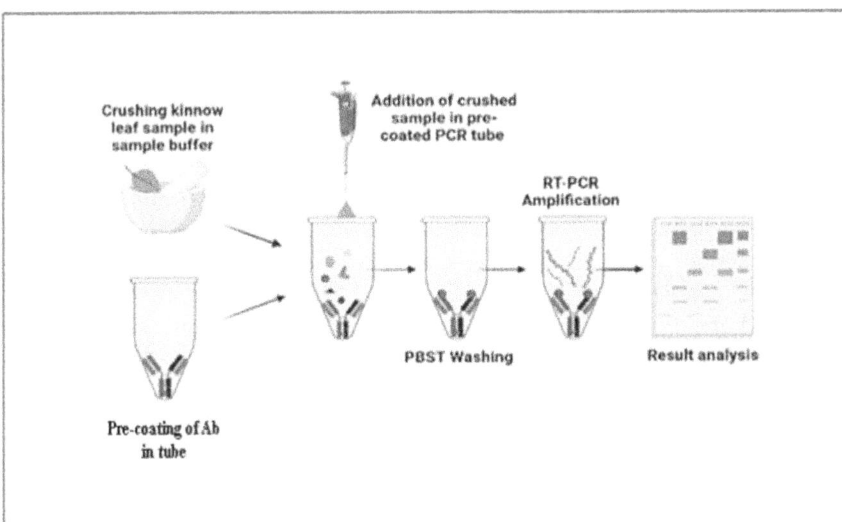

FIGURE 3.2 Diagrammatic representation of IC-RT-PCR technique for the identification of plant viruses.

it utilises the sensitivity and precision of molecular and immunological methods to identify plant viruses at very low concentration.

3.8.1 Multiplex immunocapture PCR (M-IC-PCR)

Multiplex (M)-PCR has been developed for plant viruses and enables the identification and characterisation of multiple viruses in a single PCR, followed by utilisation of microplate system for the colorimetric identification of PCR amplicons. According to (Sharman et al., 2000), the M-IC-PCR is at least 25 times more sensitive than the ELISA for the three banana viruses. This approach was effectively employed to confirm the existence of numerous other viruses such as banana streak viruses (BSV) (Javer-Higginson et al., 2014; Le Provost et al., 2006), beet mild yellowing virus (BMYV) and beet chlorosis virus (BChV) (Viganó & Stevens, 2007), potato virus Y (PVY) (Mallik et al., 2012) and dioscorea mosaic associated virus (DMaV), and yam asymptomatic virus 1 (YaV1) (Umber et al., 2020).

3.8.2 Immuno-capture Loop mediated isothermal amplification (IC-LAMP)

Immunocapture (IC) RT-LAMP (for RNA viruses)/IC-LAMP (for DNA viruses) are applicable for the detection of on-farm viruses as they coat PCR tubes with virus-specific polyclonal antibodies. In order to identify six sequences in the coat protein (CP) genes of the lily symptomless virus (LSV) and CMV, they designed two external, two internal, and two loop primer pairs, resulting in highly sensitive detection (Almasi, 2017, Y. Zhang et al., 2022). While IC-PCR is an expensive method that can only be used in well-equipped laboratories with trained personnel, but similar results can be obtained in IC-LAMP, and can be easily carried out in a water bath or temperature block without the need for a thermocycler or gel electrophoresis, making latter more sensitive than former (Dehabadi, 2013; Naganur et al., 2023). IC-RT-LAMP offered up to 100 times more sensitivity for the identification of plant viruses than IC-RT-PCR. Similarly, other stages of the procedure are easily applicable, with the exception of the primer designing process, which is quite delicate and complex. Interestingly, there are four key benefits: (1) no need for DNA extraction; (2) no need for expensive and complex instruments for amplification and detection; (3) no need for post-amplification treatment of the amplicons; and (4) a flexible, simple detection method that can be visualised with the unaided eye using a variety of visual dyes (Ahmadi et al., 2013). Unskilled labourers can easily perform this assay in the field in about an hour, and the study offers real-time on-site detection (Table 3.9) (Selvaraj et al., 2019).

3.8.3 Immunocapture Real Time PCR (IC-real-time PCR)

Immunocapture real-time PCR is a powerful molecular biology method that integrates immunology with real-time-PCR for the specific and quantitative detection of target viral nucleic acid sequences (Ling et al., 2007). The principle of this technique is to

TABLE 3.9
Plant viruses detected through IC-PCR and IC-LAMP

S.No.	Viruses	Technique	References
1	Tomato spotted wilt virus	IC-RT-LAMP	(Fukuta et al., 2004)
2	Kyuri green mottle mosaic virus	IC-RT-LAMP	(Fukuta et al., 2012)
3	Potato Leafroll virus	IC-RT-PCR and IC-RT-LAMP	(Almasi et al., 2012)
4	Tomato yellow leaf curl virus	IC-LAMP	(Dehabadi, 2013)
5	Potato virus Y	IC-RT-PCR and IC-RT-LAMP	(Almasi & Dehabadi, 2013)
6	Tomato yellow leaf curl virus	IC-PCR, and IC-LAMP	(Almasi et al., 2013)
7	Beet curly top virus	IC-PCR, and IC-LAMP	(Almasi et al., 2014)
8	Banana bract mosaic virus	IC-RT-LAMP	(J. Zhang et al., 2016)
9	Mirafiori lettuce big-vein virus	IC-RT-LAMP	(Almasi, 2017)
10	Citrus tristeza virus	IC-RT-LAMP and IC-RT-qPCR	(Selvaraj et al., 2019)
11	Lily symptomless virus and cucumber mosaic virus	IC-RT-PCR and IC-RT-LAMP	(Y. Zhang et al., 2022)
12	ToLCNDV-Rg ridge gourd isolate	IC-PCR, and IC-LAMP	(Naganur et al., 2023)
13	Peanut stripe virus	IC-PCR	(Singh et al., 2009)

first capture the viral antigen (in the form of virions) using a PCR tube coated with a specific anti-virus antibody. Then, the real-time RT-PCR reaction begins by heating the virions for dissociation to release the genomic RNA, eliminating the need for total RNA extraction, making this method highly specific and sensitive. In real-time PCR, TaqMan probe is utilised, which has 5′ terminal reporter dye 6-carboxyfluorescein (FAM) and the 3′ quencher dye 6-carboxy-tetramethyl-rhodamine (TAMRA). As the primer extends during each PCR amplification cycle, the Taq DNA polymerase's 5′–3′ exonuclease activity degrades the TaqMan probe and releases the fluorescent chromophore. At every amplification cycle, the fluorescence is detected and is directly correlated with the amount of PCR product that is synthesised. According to Yang et al. (2012), the limit of IC-real-time RT-PCR in this study was 102 times greater than that of RT-PCR for the detection of TMV in soil. Multiple advantages come with the one-step enclosed immunocapture real-time PCR (RT-PCR) method, including higher sensitivity, a lower risk of contamination, and the absence of post-PCR processing, and a reduction in the amount of time required for a reliable disease diagnosis.

3.8.4 Immunocapture Transmission Electron Microscopy (IC-TEM)

This is a specialised method that incorporates immunology and electron microscopy to study plant viruses at a very high resolution (Debrick, 1973). This method is used to study the spatial distribution and localisation of specific molecules or structures within cells or tissues. The principle involves a beam of electrons to interact with the sample, and the electron-dense markers on the antibodies provide contrast, making it possible for virus particle visualisation (Hafrén et al., 2017, Dore et al., 1988; Himmler et al., 1988; Kerlan et al., 1981; Lima & Purcifull, 1980; Luisoni et al., 1982; Milne & Lesemann, 1978; Roberts & Harrison, 1979; Van Lent & Verduin, 1986). The SOP (standard operating procedure) includes sample preparation, immunocapture, washing and staining, followed by imaging. The morphological determination of viruses is facilitated by the high level of sensitivity and specificity exhibited by the ultrathin sections in electron microscopy (Khurana & Garg, 1993). The unique aspect of IC-TEM is the use of immunocapture, which involves labelling specific molecules or structures within the sample using antibodies. Antibodies are conjugated with electron-dense markers, such as copper grids with thin-layer Parlodion or colloidal gold nanoparticles.

3.9 IMMUNOFLUORESCENCE ASSAY (IFA)

IFA is a serological technique that utilises the specific binding of fluorescently (i.e., fluorescein) labelled antibodies to viral antigens for the visualisation and detection of plant viruses. IFA enables the direct visualisation of virus-infected plant tissues or cells using fluorescence microscopy, which contains fluorescence optics that enables sensitive and specific identification of viral infections in plants (Bressan & Watanabe, 2011; Cruz et al., 1996; Dufour et al., 1989; Ko et al., 1984; Mahmood et al., 1997; Van Lent & Verduin, 1987). Potato or tobacco leaves that were infected once or twice were subjected to immunofluorescent examination utilizing 1° monoclonal antibodies conjugated with the N serotype isolates and the relevant 2° antibodies coupled with a specific fluorescent dye. The epitopes of the isolates of serotypes N and O were found using green fluorescence (FITC) and red (Texas Red), respectively (Grupa et al., 2018).

REFERENCES

Abd El-Aziz, M. (2019). Three modern serological methods to detect plant viruses. *Journal of Plant Science and Phytopathology*, *3*(3), 101–106. https://doi.org/10.29328/journal.jpsp.1001039

Ahangaran, A., Mohammadi, G. M., Habibi, M. K., Khezri, S., & Shahraeen, N. (2009). Use of rapid serological and nucleic acid- based methods for detecting the soybean mosaic virus. *Journal of Agricultural Science and Technology*, *11*, 91–97. https://doi.org/10.29328/journal.jpsp.1001039

Ahmadi, S., Almasi, M. A., Fatehi, F., Struik, P. C., & Moradi, A. (2013). Visual detection of Potato Leafroll Virus by one-step reverse transcription loop-mediated isothermal amplification of DNA with hydroxynaphthol blue dye. *Journal of Phytopathology*, *161*(2), 120–124.

Almasi, M. A. (2017). Development of a colorimetric reverse transcription loop-mediated isothermal amplification assay for the detection of Mirafiori lettuce big-vein virus. *Archives of Virology*, *162*, 2775–2780.

Almasi, M. A., & Dehabadi, S. H. (2013). Colorimetric immunocapture reverse transcription loop-mediated isothermal amplification assay for rapid detection of the Potato virus Y. *Journal of Plant Pathology and Microbiology*, *4*, 188.

Almasi, M. A., Dehabadi, S. H., & Eftekhari, Z. (2013). Immunocapture loop mediated isothermal amplification for rapid detection of tomato yellow leaf curl virus (TYLCV) without DNA extraction. *Journal of Plant Pathology and Microbiology*, *4*, 185.

Almasi, M. A., Hosseyni-Dehabadi, S. M., & Aghapour-ojaghkandi, M. (2014). Comparison and evaluation of three diagnostic methods for detection of beet curly top virus in sugar beet using different visualizing systems. *Applied Biochemistry and Biotechnology*, *173*, 1836–1848.

Almasi, M. A., Moradi, A., Nasiri, J., Karami, S., & Nasiri, M. (2012). Assessment of performance ability of three diagnostic methods for detection of Potato leafroll virus (PLRV) using different visualizing systems. *Applied Biochemistry and Biotechnology*, *168*, 770–784.

Bergervoet, J. H. W., Peters, J., van Beckhoven, J. R. C. M., van den Bovenkamp, G. W., Jacobson, J. W., & van der Wolf, J. M. (2008). Multiplex microsphere immuno-detection of potato virus Y, X and PLRV. *Journal of Virological Methods*, 149(1), 63–68.

Boonham, N., Kreuze, J., Winter, S., van der Vlugt, R., Bergervoet, J., Tomlinson, J., & Mumford, R. (2014). Methods in virus diagnostics: From ELISA to next generation sequencing. *Virus Research*, *186*, 20–31. https://doi.org/10.1016/j.virusres.2013.12.007

Bressan, A., & Watanabe, S. (2011). Immunofluorescence localisation of Banana bunchy top virus (family Nanoviridae) within the aphid vector, Pentalonia nigronervosa, suggests a virus tropism distinct from aphid-transmitted luteoviruses. *Virus Research*, *155*(2), 520–525. https://doi.org/10.1016/j.virusres.2010.12.005

Candresse, T., Lanneau, M., Revers, F., Macquaire, G., German, S., Dunez, J., Grasseau, N., & Malinovsky, T. (1994). An immunocapture PCR assay adapted to the detection and the analysis of the molecular variability of apple chlorotic leaf spot virus. *XVI International Symposium on Fruit Tree Virus Diseases*, *386*, 136–147.

Chandel, V., Rana, T., Hallan, V., & Zaidi, A. A. (2011). Detection of Prunus necrotic ring spot virus in plum, cherry and almond by serological and molecular techniques from India. *Archives of Phytopathology and Plant Protection*, *44*(18), 1779–1784. https://doi.org/10.1080/03235401003633873

Charlermroj, R., Himananto, O., Seepiban, C., Kumpoosiri, M., Warin, N., Oplatowska, M., Gajanandana, O., Grant, I. R., Karoonuthaisiri, N., & Elliott, C. T. (2013). Multiplex detection of plant pathogens using a microsphere immunoassay technology. *PloS One*, *8*(4), e62344.

Charoenvilaisiri, S., Seepiban, C., Kumpoosiri, M., Rukpratanporn, S., Warin, N., Phuangrat, B., Chitchuea, P., Siripaitoon, S., Chatchawankanphanich, O., & Gajanandana, O. (2021). Development of a triple antibody sandwich enzyme-linked immunosorbent assay for cassava mosaic disease detection using a monoclonal antibody to Sri Lankan cassava mosaic virus. *Virology Journal*, *18*(1), 100. https://doi.org/10.1186/s12985-021-01572-6

Chen, H., Ali, N., Lv, W., Shen, Y., Qing, Z., Lin, Y., Chen, B., & Wen, R. (2020). Comparison of IC-RT-PCR, dot-ELISA and indirect-ELISA for the detection of sorghum mosaic virus in field-grown sugarcane plants. *Sugar Tech*, *22*(1), 122–129. https://doi.org/10.1007/s12355-019-00738-5

Chen, Z., Liu, J., Zeng, M., Wang, Z., Yu, D., Yin, C., Jin, L., Yang, S., & Song, B. (2012). Dot immunobinding assay method with chlorophyll removal for the detection of southern rice black-streaked dwarf virus. *Molecules*, *17*(6), 6886–6900. https://doi.org/10.3390/molecules17066886

Cruz, S. S., Chapman, S., Roberts, A. G., Roberts, I. M., Prior, D. A., & Oparka, K. J. (1996). Assembly and movement of a plant virus carrying a green fluorescent protein overcoat. *Proceedings of the National Academy of Sciences*, *93*(13), 6286–6290.

Çulal KiliÇ, H., & Çat, A. (2022). Detection of bean common mosaic virus in bean seeds by immunocapture RT-PCR and DAS-ELISA methods. *Sinop Üniversitesi Fen Bilimleri Dergisi*, *7*(1), 1–7. https://doi.org/10.33484/sinopfbd.1006005

D'Hondt, L., Höfte, M., Van Bockstaele, E., & Leus, L. (2011). Applications of flow cytometry in plant pathology for genome size determination, detection and physiological status. *Molecular Plant Pathology*, *12*(8), 815–828. https://doi.org/10.1111/j.1364-3703.2011.00711.x

Darshan, K., Amrurthalakshmi, M., & Reddy, M. G. (2021). *Phytopathological Techniques Basic to Advances*.

Das, S., Agarwal, D. K., Mandal, B., Rao, V. R., & Kundu, T. (2021). Detection of the chilli leaf curl virus using an attenuated total reflection-mediated localized surface-plasmon-resonance-based optical platform. *ACS Omega*, *6*(27), 17413–17423. https://doi.org/10.1021/acsomega.1c01702

Daugrois, J.-H., Roumagnac, P., Julian, C., Filloux, D., Putra, L., Mollov, D., & Rott, P. C. E. (2023). Historical review on sugarcane streak mosaic virus that has recently emerged in Africa. *Phytopathology*, 2024 Apr; *114*(4), 668–680. https://doi.org/0.1094/PHYTO-08-23-0291-RVW

Debrick, K. S. (1973). Quantitative assay for plant viruses using serologically specific electron microscopy. *Virology*, *56*(2), 652–653.

Dehabadi, S. H. (2013). Immunocapture loop mediated isothermal amplification for rapid detection of Tomato Yellow Leaf curl Virus (TYLCV) without DNA extraction. *Journal of Plant Pathology & Microbiology*, 04(07), 1–5. https://doi.org/10.4172/2157-7471.1000185

Desbiez, C., & Lecoq, H. (1997). Zucchini yellow mosaic virus. *Plant Pathology*, *46*, 809–829.

Dhir, S., Lakshmi, V., & Hallan, V. (2021). Development of immunodiagnostics for Apple stem pitting virus and Apple mosaic virus infecting apple in India. *Indian Phytopathology*, *74*(1), 189–199. https://doi.org/10.1007/s42360-020-00305-z

Dijkstra, J., & de Jager, C. P. (1998). *Chloroplast agglutination test. Practical Plant Virology*. Springer, Berlin, Heidelberg, Springer Lab Manual.

Dore, I., Weiss, E., Altschuh, D., & Van Regenmortel, M. H. V. (1988). Visualization by electron microscopy of the location of tobacco mosaic virus epitopes reacting with monoclonal antibodies in enzyme immunoassay. *Virology*, *162*(2), 279–289.

Drygin, Y. F., Blintsov, A. N., Grigorenko, V. G., Andreeva, I. P., Osipov, A. P., Varitzev, Y. A., Uskov, A. I., Kravchenko, D. V, & Atabekov, J. G. (2012). Highly sensitive field test lateral flow immunodiagnostics of PVX infection. *Applied Microbiology and Biotechnology*, *93*, 179–189.

Dufour, O., Palloix, A., Selassie, K. G., Pochard, E., & Marchoux, G. (1989). The distribution of cucumber mosaic virus in resistant and susceptible plants of pepper. *Canadian Journal of Botany*, *67*(3), 655–660. https://doi.org/10.1139/b89-088

Dyussembayev, K., Sambasivam, P., Bar, I., Brownlie, J. C., Shiddiky, M. J. A., & Ford, R. (2021). Biosensor technologies for early detection and quantification of plant pathogens. *Frontiers in Chemistry*, *9*, 1–13. https://doi.org/10.3389/fchem.2021.636245

Eni, A. O., Hughes, J. D. A., Asiedu, R., & Rey, M. E. C. (2012). Re-evaluation of yam mosaic virus (YMV) detection methods. *Academic Journal of Plant Sciences*, *5*(1), 18–22.

Eni, A. O., Ogunsanya, P., Oviasuyi, T., & Hughes, J. d'A. (2013). Alarming increase in the incidence of Cucumber mosaic virus in cowpea (*Vigna unguiculata* (L.) Walp.) in northern Nigeria. *Archives of Phytopathology and Plant Protection*, *46*(16), 1958–1965. https://doi.org/10.1080/03235408.2013.782218

Fegla, G. I., El-Samra, I. A., Younes, H. A., & Abd El-Aziz, M. H. (2001). Plain and filter papers as solid carriers for detection of three plant viruses by dot and tissue blot immunoassays. *International Journal of Advance Agricultural Research*, *6*(4), 755–761.

Fukuta, S., Ohishi, K., Yoshida, K., Mizukami, Y., Ishida, A., & Kanbe, M. (2004). Development of immunocapture reverse transcription loop-mediated isothermal amplification for the detection of tomato spotted wilt virus from chrysanthemum. *Journal of Virological Methods*, *121*(1), 49–55.

Fukuta, S., Takeyama, K., Suzuki, M., Shichi, A., Ichikawa, K., & Nakanishi, H. (2012). Detection of Kyuri green mottle mosaic virus from soil by the immunocapture reverse transcription loop-mediated isothermal amplification reaction. *Plant Pathology Journal (Faisalabad)*, *11*(2), 51–59.

Gella, F. J., Serra, J., & G. J. (1991). Latex agglutination procedures in immunodiagnosis. *Pure and Applied Chemistry*, *63*, 1131–1134.

Gibbs, A., & Harrison, B. (1976). *Plant Virology the Principles*. Edward Arnold (Publishers) Ltd., London, 292.

Gong, H., Dusengemungu, L., Lv, P., & Igiraneza, C. (2023). Advancements in lily viruses management: Challenges and solutions in elimination and detection. *Horticulturae*, *9*(7), 790. https://doi.org/10.3390/horticulturae9070790

Grupa, A., Otulak-Kozieł, K., & Syller, J. (2018). Serological, molecular and immunofluorescent evidence for interference competition between isolates of Potato virus Y. *Plant Pathology*, *67*(9), 1997–2012.

Guo, Y., Liu, R., Liu, Y., Xiang, D., Liu, Y., Gui, W., Li, M., & Zhu, G. (2018). A non-competitive surface plasmon resonance immunosensor for rapid detection of triazophos residue in environmental and agricultural samples. *Science of the Total Environment*, *613*, 783–791.

Gutiérrez-Aguirre, I., Hodnik, V., Glais, L., Rupar, M., Jacquot, E., Anderluh, G., & Ravnikar, M. (2014). Surface plasmon resonance for monitoring the interaction of Potato virus Y with monoclonal antibodies. *Analytical Biochemistry*, *447*, 74–81.

Hafrén, A., Macia, J.-L., Love, A. J., Milner, J. J., Drucker, M., & Hofius, D. (2017). Selective autophagy limits cauliflower mosaic virus infection by NBR1-mediated targeting of viral capsid protein and particles. *Proceedings of the National Academy of Sciences*, *114*(10), E2026–E2035.

Halabi, M. H., Oladokun, J. O., Baldodiya, G. M., Borah, B. K., & Nath, P. D. (2021). Identification, prevalence and genetic diversity study of potato viruses in Northeastern states of India. *Annals of Applied Biology*, *179*(2), 185–194. https://doi.org/10.1111/aab.12691

He, W., Wu, J., Ren, Y., Zhou, X., Zhang, S., Qian, Y., Li, F., & Wu, J. (2020). Highly sensitive serological approaches for Pepino mosaic virus detection. *Journal of Zhejiang University-Science B*, *21*(10), 811–822. https://doi.org/10.1631/jzus.B2000255

Heller, F. O. (1973). DNS-Bestimmung an Keimwurzeln von Vicia faba L. mit Hilfe der Impulscytophotometrie 1. *Berichte Der Deutschen Botanischen Gesellschaft*, *86*(5–9), 437–441.

Hill, S. A. (1984). *Methods in Plant Virology*. Blackwell, Oxford.

Himmler, G., Brix, U., Steinkellner, H., Laimer, M., Mattanovich, D., & Katinger, H. W. D. (1988). Early screening for anti-plum pox virus monoclonal antibodies with different epitope specificities by means of gold-labelled immunosorbent electron microscopy. *Journal of Virological Methods*, 22(2–3), 351–357.

Iannelli, D., Barba, M., D'Apice, L., Pasquini, G., Capparelli, R., Monti, L., Parrella, G., Scala, F., & Noviello, C. (1996). Cytofluorimetric method for the detection of the cucumber mosaic virus. *Phytopathology*, *86*(9), 959–965.

Iannelli, D., D'Apice, L., Cottone, C., Viscardi, M., Scala, F., Zoina, A., Del Sorbo, G., Spigno, P., & Capparelli, R. (1997). Simultaneous detection of cucumber mosaic virus, tomato mosaic virus and potato virus Y by flow cytometry. *Journal of Virological Methods*, *69*(1–2), 137–145.

Jacobi, V., Bachand, G., Hamelin, R., & Castello, J. (1998). Development of a multiplex immunocapture RT-PCR assay for detection and differentiation of tomato and tobacco mosaic tobamoviruses. *Journal of Virological Methods*, *74*(2), 167–178. https://doi.org/10.1016/S0166-0934(98)00086-X

Jarocka, U., Wąsowicz, M., Radecka, H., Malinowski, T., Michalczuk, L., & Radecki, J. (2011). Impedimetric immunosensor for detection of plum pox virus in plant extracts. *Electroanalysis*, *23*(9), 2197–2204.

Javer-Higginson, E., Acina-Mambole, I., González, J. E., Font, C., González, G., Echemendía, A. L., Muller, E., & Teycheney, P.-Y. (2014). Occurrence, prevalence and molecular diversity of banana streak viruses in Cuba. *European Journal of Plant Pathology*, *138*, 157–166.

Jiao, K., Sun, W., & Zhang, S.-S. (2000). Sensitive detection of a plant virus by electrochemical enzyme-linked immunoassay. *Fresenius' Journal of Analytical Chemistry*, *367*, 667–671.

John, Alby, Deepa R. Chandran, and S. S. S. (n.d.). Techniques in Phytopathology. *Modern Approaches in Plant Pathology*, 207.

Kapoor, R., Mandal, B., Paul, P. K., & Jain, R. K. (2014). Simultaneous detection of potato viruses Y and X by DAC-ELISA using polyclonal antibodies raised against fused coat proteins expressed in Escherichia coli. *Journal of Plant Biochemistry and Biotechnology*, *23*(3), 332–335. https://doi.org/10.1007/s13562-013-0251-5

Kerlan, C., Mille, B., & Dunez, J. (1981). Immunosorbent electron microscopy for detecting apple chlorotic leaf spot and plum pox viruses. *Phytopathology*, *71*(4), 400–404.

Khurana, S. M. P., & Garg, I. D. (1993). New techniques for detection of viruses and viroids. *Advances in Horticulture*, *7*, 529–566.

Ko, N.-J., Zettler, F. W., Edwardson, J. R., & Christie, R. G. (1984). Light microscopic techniques for detecting orchid viruses. *VI International Symposium on Virus Diseases of Ornamental Plants 164*, 241–254.

Konate, G., & Neya, B. J. (1996). Rapid detection of cowpea aphid-borne mosaic virus in cowpea seeds. *Annals of Applied Biology*, *129*(2), 261–266. https://doi.org/10.1111/j.1744-7348.1996.tb05750.x

Krishna, G. V., Kumar, V. M., Varma, P. K., Bhavani, B., & Kumar, G. V. (2023). Detection of Sugarcane bacilliform virus in diseased sugarcane plants in Andhra Pradesh, India by serological and molecular methods. *Brazilian Journal of Microbiology*, *54*(3), 1399–1409. https://doi.org/10.1007/s42770-023-01094-z

Le Provost, G., Iskra-Caruana, M.-L., Acina, I., & Teycheney, P.-Y. (2006). Improved detection of episomal Banana streak viruses by multiplex immunocapture PCR. *Journal of Virological Methods*, *137*(1), 7–13.

Lim, M. S., Kim, S. M., & Choi, S. H. (2016). Simultaneous detection of three lily-infecting viruses using a multiplex Luminex bead array. *Journal of Virological Methods*, *231*, 34–37.

Lima, J. A. A., & Purcifull, D. E. (1980). Immunochemical and microscopical techniques for detecting blackeye cowpea mosaic and soybean mosaic viruses in hypocotyls of germinated seeds. *Phytopathology*, *70*(2), 142–147.

Lin, H.-Y., Huang, C.-H., Lu, S.-H., Kuo, I.-T., & Chau, L.-K. (2014). Direct detection of orchid viruses using nanorod-based fiber optic particle plasmon resonance immunosensor. *Biosensors and Bioelectronics*, *51*, 371–378.

Ling, K.-S., Wechter, W. P., & Jordan, R. (2007). Development of a one-step immunocapture real-time TaqMan RT-PCR assay for the broad spectrum detection of Pepino mosaic virus. *Journal of Virological Methods*, *144*(1–2), 65–72.

Luisoni, E., Milne, R. G., & Roggero, P. (1982). Diagnosis of Rice ragged stunt virus by enzyme-linked immunosorbent assay and immunosorbent electron microscopy. *Plant Disease*, *66*(10), 929–932.

Mahmood, T., Hein, G. L., & French, R. C. (1997). Development of serological procedures for rapid and reliable detection of wheat streak mosaic virus in a single wheat curl mite. *Plant Disease*, *81*(3), 250–253. https://doi.org/10.1094/PDIS.1997.81.3.250

Mahmoud Hamdy Abd El-Aziz Hosny Aly Younes. (2019). Detection of Cucumber mosaic cucumovirus in infected cowpea plants (*Vigna unguiculata* L.) from northern Egypt. *Novel Research in Microbiology Journal*, 3(2), 326–340. https://doi.org/10.21608/nrmj.2019.30612

Mallik, I., Anderson, N. R., & Gudmestad, N. C. (2012). Detection and differentiation of Potato virus Y strains from potato using immunocapture multiplex RT-PCR. *American Journal of Potato Research*, *89*, 184–191.

Mangalore, C. N., Hari D S., Bhagavan, B. V. K., Salomi, S. K. U., & Krishna, K. R. K. (2020). Pre and post evaluation of direct antigen coating – enzyme-linked immuno sorbent assay (DAC-ELISA) for the detection of BBrMV and CMV in micropropagation of banana. *Indian Journal of Plant Protection*, *48*(4), 461–473.

Milne, R. G., & Lesemann, D.-E. (1978). An immunoelectron microscopic investigation of oat sterile dwarf and related viruses. *Virology*, *90*(2), 299–304.

Miras, M., Torre, C., Gómez-Aix, C., Hernando, Y., & Aranda, M. A. (2020). Development of monoclonal antibodies against melon necrotic spot virus and their use for virus detection. *Journal of Virological Methods*, *278*, 113837. https://doi.org/10.1016/j.jviromet.2020.113837

Mumford, R., & Seal, S. (1997). Rapid single-tube immunocapture RT-PCR for the detection of two yam potyviruses. *Journal of Virological Methods*, *69*(1–2), 73–79. https://doi.org/10.1016/S0166-0934(97)00141-9

Naganur, P., Shankarappa, K. S., Mesta, R. K., Rao, C. D., Venkataravanappa, V., Maruthi, M. N., & Reddy, L. R. C. N. (2023). Detecting tomato leaf curl New Delhi virus causing ridge gourd yellow mosaic disease, and other begomoviruses by antibody-based methods. *Plants*, *12*(3), 490.

Narayanasamy, P., & Narayanasamy, P. (2011). Detection of virus and viroid pathogens in plants. *Microbial Plant Pathogens-Detection and Disease Diagnosis: Viral and Viroid Pathogens,* 3, 7–220.

Newman, A. L., Hunter, K. W., & Stanbro, W. D. (1986). The capacitive affinity sensor: A new biosensor. *Bordeaux, France: Proceedings of the Second International Meeting on Chemical Sensors.*

Noordam D. (1975). Identification of plant viruses: methods and experiments. *Cen Agric Pub Doc, Wageningen.*

Ouchterlony O. (1968). Handbook of immunodiffusion and immunoelectrophoresis. *Ann Arbor Scientific Publications, Michigan.*

Paternolli, C., Antonini, M., Ghisellini, P., & Nicolini, C. (2004). Recombinant cytochrome P450 immobilization for biosensor applications. *Langmuir*, *20*(26), 11706–11712.

Phatsaman, T., Hongprayoon, R., & Wasee, S. (2020). Monoclonal antibody-based diagnostic assays for pepper mild mottle virus. *Journal of Plant Pathology*, *102*(2), 327–333. https://doi.org/10.1007/s42161-019-00421-4

Rageshwari, S., Renukadevi, P., Malathi, V.G., P. A., & Nakkeeran S. (2017). DAC-ELISA and RT-PCR-based confirmation of systemic and latent infection by tobacco streak virus in cotton and parthenium. *Journal of Plant Pathology*, *99*(2), 1–7.

Razmi, A., Golestanipour, A., Nikkhah, M., Bagheri, A., Shamsbakhsh, M., & Malekzadeh-Shafaroudi, S. (2019). Localized surface plasmon resonance biosensing of tomato yellow leaf curl virus. *Journal of Virological Methods*, *267*, 1–7. https://doi.org/10.1016/j.jviromet.2019.02.004

Ren, R., Wang, T., Gao, L., Song, P., Yang, Y., Zhi, H., & Li, K. (2022). Development of Comprehensive Serological Techniques for Sensitive, Quantitative and Rapid Detection of Soybean mosaic virus. *International Journal of Molecular Sciences*, *23*(16), 9457. https://doi.org/10.3390/ijms23169457

Roberts, I. M., & Harrison, B. D. (1979). Detection of potato leafroll and potato mop-top viruses by immunosorbent electron microscopy. *Annals of Applied Biology*, *93*(3), 289–297.

Seepiban, C., Charoenvilaisiri, S., Warin, N., Bhunchoth, A., Phironrit, N., Phuangrat, B., Chatchawankanphanich, O., Attathom, S., & Gajanandana, O. (2017). Development and application of triple antibody sandwich enzyme-linked immunosorbent assays for begomovirus detection using monoclonal antibodies against Tomato yellow leaf curl Thailand virus. *Virology Journal*, *14*(1), 99. https://doi.org/10.1186/s12985-017-0763-z

Selvaraj, V., Maheshwari, Y., Hajeri, S., & Yokomi, R. (2019). A rapid detection tool for VT isolates of Citrus tristeza virus by immunocapture-reverse transcriptase loop-mediated isothermal amplification assay. *PLoS One*, *14*(9), e0222170.

Shahmirzaie, M., Safarnejad, M. R., Rakhshandehroo, F., Safarpour, H., Rabbani, H., Zamanizadeh, H. R., & Elbeaino, T. (2019). Production of a polyclonal antiserum against recombinant nucleocapsid protein and its application for the detection of fig mosaic virus. *Journal of Virological Methods*, *265*, 22–25. https://doi.org/10.1016/j.jviromet.2018.12.016

Sharman, M., Thomas, J. E., & Dietzgen, R. G. (2000). Development of a multiplex immunocapture PCR with colourimetric detection for viruses of banana. *Journal of Virological Methods*, *89*(1–2), 75–88. https://doi.org/10.1016/S0166-0934(00)00204-4

Singh, M. K., Chandel, V., Hallan, V., Ram, R., & Zaidi, A. A. (2009). Occurrence of peanut stripe virus on patchouli and raising of virus-free patchouli plants by meristem tip culture. *Journal of Plant Diseases and Protection*, *116*(1), 2–6. https://doi.org/10.1007/BF03356278

Sherwood, J. L., Sanborn, M. R., Keyser, G. C., & Myers, L. D. (1988). Use of monoclonal antibodies in detection of tomato spotted wilt virus. *Phytopathology*, *79*, 61–64.

Stewart, L. R., Willie, K., Wijeratne, S., Redinbaugh, M. G., Massawe, D., Niblett, C. L., Kiggundu, A., & Asiimwe, T. (2017). Johnsongrass mosaic virus Contributes to Maize Lethal Necrosis in East Africa. *Plant Disease*, *101*(8), 1455–1462. https://doi.org/10.1094/PDIS-01-17-0136-RE

Talley. J., Warren, F. H. J. B., Torrance, L., & Jones. R. A.C. (1980). A simple kit for detection of plant viruses by the latex serological test. *Plant Pathology*, *29*, 77–79.

Tatineni, S., Sarath, G., Seifers, D., & French, R. (2013). Immunodetection of Triticum mosaic virus by DAS- and DAC-ELISA using antibodies produced against coat protein expressed in Escherichia coli: Potential for high-throughput diagnostic methods. *Journal of Virological Methods*, *189*(1), 196–203. https://doi.org/10.1016/j.jviromet.2013.01.023

Teber, S., Gürcan, K., Brevet, M., Palmisano, F., Candresse, T., & Dallot, S. (2023). High genetic diversity suggests a long evolutionary history of plum pox virus-M in Turkey. *Plant Pathology*, *72*(2), 300–311.

Umber, M., Filloux, D., Gélabale, S., Gomez, R.-M., Marais, A., Gallet, S., Gamiette, F., Pavis, C., & Teycheney, P.-Y. (2020). Molecular viral diagnosis and sanitation of yam genetic resources: Implications for safe yam germplasm exchange. *Viruses*, *12*(10), 1101.

Van Lent, J. W. M., & Verduin, B. J. M. (1986). Detection of viral protein and particles in thin sections of infected plant tissue using immunogold labelling. In *Developments in Applied Biology 1. Developments and Applications in Virus Testing* (pp. 193–211).

Van Lent, J. W. M., & Verduin, B. J. M. (1987). Detection of viral antigen in semi-thin sections of plant tissue by immunogold-silver staining and light microscopy. *Netherlands Journal of Plant Pathology*, *93*, 261–272.

Van Slogteren DHM. (1955). Serological microreactions with plant viruses under paraffin oil. In *Proceedings 2nd Conference of Potato Virus Diseases. Lisse-Wageningen 1954, pp. 51–54*

Viganó, F., & Stevens, M. (2007). Development of a multiplex immunocapture-RT-PCR for simultaneous detection of BMYV and BChV in plants and single aphids. *Journal of Virological Methods*, *146*(1–2), 196–201.

Vigne, E., Garcia, S., Komar, V., Lemaire, O., & Hilly, J.-M. (2018). Comparison of serological and molecular methods with high-throughput sequencing for the detection and quantification of grapevine fanleaf virus in vineyard samples. *Frontiers in Microbiology*, *9*, 2726. doi: 10.3389/fmicb.2018.02726. PMID: 30524388; PMCID: PMC6262039

Voller, A., Bartlett, A., Bidwell, D. E., Clark, M. F., & Adams, A. N. (1976). The detection of viruses by enzyme-linked immunosorbent assay (ELISA). *Journal of General Virology*, *33*(1), 165–167. https://doi.org/10.1099/0022-1317-33-1-165

Wagih, E., Zalat, M., & Kawanna, M. (2021). Partial characterisation of two isolates of Cucumber mosaic virus (CMV). *Scientific Journal of Agricultural Sciences*, 0–0. https://doi.org/10.21608/sjas.2021.52505.1064

Walkey, D. G. A. (1985). Applied plant virology. *Heinemann, London.*

Wetzel, T., Candresse, T., Macquaire, G., Ravelonandro, M., & Dunez, J. (1992). A highly sensitive immunocapture polymerase chain reaction method for plum pox potyvirus detection. *Journal of Virological Methods*, *39*(1–2), 27–37. https://doi.org/10.1016/0166-0934(92)90122-T

Xue, F., Cornelissen, J. J. L. M., Yuan, Q., & Cao, S. (2023). Delivery of MicroRNAs by plant virus-based nanoparticles to functionally alter the osteogenic differentiation of human mesenchymal stem cells. *Chinese Chemical Letters*, *34*(1), 107448.

Yang, J.-G., Wang, F.-L., Chen, D.-X., Shen, L.-L., Qian, Y.-M., Liang, Z.-Y., Zhou, W.-C., & Yan, T.-H. (2012). Development of a one-step immunocapture real-time RT-PCR assay for detection of Tobacco mosaic virus in soil. *Sensors*, *12*(12), 16685–16694.

Zeng, C., Huang, X., Xu, J., Li, G., Ma, J., Ji, H.-F., Zhu, S., & Chen, H. (2013). Rapid and sensitive detection of maize chlorotic mottle virus using surface plasmon resonance-based biosensor. *Analytical Biochemistry*, *440*(1), 18–22.

Zhang, C., Guo, M., Dong, J., Liu, L., Zhou, X., & Wu, J. (2023). Visual and super-sensitive detection of maize Chlorotic mottle virus by dot- ELISA and Au nanoparticle-based immunochromatographic test strip. *Viruses*, *15*(7), 1607. https://doi.org/10.3390/v15071607

Zhang, J., Borth, W. B., Lin, B., Dey, K. K., Melzer, M. J., Shen, H., Pu, X., Sun, D., & Hu, J. S. (2016). Deep sequencing of banana bract mosaic virus from flowering ginger (Alpinia purpurata) and development of an immunocapture RT-LAMP detection assay. *Archives of Virology*, *161*, 1783–1795.

Zhang, J., & Zhao, J. (2019). *Chapter 5 – Immuno-Biosensor* (G. B. T.-N.-I. B. for P. A. with C. A. Li (ed.); pp. 115–137). Elsevier. https://doi.org/https://doi.org/10.1016/B978-0-12-815053-5.00005-2

Zhang, Y., Xie, Z., Wang, Y., Wang, R., Guo, Z., Qiu, Y., & He, Y. (2022). Development of immunocapture reverse transcription loop-mediated isothermal amplification assays to detect both lily symptomless virus and cucumber mosaic virus. *Annals of Applied Biology*, *180*(1), 132–139.

Zhao, C., Zhu, Q., Mu, X., Geng, C., Tian, Y., Yuan, X., Yan, Z., & Li, X. (2023). Epitope mapping and a cocktail of monoclonal antibodies to achieve full detection coverage of potato virus Y. *Plant Biotechnology Journal*, *21*(9), 1725–1727. https://doi.org/10.1111/pbi.14094

4 Molecular Diagnostics with a Special Focus on PCR-Based Approaches for the Detection of Plant Pathogens

Uzma Rashid, Sajad Un Nabi, Mohammad Irfan, Javid Iqbal Mir, Subaya Manzoor, and Sheikh Mansoor

4.1 INTRODUCTION

Every year, a multitude of biotic stressors pose a significant threat to agricultural crops, resulting in a decline in output and a deterioration in yield quality. The agricultural industry is currently facing a significant threat from plant diseases. A significant portion of the annual output of commercially significant crops is lost, reaching up to 40%, as a result of plant diseases and pests (Baldi and La Porta, 2020). Farmers frequently encounter the challenge of dealing with one or several pests or diseases, as well as the emergence of new strains of pathogens that are resistant to pesticides, all of which target the same crop. The impact of plant diseases on agricultural productivity is significant, resulting in substantial economic losses. These losses are anticipated to amount to a staggering $220 billion USD annually(FAO, 2019). The intensification of agricultural practices, such as the dense cultivation of extensive regions with monocultures, enables the swift dissemination of host specialized pathogens (Savary*et al.,* 2019; Baldi and La Porta, 2020). The exacerbation of the impact of these pathogens is attributed to the escalation of global trade, which facilitates the rapid introduction of invasive pathogens and subsequently leads to significant damage to crops and reduction in output. Furthermore, the expansion of pathogen distribution ranges as a result of human-mediated activities serves to facilitate the process of hybridization and horizontal gene transfer, ultimately resulting in the creation of novel pathogens (Gluck-Thaler and Slot, 2015). Moreover, the introduction of arthropod vectors into previously unoccupied regions can lead to the formation of

DOI: 10.1201/9781003474036-4

new relationships with both introduced and native pathogen species (Wingfield et al., 2016). The elements can potentially exert a profound economic and social influence, particularly in less developed nations (De Boer and López, 2012; Baldi and La Porta, 2020). Hence, it is imperative to mitigate the losses linked to plant diseases. It is of paramount importance to effectively mitigate these losses in a manner that is environmentally and economically sustainable.

The development of rapid, efficient, and cost-effective techniques for early detection and diagnosis of plant pathogens is imperative for the purpose of managing and, when feasible, mitigating plant diseases and the dissemination of plant pathogens to novel regions. The field of disease diagnostics plays a pivotal role in the identification and comprehension of plant diseases. The field of plant pathology employs a variety of tools and methodologies to effectively diagnose diseases in plants, enabling the correct identification of the agents responsible for these diseases and the evaluation of their effects on plant well-being. The identification of fungi and bacteria in traditional approaches is mostly focused on the observation of symptoms and the utilization of culture-based techniques. This indicates the requirement of cultivating the pathogen on specialized media under aseptic conditions, a procedure that typically spans several days or even weeks. Furthermore, the process of morphologically identifying the pathogen using microscopy necessitates the expertise of individuals who have received specialized training. Additionally, this method can be both time-consuming and arduous, since it demands a comprehensive understanding of taxonomy. The precise identification and timely detection of pathogens play a fundamental role in the treatment of diseases in many crops. The application of molecular diagnostics in the identification of plant pathogens has brought about a significant transformation in the realm of plant pathology. This is primarily due to the provision of exceptionally precise, expeditious, and sensitive techniques for detecting the existence of pathogens in plant specimens. The techniques are predicated upon the examination of the genetic material of the pathogen, commonly referred to as its DNA or RNA, in order to validate its specific classification. The utilization of molecular technologies for diagnosing plant diseases presents diagnosticians with numerous advantages in comparison to conventional methods. For instance, this technology facilitates the differentiation of species that share similar physical characteristics and the early detection of infection prior to the manifestation of symptoms. Molecular instruments possess capabilities that extend beyond enhancing the efficiency, precision, and expediency of diagnostic procedures. The utilization of shared technology foundations presents supplementary advantages, particularly in situations where resources are constrained and the upkeep of conventional functionality becomes challenging. Over the course of the last three decades, significant advancements have been made in the field of molecular detection technologies for plant diseases. One significant development was the emergence of antibody-based detection methods, specifically monoclonal antibodies and enzyme-linked immunosorbent assay (ELISA). This methodology constituted a significant milestone in the fields of virology and bacteriology due to its ability to expedite the identification and detection of pathogens (Miller and Martin, 1988). However, it should be noted that the manufacturing process of monoclonal antibodies can be costly, and there have been instances where issues regarding low specificity have been documented (Martinelli *et al.,* 2015). Subsequently, DNA-based methodologies, such as the polymerase chain reaction (PCR), emerged and demonstrated the ability to amplify the original target DNA by several millionfold (Mullis and Faloona, 1987). In recent decades, an increasing number of laboratories have embraced DNA-based techniques for the purpose of pathogen detection and identification. These techniques primarily involve the utilization of PCR and its several iterations, including quantitative PCR (qPCR), nested PCR, multiplex PCR, and digital PCR (ddPCR) (Martinelli et al., 2015). These approaches typically exhibit a high degree of specificity, rapidity, and cost-effectiveness. In recent years, there has been a proliferation of emerging technologies in the field of DNA amplification that utilize isothermal amplification techniques. These include loop-mediated isothermal amplification (LAMP), recombinase polymerase amplification (RPA), and helicase-dependent amplification (HDA), among others (Lau and Botella, 2017). These technologies have shown promise in addressing certain limitations associated with PCR-based methods (Zhong and Zhao, 2018). All of these methodologies, albeit being grounded on distinct principles, possess the primary attribute of enhancing the quantity of target DNA at a consistent temperature, hence obviating the necessity for a thermal cycler.

This chapter examines molecular diagnostic techniques utilized in the identification of plant diseases, with a particular emphasis on PCR-based methods. PCR-based methods are now the most prevalent and dependable means of pathogen identification.

4.2 MOLECULAR DIAGNOSTIC TECHNIQUES FOR THE DIAGNOSIS OF PLANT PATHOGENS

The utilization of molecular diagnostic techniques has significantly transformed the process of identifying plant diseases, facilitating precise and expeditious diagnosis. Theseprocedures are dependent on the examination of distinct DNA or RNA sequences that are exclusive to the particular disease under investigation. A wide range of molecular techniques are progressively emerging as significant resources in many facets of pathogen diagnostics. The techniques encompass immunological methods, nucleic acid-based probe technology, and PCR technology. Nucleic acid sequences, specifically DNA or RNA, serve as highly effective molecular targets for the detection and identification of microorganisms, including those that are pathogenic. These techniques can be applied to detect and identify viral, fungal, and bacterial pathogens. The identification of a specific genetic sequence exclusive to the target pathogen can be achieved using many methods, including PCR, isothermal amplification techniques, and hybridization-based techniques. There are numerous approaches available for the detection of plant diseases, with nucleic acid-based molecular techniques being the most widely employed methods (Figure 4.1). In the following discussion, we will go into these techniques in greater depth, with a particular emphasis on PCR-based methods. A common requirement for several nucleic acid-based procedures, particularly those using PCR, is the necessity of obtaining highly pure DNA (or RNA) from the target pathogen present in the sample. This is crucial due to the sensitivity of most nucleic acid-based tests to potential inhibitors that could be coextracted alongside the target pathogen.

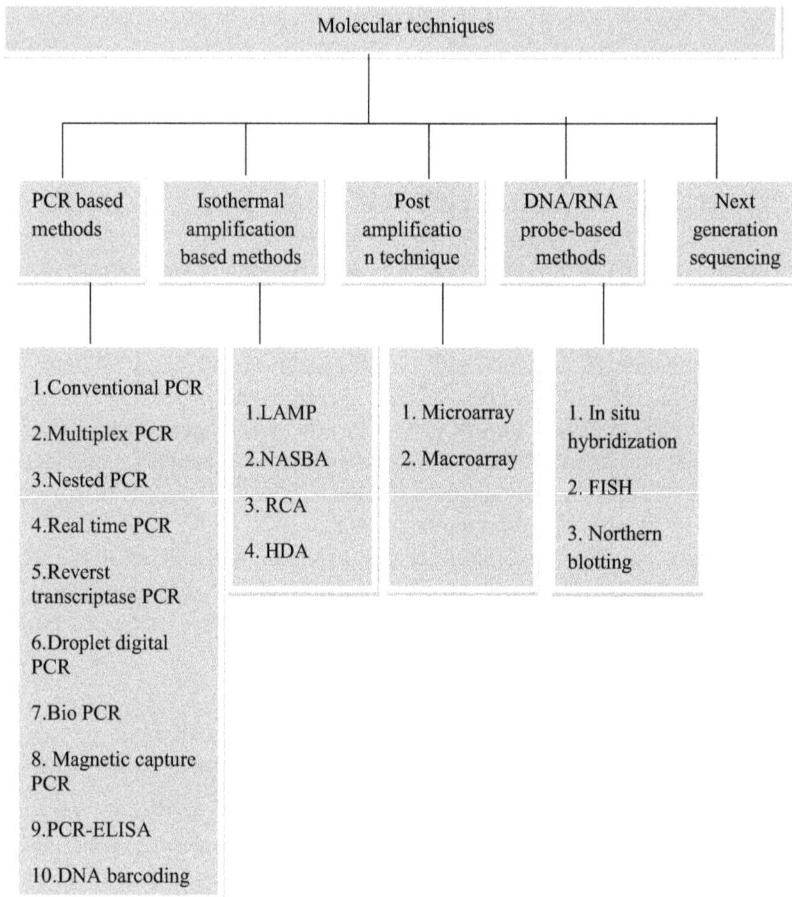

FIGURE 4.1 Flowchart of molecular techniques for diagnosis of plant pathogens.

4.2.1 PCR Amplification Methods

The utilization of PCR allows for the targeted amplification of DNA sequences specific to a particular disease, hence facilitating the detection of even minuscule quantities of the desired organism. PCR provides a high level of sensitivity and specificity by employing primers that are specifically designed to complement the genetic material of the pathogen. In addition, the utilization of real-time PCR, also known as qPCR, enables the quantitative evaluation of pathogen burden, hence facilitating informed decisions regarding disease management. The utilization of PCR-based techniques has significantly accelerated the process of diagnosing plant diseases, resulting in prompt interventions and effectively protecting agricultural crops from possibly catastrophic consequences. The pathogens identified using several PCR techniques have been listedin Table 4.1.

TABLE 4.1
PCR-based approaches to diagnose pathogenic fungi in different crops

Assay	Host	Disease	Diagnosed fungi	Target gene[a]
End-point PCR	Capsicum assamicum	Leaf spot	*Cercospora tezpurensis sp. nov.*	ACT, CAL, HIS and TEF-1a
End-point PCR	Blueberry	Leaf and fruit spot	*Exobasidium maculosum*	LSU-rDNA
End-point PCR	Cannabis sativa	Hemp powdery mildew	*Golovinomyces cichoracearum sensu lato*	ITS
End-point PCR	Cannabis sativa	Hemp leaf spot	*Cercospora cf. flagellaris*	ITS, TEF-1a, CAL, HIS and ACT
End-point PCR	Macadamia	Leaf spot	*Neopestalotiopsis clavispora and Colletotrichum siamense*	ITS, TUB2, TEF-1a, ACT and GAPDH
Nested PCR	Wheat	Stripe rust	*Puccinia striiformis f. sp. tritici*	PSR
Nested PCR	Strawberry	Crown rot	*Phytophthora cactorum*	ITS
Nested PCR	Dioscorea spp.	Greater yam anthracnose	*Colletotrichum gloeosporioides*	ITS
Nested PCR	Pomegranate	Twig blight and crown rot	*Pilidiella granati*	SSU-rDNA
Multiplex PCR	Maize	Stalk rot and ear rot	*Fusarium Verticillioides and F. subglutinans*	gaoB
Multiplex PCR	Apple	Bull's eye rot	*Neofabraea alba, N. perennans and N. keinholzii*	TUB2
Multiplex PCR	Musa spp.	Dessert/beer bananas	*Fusarium oxysporum f. sp. cubense lineage VI strains*	TEF-1a and RPC2
Quantitative PCR	Cucurbits	Gummy stem blight	*Didymella bryoniae*	RAPD
Quantitative PCR	Barely	Ramularia leaf spot	*Ramularia collo-cygni*	Not mentioned
Quantitative PCR	Tobacco	Target spot	*Rhizoctonia solani*	ITS
Quantitative PCR	Rice	Rice blast	*Magnaporthe oryzae*	18S-28S rDNA
Quantitative PCR	Brassica napus	Wilt and stem stripe	*Verticillium longisporum*	TUB2

(*continued*)

TABLE 4.1 (Continued)
PCR-based approaches to diagnose pathogenic fungi in different crops

Assay	Host	Disease	Diagnosed fungi	Target gene[a]
Quantitative PCR	Wheat	Tan (yellow) spot and Septoria blotch	*Pyrenophora tritici-repentis and Parastagonospora nodorum*	ToxA
Quantitative PCR	Cereals	Foot and root rot and Fusarium head blight	*Fusarium culmorum*	COX2
Quantitative PCR	Pineapple	Fusariosis	*Fusarium guttiforme*	TEF-1a and TUB2
End-point PCR and quantitative PCR	Apple	Speck rot and Sphaeropsis rot	*Phacidiopycnis washingtonensis and Sphaeropsis pyriputrescens*	ITS
End-point PCR and quantitative PCR	Citrus spp.	Citrus black spot	*Guignardia citricarpa*	ITS

4.2.1.1 Conventional PCR

The advent of PCR has had a profound impact on the study of plant pathology. This technique enables the synthesis of a targeted segment of DNA in million copies by employing a series of alternate cycles of denaturation, annealing, and elongation, facilitated by the use of specialized primers. PCR, which was pioneered by Kary Mullis in the mid-1980s, stands as a significant milestone in the field of molecular biology. Technological advancements in PCR have enabled the precise identification and characterization of plant diseases, thereby mitigating the limitations associated with conventional methodologies. The PCR method necessitates the utilization of many chemicals and constituents, encompassing the DNA template, Taq polymerase, two primers, deoxynucleotide triphosphates (dNTPs), buffer solution, as well as bivalent cations (Mg2+ or Mn2+) and monovalent cation (K+). Effective primer design facilitates the amplification of a distinct DNA fragment that is exclusive to the pathogen being targeted. The confirmation of the presence of the target pathogen is achieved by detecting a PCR fragment of the anticipated size (Shen, 2023. In a typical conventional polymerase chain reaction (cPCR) setup, the evaluation of amplification is commonly performed using an end-point detection method, such as agarose gel electrophoresis. Nevertheless, it is worth noting that the sensitivity of PCR may be affected by factors such as the method used for nucleic acid extraction and the variability of the target DNA (Ali et al., 2017). In order to address this concern in plants, the cetyl trimethyl ammonium bromide technique is commonly employed, along with the incorporation of specific chemical and enzymatic interventions. PCR is capable of amplifying minute quantities of

DNA owing to its exceptional sensitivity. Consequently, it is extensively employed in the identification of pathogens characterized by an extended period of latency between the initial infection and the manifestation of disease symptoms. According to Poisson statistics, the theoretical capability of the technique is to amplify as few as threecopies of the target nucleic acids. However, in practice, the limit of detection is heavily influenced by various factors, including the sample type, efficiency of DNA extraction, and efficiency of amplification. These factors, in turn, are affected by the PCR set-up and primer design (Kralik and Ricchi, 2017). The heightened sensitivity enables the detection of cells that are present in low quantities, have poor growth rates, or are unable to be cultured. The accuracy and precision of PCR are primarily contingent upon the meticulous design of primers that exhibit a high degree of selectivity. This selectivity, in turn, is predicated upon the accessibility of genetic information pertaining to the specific microorganism being targeted. The design of primers necessitates meticulous attention to ensure selective amplification of desired genetic sequences while minimizing the occurrence of false-positive outcomes. Therefore, it is imperative to assess and validate the sensitivity and specificity of a PCR assay on a case-by-case basis. However, conventional PCR is subject to some limitations, namely its vulnerability to contamination and its lack of robustness. This approach exhibits a notably low resolution, around ten-fold, due to the limitations of agarose gel in effectively resolving variations in yield. Nevertheless, as compared to conventional techniques for pathogen detection in plants, it exhibits greater specificity and sensitivity. Currently, the utilization of traditional PCR continues to be prevalent in the detection of plant diseases. Azizi et al. (2019a,b) utilized a conventional PCR technique to identify *Pantoeaananatis* and *P. stewartii* subsp. *indologenes* as the causal agents responsible for the occurrence of bacterial leaf blight (BLB) disease in rice in Malaysian. Similarly, in the region of Southern Karnataka, India, the PCRtechnique was utilized to analyze and classify the *Ralstoniasolanacerum* isolates responsible for inducing bacterial wilt in tomato plants (Shweta *et al.,* 2018). In addition, Murugan *et al.* (2020) effectively discovered *Fusarium oxysporum* f.sp. *lycopersici* (Fol) isolates from tomatoes that were contaminated using PCR.

One of the primary drawbacks associated with this approach is the considerable susceptibility to contamination, leading to an extended duration for obtaining the outcome. This approach facilitates the efficient dispersion of amplicons across the environment. The establishment of work disciplines and the isolation of the work environment from external factors are crucial measures to address the issue of false-positive outcomes. Despite the initial hypothesis suggesting that PCR would have a detection limit of less than 10 target sequences per sample, researchers have encountered challenges in developing tests that consistently achieve this level of detection across several methods. The current detection level of the approach is far lower than the threshold, and its sensitivity has been diminished due to the interference of multiple causes. Several factors significantly influence the outcomes, including the lack of a standardized genetic purification procedure, the utilization of low-quality materials, and the inability to effectively optimize the test. Nevertheless, as working conditions undergo alterations, the sensitivity of the tests diminishes.

Hence, it is important to take into account the ongoing quality control of the facilities being employed (Hajia, 2018).

4.2.1.2 Multiplex PCR

Multiplex PCR (M-PCR) is a modified version of the PCR technique, wherein multiple target sequences are amplified simultaneously within a single reaction tube utilizing multiple primer pairs. The approach holds significant importance in the field of plant pathology, particularly in cases where plants become infected with many pathogens. Multiple fragments specific to target pathogenic fungus have beenamplified simultaneously and identified based on their molecular sizes on agarose gels. In addition to its practical applications, the PCR technique has the potential to significantly reduce both time and effort expended in various scientific endeavors. The successful co-amplification of several targets in a single reaction is contingent upon the compatibility of the PCR primers employed in the reaction. In order for the amplification process to occur effectively, it is crucial that all primers utilized in the reaction exhibit comparable melting temperatures (Tm). This ensures that they can adequately bind to and detach from complementary DNA sequences at approximately equivalent temperatures. Consequently, each amplification may progress consistently at the designated temperature. The execution of this method would be hindered if the annealing of one primer set coincided with the dissociation of another primer set from its target. Hence, it is imperative to carefully choose primers with melting temperatures (Tms) that are closely aligned, differing only by a few degrees Celsius. Each amplification process occurs autonomously, without being influenced by the other processes, as long as none of the reagents are present at quantities that limit the rate of reaction. Additionally, each specific amplification product, known as an amplicon, is produced without hindrance. It is advisable to select primers that define amplicons within a similar size range (100-500 bp) to ensure efficient and equitable synthesis of each primer. In addition to the amplification stage, it is essential for every M-PCR test to incorporate a detection step that is capable of accurately detecting each amplicon. The separation and identification of distinct amplicons of varying sizes can be achieved using gel electrophoresis, or alternatively, through hybridization with DNA probes that are specific to the target sequences. Detection of the separated or hybridized amplicons can be accomplished using techniques such as spectrophotometry, fluorometry, autoradiography, or chemiluminescence.

Despite the substantial amount of effort required for the initial construction of a multiplex PCR reaction, its validation subsequently enhances the diagnostic capabilities of the approach by saving time, effort, and expensive reagents for each PCR run. One of the disadvantages associated with multiplex PCR, however, is its increased susceptibility to nonspecific DNA amplification, leading to the occurrence of false-positive outcomes. This is mostly attributed to the utilization of several primer pairs (Lau and Botella, 2017). In addition, it is worth noting that the utilization of multiplexing techniques has the potential to undermine the sensitivity of the amplification process. This is mostly because a specific target is typically amplified with more efficiency, so overshadowing the amplification of other targets. Certain constraints have been addressed through enhancements in the nucleic acid extraction techniques, improvements in the quality of the resulting products, and modifications to various parameters of the PCR technology. The utilization of magnetic nanobeads, dual priming oligonucleotide (DPO) primers, or a nested reaction has been essential in enhancing the levels of specificity and sensitivity. Additional constraints associated with the multiplex PCR method include the utilization of agarose gel-based detection for distinguishing amplicons of certain sizes, which may introduce potential optical inaccuracies. Furthermore, there is a need to verify that the amplified DNA fragments indeed belong to the intended target sequences. Agarose gel usage has been circumvented by employing alternate methodologies, such as the utilization of species-specific biotinylated probes within streptavidin-coated microtiter wells (Pallás*et al.*, 2018). The multiplex PCR technology has been employed for the concurrent identification of fungal pathogens, including *Fusarium oxysporumand Bacillus cactivora*, *P. nicotinae*, and *P. cactum,* which are known to induce disease in grafted Cacti (Ji Cho *et al.*,2016).

4.2.1.3 Nested PCR

Nested PCR is a variant of the PCR technique that was developed to enhance the sensitivity and specificity of the amplification process. This approach serves to reduce the occurrence of nonspecific amplification of the DNA template. Nested PCR is a molecular technique that employs two distinct sets of primers and two consecutive PCR amplification processes. The initial reaction entails the utilization of a primary set of primers, commonly referred to as outer primers. These outer primers are designed to encompass the target sequence, as well as an additional flanking sequence that extends beyond both ends of the target sequence. Following the initial reaction, a further reaction is conducted on the resultant products of the first PCR using a distinct set of primers known as inner primers or nested primers. These primers are designed to specifically bind to the target sequence and are positioned within the amplified region generated by the first PCR. The reduction in nonspecific binding is attributed to the fact that in the subsequent reaction, the majority of the amplicons generated from the initial reaction consist primarily of the target sequence and its adjacent sequences. The employed technique often yields a heightened level of detection sensitivity which is a result of the elevated total cycle count, while the enhanced specificity is a consequence of the second primer set annealing to sequences generated during the initial round. The high sensitivity of nested polymerase chain reaction (nPCR) renders it particularly advantageous in identifying pathogens with low concentrations. This is demonstrated in the nPCR assay designed to detect *Phytoplasmas, Phytophthora*, and citrus tristeza virus in diverse plant samples (Venbrux *et al.,* 2023; Nair and Manimekalai, 2021). Nevertheless, the likelihood of carryover contamination in the reaction is generally heightened as a result of further handling of amplicon products. To mitigate carryover, it is advisable to physically segregate distinct components of the process, ideally by allocating them to separate rooms. Alternatively, a technique known as single-tube nested PCR (STNPCR) can be employed, wherein both sets of primers are introduced into the initial reaction vessel and an extended PCR is conducted. The detection of amplicons from nested PCR assays follows the same methodology as that of PCR described before. The primary objective of nested PCR is to enhance the sensitivity of the assay by conducting a subsequent amplification of the target DNA from a template that has been previously enriched through the initial PCR. Nontarget sequences that were amplified in a nonspecific manner during the first PCR are not subjected to re-amplification in the subsequent reaction. This is because these sequences are unlikely to have the specific binding sites required by the internal primers used in the second PCR.

4.2.1.4 Real-time PCR

Quantitative polymerase chain reaction (qPCR), also known as real-time PCR, functions based on the same fundamental principle as conventional PCR. However, the key distinction lies in the fact that the amplified DNA is quantitatively assessed in real time throughout the PCR reaction, as opposed to being detected at the conclusion of the reaction. The RT-PCR technology has garnered significant attention from researchers worldwide and gives a pioneering alternative to conventional molecular techniques. This application is considered to be highly potent in terms of its rapidity and sensitivity in detection, hence minimizing the potential for cross-contamination (White *et al.* 2011). In addition, the technique has the capability to identify the little quantity of DNA pathogens present in both infected plant tissues and insect vectors (Crosslin*et al.,* 2006). Real-time PCR employs the fundamental principles of PCR, with the distinction that the advancement of the reaction is continuously observed through the utilization of a camera or detector, providing real-time feedback. This novel methodology is different from conventional PCR, wherein the detection of the reaction product occurs at completion of the reaction. Various strategies are employed to enable the monitoring of the progression of a PCR. Every method employs a specific type of fluorescent label that forms a bond with the DNA. The process of real-time PCR involves the utilization of a thermal cycler that has the capability to expose each sample to a specific wavelength of light. This allows for the identification of the amplified DNA fragment using oligonucleotide probes that generate fluorescence during the amplification process (Figure 4.2). The PCR methodology involves a cycle wherein the level of fluorescence released during amplification is directly correlated with the quantity of amplified DNA/RNA within the sample. The phrase used to describe this cycle is known as the quantitation cycle (Cq) or cycle threshold (Ct). A probe, such as the TaqMan probe, is specifically engineered to bind to the target sequence located between the conventional forward and reverse primers. The probe is tagged with a reporter fluorochrome at the 5' terminus, while a quencher fluorochrome is attached at the 3' terminus. The design of the probe is intended to possess a melting Tm that is greater than that of the primers. It is crucial for the probe to achieve complete hybridization, with a 100% binding efficiency, during the extension phase in order to ensure the success of the test. The presence of fluorochromes on the probe leads to the inhibition of fluorescence by the reporter molecule, facilitated by the quencher. Nevertheless, it should be noted that during the process of primer extension by Taq polymerase, the inherent 5' to 3' nuclease activity of Taq leads to the degradation of the probe, resulting in the liberation of the reporter fluorochrome. The quantity of fluorescence emitted during the process of amplification is directly proportional to the quantity of product created in each cycle.

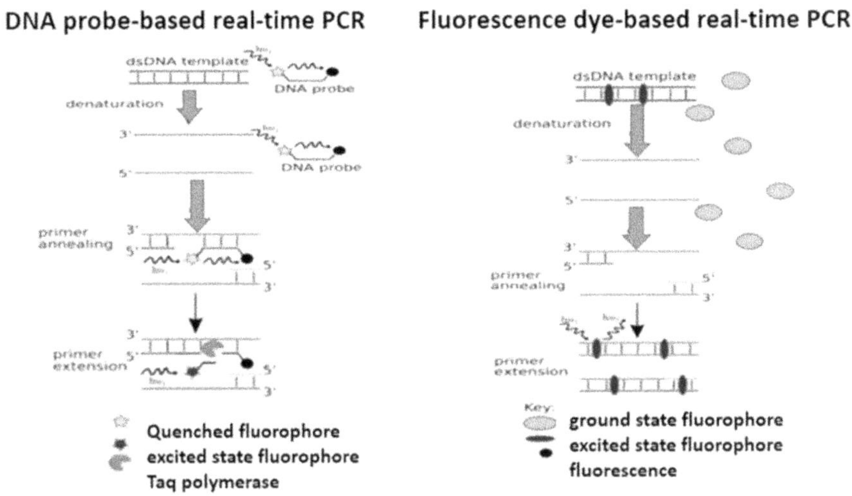

FIGURE 4.2 Diagramatic representation of real time PCR.

Currently, real-time PCR is widely regarded as the preferred and most reliable technique for the identification and quantification of plant pathogens. The utilization of real-time PCR enables the precise and dependable quantification of target pathogen DNA in diverse environmental samples, encompassing host tissues, soil, water, and air. Consequently, this technique presents novel avenues for research in the areas of diagnosis, determination of inoculum threshold levels, as well as the investigation of epidemiology and interactions between hosts and pathogens. The practical application of real-time PCR in the diagnosis of plant diseases is highly flexible. Real-time PCR enables the identification and detection of target pathogens, as well as the quantification of their presence in a given sample.

The utilization of quantitative (qPCR) in the detection of plant pathogens presents numerous benefits. The enhanced sensitivity of the technique compared to conventional PCR can be attributed to two primary factors. Firstly, instrumental fluorescence measurements offer greater sensitivity compared to the visualization of a DNA fragment through gel electrophoresis. Secondly, the targets of qPCR are typically shorter in length, ranging from 70 to 150 base pairs, which allows for more efficient amplification. The heightened sensitivity of this instrument renders it a great asset in the early identification of pathogens, even prior to the manifestation of disease signs. Furthermore, the absence of gel electrophoresis as a need contributes to expedited analysis time and enhances the potential for automation. In addition, the presence and abundance of pathogenic RNA viruses can be determined by the inclusion of a reverse transcription process. Nevertheless, the most significant advantage lies in the potential for quantifying the pathogen. This allows for the establishment of action thresholds in the field, specifically referring to the point at which treatment becomes necessary due to pathogen levels. Consequently, this approach reduces the frequency of chemical pesticide application, resulting in a disease control strategy that is both more effective and sustainable. While quantification is feasible, the assays are predominantly utilized for qualitative objectives, specifically determining the presence or absence of certain microorganisms. Numerous instances of qPCR techniques utilized for the detection of plant pathogens have been elucidated in scholarly works, highlighting their superiority over cPCR procedures (Narayanasamy, 2010). Verdecchia et al. (2021) reported the development of a qPCR assay for the purpose of detecting *P. cactorum* in strawberry samples. The assay demonstrated a notable level of specificity and sensitivity, enabling the detection of as few as 10 zoospores per gram of plant material. In the realm of bacterial plant pathogens, numerous qPCR techniques have been devised. For instance, detection of rhizogenic *Agrobacterium* strains in irrigation water of tomato greenhouses showed a sensitivity of 1 colony-forming unit per milliliter (CFU/ml) of water (Bosmans *et al.*, 2016).

4.2.1.5 Reverse transcription PCR

Reverse-transcription polymerase chain reaction (RT-PCR) is a highly sensitive technique that allows for the detection and quantification of RNA molecules, even at the level of a solitary cell. RT-PCR is a molecular technique utilized to convert RNA molecules into their complementary DNA (cDNA) sequences by the action of reverse transcriptases. Subsequently, the newly created cDNA is amplified using normal PCR protocols. The method employed for investigating gene expression, commonly referred to as RT-PCR, derives its name from the involvement of reverse transcriptase (RT) enzyme in the production of first-strand cDNA. cDNA exhibits greater stability compared to RNA due to its resistance to destruction by RNase enzymes. In the process of RT-PCR, the initial RNA molecules are subsequently destroyed, leading to the production of double-stranded DNA (dsDNA). Following this, the PCR amplification occurs in the conventional fashion. RT-PCR is a molecular biology technique that involves a sequential two-step process. The normalization of samples holds significant importance in the context of RT-PCR. The efficiency of first-strand cDNA synthesis plays a crucial role in determining the outcome of this approach, hence making it a key factor in its success or failure. Due to this rationale, it is more advantageous from a strategic standpoint to generate a substantial cDNA pool, from which separate aliquots can be extracted for individual applications, rather than repeatedly doing the same cDNA synthesis operation. The development of effective primers requires the careful consideration and optimization of several key factors, including template specificity, thermodynamic stability in relation to template base-pairing, and the ability of one primer to work in conjunction with other primers to facilitate RT-PCR. The most appropriate way to characterize the potential cooperative behavior of one or more pairs of oligonucleotide primers is by considering the Tm of each individual primer involved. The Tm, or melting temperature, refers to the temperature at which half of the potential annealing events between the primer and template have taken place, while the other half are still pending.

RT-PCR is widely regarded as the preeminent molecular technique employed for the identification of plant viruses owing to its exceptional sensitivity and specificity. The quantification of *Fusarium graminearum* fungi, which are responsible for causing the Fusarium ear blight (FEB) disease in cereals including wheat, rye, barley, oat, and maize, has been conducted using the RT-PCR technique (Brown *et al.*, 2011).

4.2.1.6 Digital droplet polymerase chain reaction (ddPCR)

Droplet Digital PCR (ddPCR) is widely recognized as the third generation of quantitative PCR. The process operates by employing water-in-oil emulsions to partition samples into droplets, enabling the identification and quantification of their genetic material. ddPCR represents a subsequent iteration of the PCR technique, offering the capability to precisely quantify nucleic acid targets present in a given sample. The ability of ddPCR to precisely identify and quantify targets that are present in low abundance has resulted in its rapid expansion in the field of pathogen identification. The methodology involves the segregation of DNA molecules inside a specimen into approximately 20,000 miniscule droplets composed of water-in-oil emulsions. Ideally, each individual droplet is expected to have either zero or a single copy of template DNA (Chen *et al.,* 2021). Each droplet will function as an independent vessel for a PCR, wherein a specific DNA region associated with the target pathogen is amplified. The utilization of fluorescent probes or intercalating dyes facilitates the identification of the occurrence of a PCR reaction. The droplets that are produced are thereafter individually transferred through a microfluidics device in order to assess their fluorescence levels (Zhao *et al.,* 2016). The quantification of template DNA in the original sample can be achieved by utilizing Poisson statistics, wherein the count of droplets containing an amplicon serves as an indicator (Chen *et al.,* 2021). When comparing the ddPCR technique to qPCR, several advantageous features can be identified. Firstly, ddPCR enables absolute quantification by utilizing the principles of sample partitioning in uniform droplets and applying Poisson statistics correction for multiple target molecules per droplet. This approach effectively addresses challenges related to normalization and calibration. Secondly, ddPCR has demonstrated enhanced precision, robustness, and sensitivity in the detection of low target copies. Thirdly, ddPCR exhibits relative insensitivity to potential PCR inhibitors. Fourthly, ddPCR allows for the measurement of the absolute number of microRNA copies per microliter of reaction, providing confidence intervals, a wide dynamic range, and high throughput capabilities. Fifthly, ddPCR is user-friendly, resembling the ease of use associated with conventional RT-qPCR. Lastly, ddPCR has the potential to exhibit superior diagnostic performance compared to conventional RT-qPCR. Several recent studies have presented novel ddPCR methodologies for the detection of plant pathogens such as *Xylella fastidiosa* (Dupas et al., 2019), *Acidovoraxcitrulli* (Cubero *et al.,* 2024, and *Tilletiacontroversa* (Liu *et al.,* 2020), and an RT-ddPCR method was used for the detection of peach latent mosaic viroids extracted from infected peach leaves (Lee *et al.,* 2021). The utility of ddPCR extends to multiplexing applications, but the existing literature on plant pathogen detection provides limited examples in this regard. Notably, Maheshwari et al. (2021)proposed a multiplex ddPCR test for the detection *of CandidatusLiberibacter asiaticus* and *Spiroplasmacitri.* Collectively, these investigations have demonstrated that ddPCR is a highly sensitive and reliable method that holds significant value in the surveillance of low titer pathogens inside intricate sample matrices.

4.2.1.7 Bio-PCR

The Bio-PCR assay is a variant of the end-point PCR technique that incorporates a pre-assay incubation step in a sample affected by a disease. This incubation step is employed to enhance the biomass of the causal agent. The primary purpose of this strategy is to enhance the concentration of certain pathogens by cultivating them in a growth medium that inhibits the growth of nontarget microorganisms. This approach has proven to be successful in the identification of seed-borne fungal pathogens (Kumar *et al.,* 2020). The advantages of utilizing endpoint PCR techniques include enhanced sensitivity, effective eradication of PCR inhibitors, and the ability to detect viable cells, hence mitigating the occurrence of false positive results. The approach in question is subject to certain limitations, namely its time-consuming nature and the associated expenditures paid when employing selective media for the experiment. The assay's reliability may be impaired by the action exerted, such as by bacteria during the cultivation of fungal biomass that is appropriate for detection. Hence, the drawbacks are associated with the escalation in test expenses, the temporal aspect, and the potential for environmental contamination. Fungal pathogens, specifically *Alternaria radicina* and A. *dauci*, which were transmitted through infected Alternaria alternata seeds, were identified utilizing ITS-specific rDNA primers through the utilization of a deep-freeze-blotter method in the Bio-PCR test (Konstantinova *et al.,* 2002).

4.2.1.8 Magnetic-capture hybridization PCR

Magnetic-capture hybridization PCR (MCH-PCR) involves a process of DNA isolation that includes a purification step involving the hybridization of single-stranded DNA (ssDNA) probes with magnetic beads. This is then followed by the amplification of target DNA sequences using PCR. The primary objective of developing this PCR test was to address the issue of PCR inhibitors present in plant extracts during the DNA isolation process. The magnetic beads employed in this study are coated with a biotinylated oligonucleotide that exhibits specificity towards a DNA region associated with the fungal pathogen under investigation (Botella, 2022). The process of hybridizing dsDNA with magnetic beads facilitates the isolation of the complex from inhibitory substances. The performance of MCH and real-time based PCR tests was assessed for the detection of two cucurbit seed pathogens, *Acidovoraxavenae* subsp. *Citrulli* and *Didymellabryoniae,* which are responsible for the occurrence of bacterial fruit blotch and gummy stem blight, respectively. This assay was able to detect both tested pathogens in cucurbit seed samples simultaneously (Ha *et al.,* 2009). The process of designing a capture probe for use in MCH-PCR entails the careful selection of an oligonucleotide probe sequence derived from highly conserved regions of fungal pathogens. The chosen sequence can be analyzed in silico for specificity by employing the BLAST algorithm. The probe is labeled with biotin at its 5' end in order to facilitate its binding to streptavidin-coated magnetic beads. The utilization of MCH-PCR has the potential to reduce the overall detection time, enhance the sensitivity of PCR, and eliminate a significant portion of the inhibitors present in the amplification reaction, as well as excessive nontarget DNA (Amagliani*et al.,* 2006).

4.2.1.9 DNA barcoding

DNA barcoding involves the utilization of a specific piece of DNA to identify the species of all eukaryotic life domains, employing molecular diagnostic methods. Barcode areas are ubiquitously present in target lineages. The observed DNA sequence variation is sufficient to discern distinct species. DNA barcoding is a widely employed technique that utilizes standardized sequences spanning 500-800 base pairs to facilitate species identification. These sequences serve as markers that are applicable across a diverse array of taxonomic groups. The PCR-based method is valuable for identifying microbiological organisms, including fungi, bacteria, and algae. It serves the purpose of recognizing cultured species as well as identifying species from uncultured taxa in natural environments. The efficacy of species identification by barcode analysis is contingent upon the quantity of representative specimens available inside the database. DNA barcoding enables the identification of fungi during life-cycle stages that are unsuitable for morphological identification. The effectiveness of barcoding is predicated upon the premise that genetic differences within a given species are far smaller in magnitude as compared to the variances observed between different species. Fungi that are not amenable to morphological identification and discovery can be effectively recognized and discovered through the utilization of DNA barcoding techniques at the molecular level. The complete operational procedure of the DNA barcoding technology. DNA barcoding offers numerous scientific benefits, such as the ability to identify species at any life stage, the potential to discover new species through phylogenetic analysis of nucleic acid sequences, insights into the diversification of life, and the promotion of DNA sequencing tools for biodiversity research. Barcodes that are reliable offer a uniform and credible means of species identification, which becomes valuable in the assessment of biodiversity and environmental research. The process of selecting a barcode region is a complex task that presents challenges in the pursuit of designing optimal and universally applicable DNA markers for PCR amplification. The molecular systematics of fungi heavily rely on the examination of the nuclear ribosomal RNA (rRNA) cistron, which comprises the small (18S) and large (28S) subunits. The amplification of the Internal Transcribed Spacer (ITS) region can be achieved from a diverse range of fungal species through the utilization of a restricted number of primers. Furthermore, research has revealed that variations across different species tend to be more pronounced compared to variations within a single species. The ITS region is commonly employed as a universal barcode region in several studies (Schoch *et al.*, 2012; Li *et al.*, 2011). The considerable degree of variability exhibited by the ITS region renders it a robust and credible contender for fungal DNA barcoding. In the realm of taxonomy, the efficacy of primary barcodes in detecting harmful fungal species is frequently found to be inadequate, prompting the ongoing establishment of secondary barcodes. The proposed approach has promise as a highly accurate and practical means of identifying phytopathogenic fungus through DNA sequence analysis. The barcoding markers utilized for fungal DNA barcoding include ITS, SSU, LSU EF-1α, and RPB.

4.2.1.10 PCR ELISA

The early 1990s saw the development of a technique known as PCR-ELISA, which combines traditional PCR and enzyme-linked immunosorbent assay (ELISA) (Nutman*et al.*, 1994). The experimental procedure entails the process of combining the labeled PCR product with an immobilized probe on an ELISA plate, followed by the introduction of an enzyme conjugate and a substrate. This allows for the analysis of the captured PCR product using a photometric measurement. The technology under consideration is a robust molecular diagnostic method that integrates the high sensitivity of PCR with the exceptional specificity of ELISA. The PCR-ELISA procedure typically commences with the amplification of a specific DNA sequence of interest through the utilization of PCR, mirroring the approach employed in conventional PCR techniques. Instead of employing gel electrophoresis as a means of viewing the PCR product, the amplified DNA is affixed to a microtiter plate or a solid surface. Subsequently, the utilization of ELISA is implemented for the purpose of detecting the DNA that has been immobilized. The methodology entails the introduction of antibodies or probes that are specifically labeled with enzymes, which exhibit affinity towards the targeted DNA sequences. The enzymatic reaction between the enzyme and substrate produces a detectable signal, typically in the form of a change in color or luminosity, thusindicating the presence of the amplified DNA. The methodology has demonstrated efficacy in the identification of viruses (Shindo*et al.*, 1994), bacteria (Daly *et al.*, 2002), and fungi (Bailey *et al.*, 2002; Somai*et al.*, 2002) with enhanced specificity compared to traditional PCR. Nevertheless, despite its notable selectivity, the assay has the potential to yield false-positive outcomes when identifying plant pathogens due to the occurrence of hybridization between the PCR amplified DNA and the ELISA probe of different species.

4.2.2 Methods based on isothermal amplification

Isothermal amplification techniques employ amplification mechanisms that do not necessitate thermal cycling equipment, but rather depend on the utilization of strand-displacing DNA polymerases. Heating blocks can be utilized as an alternative to thermal cyclers for the execution of experiments (Ivanov *et al.*, 2021). One further benefit of isothermal amplification methods is their independence from the need for DNA of high purity. A number of detection techniques employ isothermal amplification, including Nucleic acid sequence-based amplification (NASBA), loop-mediated isothermal amplification (LAMP), recombinase polymerase amplification (RPA), and helicase-dependent amplification (HDA) (Becherer et al., 2020), as discussed below.

4.2.2.1 Loop-mediated isothermal amplification (LAMP)

The LAMP technique is a molecular biology approach that employs a single-tube reaction system and an isothermal polymerase enzyme to achieve DNA amplification at a specific temperature. During the LAMP test, a set of four to six diverse primers is employed to specifically bind to six to eight unique primers located on the target gene. The primer set consists of four primers: two outer primers, F3 and B3, and two inner primers, the Forward Inner Primer (FIP) and the Backward Inner Primer (BIP). Additionally, the primer set includes loop forward and loop backward primers. The utilization of the Bst DNA polymerase enzyme is employed in the isothermal reaction. The utilization of LAMP assays has emerged as a contemporary development in the field of plant pathogen detection. Ristaino*et al.* (2020) developed a LAMP assay for the detection of *Phytophthora infestans*, a pathogen causing late blight in potatoes. The developed technique was compared to conventional PCR, real-time LAMP, and ddPCR. The sensitivity of the SYBR Green real-time LAMP tests was shown to be tenfold higher compared to conventional PCR methods. Additionally, these assays demonstrated specificity for detecting *P. infestans* in both potato and tomato samples. In order to detect the presence of *Pseudomonas syringae*, a research team developed the LAMP approach. The most recent development in the identification of *Phytophthora agathidicida*, an oomycete that causes root rot in kauri trees, is the use of the LAMP assay (Winkworth*et al.*, 2020).

4.2.2.2 Nucleic acid sequence-based amplification (NASBA)

The NASBA system is a highly sensitive and isothermal amplification method specifically developed for the detection of RNA but can sometimesbe used to amplify DNA. This phenomenon is commonly referred to as self-sustained sequence replication (3SR). The utilization of NASBA offers numerous advantages in comparison to alternative methods of mRNA amplification. The amplification of nucleic acid sequences can be achieved by the action of three enzymes, resulting in the generation of more than 109 copies within a time frame of one and a half hours. The reaction necessitates the utilization of three enzymes, namely AMV-RT for the purpose of reverse transcription and the acquisition of double-stranded cDNA, RNase H for the hydrolysis of the RNA fragment of the hybrid molecule DNARNA, and T7 RNA polymerase for the generation of a substantial quantity of anti-sense, single-strand RNA transcripts that correspond to the initial RNA target. The achievement of this can be facilitated through the utilization of two distinct primers, with one of them including at 5'end T7 promoter, NTPs, and dNTPs. The NASBA technique involves an isothermal reaction conducted at a temperature of 41°C. This specific temperature selection eliminates the requirement for a thermal cycler and enables the efficient development of point-of-test devices. One significant advantage offered by NASBA is the generation of single-stranded RNA amplicons. These amplicons can be utilized immediately in further amplification cycles or can be examined for detection purposes without the need for denaturation or strand separation. The application of this technique has been utilized in the detection of plant viruses, including Apple stem pitting virus (Klerks*et al.*, 2001), PPV (Olmos *et al.*, 2007a), Potato virus Y, ArMV, and the bacteria *C. Michiganensis* subsp. *sepedonicus* and *R. Solanacearum* (Lopez et al., 2009).

4.2.2.3 Helicase-Dependent Amplification (HDA)

The technique known as helicase-dependent amplification(HDA) was developed by New England Biolabs in 2004 as an isothermal method. This isothermal approach bears resemblance to the conventional PCR method, butwith the distinction of not need heat denaturation for the separation of double-stranded DNA. Moreover, this technique facilitates the annealing of primers to their complementary target sequences. The process of HDA involves the utilization of DNA helicase to produce ssDNA, which is then used for the annealing of primers. This is followed by primer extension under isothermal conditions. In order to impede the reassociation of ssDNA into dsDNA, the reaction incorporates the inclusion of single-stranded binding protein (SSB) and MutL endonuclease. The HDA method often produces PCR amplicons that can be detected for subsequent analysis within a time frame of 60 minutes. The utilization of HDA has gained significant popularity as an isothermal approach owing to its straightforward reaction steps. While the process of amplifying target sequences in LAMP shares the same idea as PCR, it offers a simpler procedure by eliminating the requirement for several temperature cycling steps. One significant limitation of this approach is the necessity for intricate optimization procedures in order to achieve synchronized enzyme activity between the helicase and DNA polymerase. Furthermore, the inclusion of SSB and MutL, both of which play a crucial role in preventing ssDNA from recombining to produce dsDNA, has the potential to exert a substantial impact on the ultimate outcomes. The utilization of HDA for the purpose of detecting plant pathogens remains constrained, with its application thus far being mostly focused on the identification of specific pathogens such as citrus leprosis virus C, tobacco mosaic virus (Corona *et al.*, 2010), tomato spotted wilt virus (Barreda *et al.*, 2018), and several species of *Phytophthora* (Strachan *et al.*, 2022).

4.2.2.4 Rolling circle amplification (RCA)

Rolling circle amplification (RCA) is a molecular technique that utilizes DNA or RNA polymerase enzymes to produce ssDNA or RNA molecules in an isothermal environment. The essential components needed for conducting an RCA assay include DNA polymerase, a homologous buffer, a short DNA/RNA primer, a circular template, and dNTPs. The utilization of phi29 DNA polymerase, which possesses strand displacement activity, is crucial for the amplification of DNA in this experiment. This polymerase facilitates the extension of one or more primers that anneal to a circular DNA template (van Emmerik *et al.*, 2020). The strand displacement process facilitates the displacement of a previously manufactured DNA molecule by a newly synthesized DNA template, resulting in the release of ssDNA. A series of strand displacement events results in the generation of long ssDNA molecules containing a variable number of tandem repeats (ranging from 100 to 1,000) of the initial targeted sequence. Four padlock probes, namely PLP-Nm, PLP-Np, PLP-Nk, and PLP-Nv, were specifically constructed to target TEF-1a segments for the purpose of identifying *Neofabraeamalicorticis*, *N. perennans*, *N. kienholzin,* and *N. vangabunda*, a pathogenic fungus that is responsible for causing bull's eye rot in apple fruit. The researchers in the study conducted an RCA test to diagnose the pathogens. The assay demonstrated efficacy and sensitivity in monitoring the pathogens within the quarantine sector (Lin *et al.,* 2018). The advantages of the RCA test include simplicity, efficiency, and the absence of temperature cycling equipment. Additionally, this technique can be employed for the examination of gene expression, single nucleotide polymorphism, mRNA splicing, and post-translational modification of protein molecules.

4.2.3 Post-amplification techniques

4.2.3.1 Microarrays

The utilization of microarray DNA chip technology enables the simultaneous analysis of several mRNAs, facilitating the observation of alterations in gene expression. DNA microarrays are widely recognized as a significant instrument in the realm of gene expression analysis. This technology facilitates the monitoring of vast quantities of RNA molecules that are generated within numerous diverse cell types. The production of complementary DNA copies of RNA transcripts present in a tissue sample is achieved by a reverse transcription procedure. The subsequent step is the deposition of these complementary DNA copies onto slides, employing one or more color-coding systems that align with gene expression algorithms' patterns. The utilization of DNA microarrays is anticipated to expand in the realm of plant disease detection and diagnosis due to the substantial increase in information resulting from the sequencing of plant pathogens. The development of DNA microarrays was based on the fundamental principle that complementary base sequences have the ability to hybridize and generate base pairs. In order to locate a particular complementary DNA probe on the array, it is necessary for the DNA of the target to have a binding interaction with a specific complementary DNA probe. Each of these probes is comprised of many cDNA molecules or oligonucleotides that are designed to target a certain gene, DNA sequence, or RNA sequence. The identification of nucleic acid samples can be achieved by the use of base-pairing principles, which rely on DNA and RNA sequences. The DNA probe is applied onto a solid substrate, and a small volume of a few nanoliters is utilized to generate each array.

The microarray approach is commonly referred to as destructive testing due to the necessity to physically disrupt cells, in order to access their gene expression patterns. It is important to note that the destruction of mRNA might lead to the generation of inaccurate microarray data. Singh and Kumar (2013) employed microarray technique to identify the *Aspergillus candida* species. A microarray assay (ArrayTube) was employed to detect different sugar beet root rot pathogens, namely *Aphanomyces cochlioides*, *Botrytis cinerea*, and *Penicillium expansum*, utilizing ITS, TEF-1α, and 16S rDNA, and demonstrated high probe performance.

4.2.3.2 Macroarrays

Macroarray, also known as DNA array hybridization or reverse dot plot, is a technique utilized in molecular biology and genetics research. This approach leverages the inherent sensitivity of DNA amplification techniques, eliminating the need for radioisotopes. The utilization of macroarray has gained widespread acceptance as a prominent molecular tool for conducting diagnostic and epidemiological investigations in an expanding array of laboratories worldwide. The functioning of this process is predicated upon the concurrent amplification of the associated species via PCR. In addition, this method concurrently examines many amplified sequences inside a single hybridization experiment. This assay involves the integration of PCR amplification and hybridization techniques, resulting in a significant enhancement in sensitivity by up to 1000-fold or more compared to the utilization of PCR alone. The procedure has a much shorter turnaround time of 1-2 days in contrast to the radioactive culture methods, which necessitate a completion period of 2-8 weeks. This technique involves the construction of internal probes with the purpose of distinguishing between various species. The probes are affixed to the support of the nylon membrane. The oligose molecules are affixed to the membrane in a permanent manner through the process of UV cross-linking. Subsequently, the PCR-amplified products are subjected to hybridization with a set of species-specific probes that are immobilized on the strips (Tsui *et al.*, 2011). The utilization of macroarray in conjunction with PCR has been employed for the purpose of detecting and classifying *myobacterium* at the species level (Leinberger*et al.* 2005). The utilization of macroarray technology presents a dependable and efficient approach for the diagnosis of pathogens, especially in scenarios where the sample comprises a multitude of pathogenic agents.

4.2.4 DNA/RNA PROBE-BASED TECHNIQUES (SEQUENCING-INDEPENDENT METHODS)

4.2.4.1 In situ hybridization

The technique of in situ hybridization (ISH) is employed to identify and localize messenger RNA (mRNA) molecules within a fixed sample. The primary phase of this test involves the design of an antisense single-stranded RNA probe with the objective of binding to the target mRNA, specifically the sequence of interest. In addition, alternative options for probe selection include the utilization of synthetic oligonucleotide probes and cDNA probes (Jensen, 2014). In general, radioactive isotopes like as 35S, 125I, and 32P are commonly employed for the labeling of probes due to their high sensitivity and ease of quantification, facilitating their identification. Various nonisotopic probes can be employed for probe labeling, including biotin, digoxigenin, tyramide, alkaline phosphatase, and bromodeoxyuridine. Signal detection can be achieved by several methods, including photography, autoradiography using X-ray film, liquid emulsion, and microscopic techniques. The rust fungi *Puccinia horiana* isolate PA-11, *Uromyces transversalis* isolate CA-07, and *Phakopsorapachyrhizi* isolate Taiwan 72-1 were identified as rust pathogens in *Chrysanthemum*×*morifolium*, *Gladiolus*×*hortulanus*, and *Glycine max*, respectively (Ellison *et al.*, 2016). The researchers employed the ISH technique to differentiate these pathogens in their respective plant tissues that had been preserved using paraffin fixation. ISH facilitates the optimal utilization of challenging-to-acquire tissue samples. However, the primary constraints associated with ISH pertain to its cost, the potential hazards posed by radioactive probes, and the challenges encountered in identifying targets with low DNA and RNA quantities.

4.2.4.2 Fluorescence in situ hybridization (FISH)

Fluorescence in situ hybridization (FISH) represents a very new and pioneering technological advancement within the field of plant disease detection. The key attributes of FISH that render it highly advantageous are its reproducibility, sensitivity, specificity, accuracy, and efficiency. This method is well recognized and employed for the identification of bacteria and fungus in plants. FISH assays encompass the identification of DNA or RNA sequences inside cellular or tissue samples by the utilization of DNA or RNA probes, which are labeled either directly or indirectly with fluorochromes. In conventional FISH techniques, oligonucleotide probes that are tagged with a single fluorescent molecule are used to hybridize with the rRNA present in microbial cells. Subsequently, the cells that have been stained are observed using wide field epifluorescence or confocal laser scanning microscopy techniques (Lukumbuzya et al., 2019). Cytogenetics plays a crucial role in this procedure as it facilitates the identification and precise localization of certain DNA sequences within the genome. The probes utilized in this study are labeled with haptene molecules, as they possess specific binding properties that enable them to selectively bind to sequences exhibiting a high degree of similarity. Following the invasion of pathogens, plants will exhibit the presence of pathogen-specific rRNA sequences. The detection of this particular information sent by RNA can be achieved by the utilization of the FISH technique. The occurrence of Southern blight in tomato plants can be attributed to the presence of *Sclerotium rolfsii*, a pathogenic microorganism that resides in the soil. The presence of soil smears was observed throughout the DNA isolation process, utilizing a concentration of 0.06 pgμl−1 of S. The presence of *S. rolfsii* was successfully identified through the utilization of the FISH technique, employing an oligonucleotide probe that was tagged with cyanine dyes Cy3 and Cy5 (Milner *et al.*, 2019).

4.2.4.3 Northern Blot technique

The Northern blot, alternatively referred to as the RNA blot, is employed to facilitate the transfer of RNA molecules onto a suitable substrate for the purpose of detecting and identifying pathogenic fungus. The Northern blot technique is analogous to the Southern blot technique, with the key distinction being the utilization of RNA material rather than DNA. Initially, it is important to purify the RNA extracted from each tissue in order to facilitate the analysis of gene expression pertaining to the gene of interest. Subsequently, the RNA material is put onto an agarose gel. Subsequently, the gel is subjected to an electric current, which facilitates the migration of RNA molecules towards the lower region of the gel. The velocity of smaller RNA molecules is greater in comparison to that of bigger RNA molecules. Subsequently, the fragmented RNA molecules are transferred onto a specialized filter paper, ensuring that each individual RNA molecule retains its spatial arrangement in relation to all other molecules. Subsequently, the filter is subjected to radioactive probes in order to facilitate hybridization with complimentary sequences. Subsequently, the sample is subjected to filtration in order to facilitate the process of film development during autoradiography. In conclusion, the presence of a band on the autoradiograph indicates successful hybridization of the probe with an RNA fragment immobilized on the filter. The Northern blot technique is advantageous for investigating gene expression. The relative positioning of bands on the blot provides information about the size of the RNA molecules. Knowing the size of RNA can offer an estimate of the coding capacity of transcripts and the corresponding size of the encoded protein. One of the primary drawbacks associated with the blotting approach is the potential risk of mRNA degradation occurring during the process of electrophoresis. The identification of *Magnaporthe grisea* in rice plants was accomplished using the utilization of real-time PCR and Northern blotting techniques (Qi and Yang in 2002).

4.2.5 Next-generation sequencing

Next-generation sequencing (NGS), also known as high-throughput sequencing (HTS), represents a novel diagnostic technique. The advancement of NGS technologies has facilitated the emergence of novel methodologies for the detection and identification of phytopathogens. The primary steps required in DNA-based NGS include the isolation and fragmentation of DNA, library preparation, massive parallel sequencing, bioinformatics analysis, and variant/mutation annotation and interpretation (Mansoor *et al.,* 2024). Several sophisticated sequencing technologies are routinely used in HTS, including massively parallel signature sequencing, pyrosequencing, polony sequencing, and sequencing by oligonucleotide ligation detection (SOLID). RNA-sequencing (RNA-Seq) provides enhanced coverage and improved resolution of the dynamic characteristics of the transcriptome. The Illumina HiSeq platform is widely regarded as the most versatile next-generation sequencing (NGS) platform for RNA-Seq applications, and it has become the benchmark for NGS technology. RNA-Seq, a NGS technique, has the potential to expedite the detection and characterization of fungal plant infections responsible for emerging illnesses. In this study, a methodology employing Illumina MiSeq technology was developed for the purpose of identifying the novel fungal pathogen, *Calonectriapseudonaviculata,* responsible for Sarcococca blight in ornamental plants. The strategy utilized whole-genome sequencing techniques. The utilization of datasets derived from population genomics that are based on next-generation sequencing techniques can be leveraged to recover genetic variations, such as single nucleotide polymorphisms, insertions and deletions (INDELS), and structural variations (Potgieter *et al.,* 2020).

CONCLUSION

In conclusion, molecular diagnostics, particularly nucleic acid and PCR-based approaches, have revolutionized the detection of plant pathogens. The precision, sensitivity, and speed offered by PCR and other molecular techniques have significantly enhanced our ability to identify and manage plant diseases. PCR allows for the specific amplification of target DNA, enabling the detection of pathogens even at low concentrations. This level of accuracy is crucial for early and reliable diagnosis, leading to more effective disease control measures.

The versatility of PCR-based methods, including real-time PCR and multiplex PCR, enables simultaneous detection of multiple pathogens in a single assay. This not only streamlines diagnostic processes but also provides a comprehensive understanding of the pathogenic landscape within plant populations. Additionally, the development of novel molecular markers and the integration of high-throughput sequencing technologies have expanded the scope of molecular diagnostics, allowing for the identification of new and emerging plant pathogens. Furthermore, the adoption of molecular diagnostic tools has facilitated the implementation of rapid response strategies, reducing the potential economic and environmental impacts of plant diseases. Early detection through PCR-based and other nucleic acid-based approaches such as LAMP and microarray techniques, etc., enable timely and targeted interventions, such as quarantine measures, precision application of fungicides, and the development of resistant plant varieties. As we continue to face challenges posed by emerging plant pathogens and global changes in agriculture, the ongoing advancements in molecular diagnostics, especially PCR technologies, will play a pivotal role in enhancing our capacity to monitor, manage, and mitigate the impact of plant diseases. This underscores the importance of ongoing research and technological innovation in the field of molecular diagnostics for plant pathogens, ensuring sustainable and resilient agricultural practices in the face of evolving challenges.

REFERENCES

Ali, N., Rampazzo, R. D. C. P., Costa, A. D. T., & Krieger, M. A. (2017). Current nucleic acid extraction methods and their implications to point-of-care diagnostics. *BioMed Rresearch International*, *2017*(1), 9306564.

Amagliani, G., Omiccioli, E., Del Campo, A., Bruce, I. J., Brandi, G., & Magnani, M. (2006). Development of a magnetic capture hybridization-PCR assay for Listeria monocytogenes direct detection in milk samples. *Journal of Applied Microbiology*, *100*(2), 375–383.

Azizi, M. M. F., Ismail, S. I., Hata, E. M., Zulperi, D., Ina-Salwany, M. Y., & Abdullah, M. A. F. (2019a). First report of Pantoeastewartii subsp. indologenes causing leaf blight on rice in Malaysia. *Plant Disease*, *103*(6), 1407–1407.

Azizi, M. M. F., Zulperi, D., Rahman, M. A. A., Abdul-Basir, B., Othman, N. A., Ismail, S. I., ... & Abdullah, M. A. F. (2019b). First report of Pantoeaananatis causing leaf blight disease of rice in Peninsular Malaysia. *Plant Disease*, *103*(8), 2122.

Bailey, A. M., Mitchell, D. J., Manjunath, K. L., Nolasco, G., & Niblett, C. L. (2002). Identification to the species level of the plant pathogens Phytophthora and Pythium by using unique sequences of the ITS1 region of ribosomal DNA as capture probes for PCR ELISA. *FEMS Microbiology Letters*, *207*(2), 153–158.

Baldi, P., & La Porta, N. (2020). Molecular approaches for low-cost point-of-care pathogen detection in agriculture and forestry. *Frontiers in Plant Science*, *11*, 570862.

Barreda-García, S., Miranda-Castro, R., de-Los-Santos-Álvarez, N., Miranda-Ordieres, A. J., & Lobo-Castañón, M. J. (2018). Helicase-dependent isothermal amplification: a novel tool in the development of molecular-based analytical systems for rapid pathogen detection. *Analytical and Bioanalytical Chemistry*, *410*, 679-693.

Becherer, L., Borst, N., Bakheit, M., Frischmann, S., Zengerle, R., & vonStetten, F. (2020). Loop-mediated isothermal amplification (LAMP)–review and classification of methods for sequence-specific detection. *Analytical Methods*, *12*(6), 717–746.

Bosmans, L., Paeleman, A., Moerkens, R., Wittemans, L., Van Calenberge, B., Van Kerckhove, S., ... & Lievens, B. (2016). Development of a qPCR assay for detection and quantification of rhizogenic Agrobacterium biovar 1 strains. *European Journal of Plant Pathology*, *145*, 719–730.

Botella, J. R. (2022). Point-of-care DNA amplification for disease diagnosis and management. *Annual Review of Phytopathology*, *60*(1), 1–20.

Brown, N. A., Bass, C., Baldwin, T. K., Chen, H., Massot, F., Carion, P. W., ... & Hammond-Kosack, K. E. (2011). Characterisation of the Fusarium graminearum-wheat floral interaction. *Journal of Pathogens*, 2011(1), 626345

Chen, B., Jiang, Y., Cao, X., Liu, C., Zhang, N., & Shi, D. (2021). Droplet digital PCR as an emerging tool in detecting pathogens nucleic acids in infectious diseases. *Clinica Chimica Acta*, *517*, 156–161.

Corona, F. M. O., Arif, M., Caasi, D., & Daniels, J. (2010). Application of helicase dependent amplification, 5'non-complementary A/T rich sequences and optimal thermodynamics during primer design for increased isothermal PCR detection. In *Conferenceof the International Organanization of Citrus Virologists*, *18*, 70.

Crosslin, J.M., Vandemark, G.J., & Munyaneza, J.E. (2006). Development of a real-time, quantitative PCR for detection of the Columbia basin potato purple top phytoplasma in plants and beet leafhoppers. *Plant Disease*, *90*, 663–667.

Cubero, J., Zarco-Tejada, P. J., Cuesta-Morrondo, S., Palacio-Bielsa, A., Navas-Cortés, J. A., Sabuquillo, P., & Garita-Cambronero, J. (2024). New Approaches to Plant Pathogen Detection and Disease Diagnosis. *Phytopathology®*, PHYTO-10.

Daly, P., Collier, T., & Doyle, S. (2002). PCR-ELISA detection of Escherichia coli in milk. *Letters in Applied Microbiology*, *34*(3), 222–226.

De Boer, S. H., & López, M. M. (2012). New grower-friendly methods for plant pathogen monitoring. *Annual Review of Phytopathology*, *50*, 197–218.

Dupas, E., Legendre, B., Olivier, V., Poliakoff, F., Manceau, C., & Cunty, A. (2019). Comparison of real-time PCR and droplet digital PCR for the detection of Xylella fastidiosa in plants. *Journal of Microbiological Methods*, *162*, 86–95.

Ellison, M. A., McMahon, M. B., Bonde, M. R., Palmer, C. L., & Luster, D. G. (2016). In situ hybridization for the detection of rust fungi in paraffin embedded plant tissue sections. *Plant Methods*, *12*, 1–11.

FAO. (2019). New standards to curb the global spread of plant pests and diseases. *Web Page of the Food and Agriculture Organization of the United Nations.*

Gluck-Thaler, E., & Slot, J. C. (2015). Dimensions of horizontal gene transfer in eukaryotic microbial pathogens. *PLoSPathogens*, *11*(10), e1005156.

Ha, Y., Fessehaie, A., Ling, K. S., Wechter, W. P., Keinath, A. P., & Walcott, R. R. (2009). Simultaneous detection of Acidovoraxavenae subsp. citrulli and Didymellabryoniae in cucurbit seedlots using magnetic capture hybridization and real-time polymerase chain reaction. *Phytopathology*, *99*(6), 666–678.

Hajia, M. (2018). Limitations of different PCR protocols used in diagnostic laboratories: a short review. *Modern Medical Laboratory Journal*, *1*(1), 1–6.

Ivanov, A. V., Safenkova, I. V., Zherdev, A. V., & Dzantiev, B. B. (2021). The potential use of isothermal amplification assays for in-field diagnostics of plant pathogens. *Plants*, *10*(11), 2424.

ji Cho, H., Hong, S. W., Kim, H. J., & Kwak, Y. S. (2016). Development of a multiplex PCR method to detect fungal pathogens for quarantine on exported cacti. *The Plant Pathology Journal*, *32*(1), 53.

Klerks, M. M., Leone, G., Lindner, J. L., Schoen, C. D., & Van den Heuvel, J. F. J. M. (2001). Rapid and sensitive detection of Apple stem pitting virus in apple trees through RNA amplification and probing with fluorescent molecular beacons. *Phytopathology*, *91*(11), 1085–1091.

Konstantinova, P., Bonants, P. J., Van Gent-Pelzer, M. P., & Van Der Zouwen, P. (2002). Development of specific primers for detection and identification of Alternaria spp. in carrot material by PCR and comparison with blotter and plating assays. *Mycological Research*, *106*(1), 23–33.

Kralik, P., & Ricchi, M. (2017). A basic guide to real time PCR in microbial diagnostics: definitions, parameters, and everything. *Frontiers in Microbiology*, *8*, 108.

Kumar, R., & Gupta, A. (Eds.). (2020). *Seed-borne Diseases of Agricultural Crops: Detection, Diagnosis & Management.* Springer Singapore.

Lau, H. Y., & Botella, J. R. (2017). Advanced DNA-based point-of-care diagnostic methods for plant diseases detection. *Frontiers in Plant Science*, *8*, 2016.

Lee, H. J., Cho, I. S., Ju, H. J., & Jeong, R. D. (2021). Development of a reverse transcription droplet digital PCR assay for sensitive detection of peach latent mosaic viroid. *Molecular and Cellular Probes*, *58*, 101746.

Leinberger, D. M., Schumacher, U., Autenrieth, I. B., & Bachmann, T. T. (2005). Development of a DNA microarray for detection and identification of fungal pathogens involved in invasive mycoses. *Journal of Clinical Microbiology*, *43*(10), 4943–4953.

Li, D. Z., Gao, L. M., Li, H. T., Wang, H., Ge, X. J., ... & Duan, G. W. (2011). Comparative analysis of a large dataset indicates that internal transcribed spacer (ITS) should be incorporated into the core barcode for seed plants. *Proceedings of the National Academy of Sciences*, *108*(49), 19641–19646.

Lin, H., Jiang, X., Yi, J., Wang, X., Zuo, R., Jiang, Z., ... & Zhou, E. (2018). Molecular identification of Neofabraea species associated with bull's-eye rot on apple using rolling-circle amplification of partial EF-1α sequence. *Canadian Journal of Microbiology*, *64*(1), 57–68.

Liu, J., Li, C., Muhae-Ud-Din, G., Liu, T., Chen, W., Zhang, J., & Gao, L. (2020). Development of the droplet digital PCR to detect the teliospores of TilletiacontroversaKühn in the soil with greatly enhanced sensitivity. *Frontiers in Microbiology*, *11*, 4.

López, M. M., Llop, P., Olmos, A., Marco-Noales, E., Cambra, M., & Bertolini, E. (2009). Are molecular tools solving the challenges posed by detection of plant pathogenic bacteria and viruses?. *Current issues in molecular biology*, *11*(1), 13-46.

Lukumbuzya, M., Schmid, M., Pjevac, P., & Daims, H. (2019). A multicolor fluorescence in situ hybridization approach using an extended set of fluorophores to visualize microorganisms. *Frontiers in Microbiology*, *10*, 1383.

Maheshwari, Y., Selvaraj, V., Godfrey, K., Hajeri, S., & Yokomi, R. (2021). Multiplex detection of "CandidatusLiberibacter asiaticus" and Spiroplasmacitri by qPCR and droplet digital PCR. *PloS One*, *16*(3), e0242392.

Martinelli, F., Scalenghe, R., Davino, S., Panno, S., Scuderi, G., Ruisi, P., Villa, P., et al. (2015). Advanced methods of plant disease detection. A review. *Agronomy for Sustainable Development*, 35, 1–25.

Miller, S. A., & Martin, R. R. (1988). Molecular diagnosis of plant disease. *Annual Review of Phytopathology*, *26*(1), 409–432.

Milner, H., Ji, P., Sabula, M., & Wu, T. (2019). Quantitative polymerase chain reaction (Q-PCR) and fluorescent in situ hybridization (FISH) detection of soilborne pathogen Sclerotium rolfsii. *Applied Soil Ecology*, *136*, 86–92.

Mullis, K. B., & Faloona, F. A. (1987). Specific synthesis of DNA in vitro via a polymerase-catalyzed chain reaction. *Methods in Enzymology* (Vol. 155, pp. 335–350).

Murugan, L., Krishnan, N., Venkataravanappa, V., Saha, S., Mishra, A. K., Sharma, B. K., & Rai, A. B. (2020). Molecular characterization and race identification of Fusarium oxysporum f. sp. lycopersici infecting tomato in India. *3 Biotech*, *10*, 1–12.

Nair, S., & Manimekalai, R. (2021). Phytoplasma diseases of plants: Molecular diagnostics and way forward. *World Journal of Microbiology and Biotechnology*, *37*(6), 102.

Narayanasamy, P. (2010). *Microbial Plant Pathogens-Detection and Disease Diagnosis: Viral and Viroid Pathogens, Vol. 3*. Springer Science & Business Media.

Nutman, T. B., Zimmerman, P. A., Kubofcik, J., & Kostyu, D. D. (1994). A universally applicable diagnostic approach to filarial and other infections. *Parasitology Today*, *10*(6), 239–243.

Olmos, A., Bertolini, E., & Cambra, M. (2007). Isothermal amplification coupled with rapid flow-through hybridisation for sensitive diagnosis of Plum pox virus. *Journal of Virological Methods*, *139*(1), 111–115.

Pallás, V., Sánchez-Navarro, J. A., & James, D. (2018). Recent advances on the multiplex molecular detection of plant viruses and viroids. *Frontiers in Microbiology*, *9*, 2087.

Potgieter, L., Feurtey, A., Dutheil, J. Y., & Stukenbrock, E. H. (2020). On variant discovery in genomes of fungal plant pathogens. *Frontiers in Microbiology*, *11*, 626.

Qi, M., & Yang, Y. (2002). Quantification of Magnaporthe grisea during infection of rice plants using real-time polymerase chain reaction and northern blot/phosphoimaging analyses. *Phytopathology*, *92*(8), 870–876.

Mansoor, S., Hamid, S., Tuan, T. T., Park, J. E., & Chung, Y. S. (2024). Advance computational tools for multiomics data learning. *Biotechnology Advances*, 108447.

Ristaino, J. B., Saville, A. C., Paul, R., Cooper, D. C., & Wei, Q. (2020). Detection of Phytophthora infestans by loop-mediated isothermal amplification, real-time LAMP, and droplet digital PCR. *Plant Disease*, *104*(3), 708–716.

Savary, S., Willocquet, L., Pethybridge, S. J., Esker, P., McRoberts, N., & Nelson, A. (2019). The global burden of pathogens and pests on major food crops. *Nature Ecology & Evolution*, *3*(3), 430–439.

Schoch, C. L., Seifert, K. A., Huhndorf, S., Robert, V., Spouge, J. L., Levesque, C. A., ... & White, M. M. (2012). Nuclear ribosomal internal transcribed spacer (ITS) region as a universal DNA barcode marker for Fungi. *Proceedings of the National Academy of Sciences*, *109*(16), 6241–6246.

Shen, C. H. (2023). *Diagnostic Molecular Biology*. Elsevier.

Shindo, M., Di Bisceglie, A. M., Silver, J., Limjoco, T., Hoofnagle, J. H., & Feinstone, S. M. (1994). Detection and quantitation of hepatitis C virus RNA in serum using the polymerase chain reaction and a colorimetric enzymatic detection system. *Journal of Virological Methods*, *48*(1), 65–72.

Shweta, H. M., Kumar, M. P., Teli, K., Kunduru, B., & Shekar, B. C. (2018). Isolation, identification and molecular characterization of Ralstoniasolanacerum isolates collected from Southern Karnataka. *Journal of Applied and Natural Science*, *10*, 886–893.

Singh, A., & Kumar, N. (2013). A review on DNA microarray technology. *International Journal of Current Research and Review*, *5*(22), 1.

Somai, B. M., Keinath, A. P., & Dean, R. A. (2002). Development of PCR-ELISA for detection and differentiation of Didymellabryoniae from related Phoma species. *Plant Disease*, *86*(7), 710–716.

Strachan, S., Bhuiyan, S. A., Thompson, N., Nguyen, N. T., Ford, R., & Shiddiky, M. J. (2022). Latent potential of current plant diagnostics for detection of sugarcane diseases. *Current Research in Biotechnology*, *4*, 475–492.

Szemes, M., & Schoen, C. D. (2003). Design of molecular beacons for AmpliDet RNA assay—characterization of binding stability and probe specificity. *Analytical Biochemistry*, *315*(2), 189–201.

Tsui, C. K., Woodhall, J., Chen, W., Lévesque, C. A., Lau, A., Schoen, C. D. & de Hoog, G. S. (2011). Molecular techniques for pathogen identification and fungus detection in the environment. *IMA Fungus*, *2*, 177–189.

van Emmerik, C. L., Gachulincova, I., Lobbia, V. R., Daniëls, M. A., Heus, H. A., Soufi, A., ... & van Ingen, H. (2020). Ramified rolling circle amplification for synthesis of nucleosomal DNA sequences. *Analytical Biochemistry*, *588*, 113469.

Venbrux, M., Crauwels, S., & Rediers, H. (2023). Current and emerging trends in techniques for plant pathogen detection. *Frontiers in Plant Science*, *14*, 1120968.

Verdecchia, E., Ceustermans, A., Baets, D., Ferreira, J., Bonants, P., Melis, P., ... & Lievens, B. (2021). Quantitative PCR for detection and quantification of Phytophthora cactorum in the cultivation of strawberry. *European Journal of Plant Pathology*, *160*(4), 867–882.

White, S., Schultz, T., & Enuameh, Y. A. K. (2011). *Synthesizing Evidence of Diagnostic Accuracy*. Philadelphia: Lippincott Williams & Wilkins.

Wingfield, M. J., Garnas, J. R., Hajek, A., Hurley, B. P., de Beer, Z. W., & Taerum, S. J. (2016). Novel and co-evolved associations between insects and microorganisms as drivers of forest pestilence. *Biological Invasions*, *18*, 1045–1056.

Winkworth, R. C., Nelson, B. C., Bellgard, S. E., Probst, C. M., McLenachan, P. A., & Lockhart, P. J. (2020). A LAMP at the end of the tunnel: A rapid, field deployable assay for the kauri dieback pathogen, Phytophthora agathidicida. *PLoS One*, *15*(1), e0224007.

Zhao, Y., Xia, Q., Yin, Y., & Wang, Z. (2016). Comparison of droplet digital PCR and quantitative PCR assays for quantitative detection of Xanthomonas citri subsp. citri. *PloS One*, *11*(7), e0159004.

Zhong, J., & Zhao, X. (2018). Isothermal amplification technologies for the detection of foodborne pathogens. *Food Analytical Methods*, *11*, 1543–1560.

5 Droplet Digital PCR in Fungal Pathogen Detection

Mohammad Irfan, Uzma Rashid, Javid Iqbal Mir, Diaa Abd El Moneim, and Sheikh Mansoor

5.1 INTRODUCTION

Accurate and rapid detection of fungal pathogens is crucial for the timely diagnosis and effective treatment of invasive fungal infections(Ellis, 2002). Traditional methods for fungal pathogen detection, such as culture-based techniques and microscopy, can be time-consuming and have limited sensitivity. Molecular approaches, like real-time quantitative PCR (qPCR), have improved the detection of fungal DNA, but they can still be susceptible to inhibition and require careful optimization of assays (Kumar et al., 2016; Cheng et al., 2020). Droplet digital PCR (ddPCR) is an emerging technology that offers several advantages over conventional qPCR for the detection and quantification of fungal pathogens. In ddPCR, the sample is partitioned into thousands of nanoliter-sized droplets, each of which is individually analyzed for the presence of target DNA sequences (McDermott et al., 2013). This digital amplification approach provides absolute quantification of target nucleic acids without the need for standard curves, making it less susceptible to inhibition and more precise than traditional qPCR (Lei et al., 2021). Recent studies have demonstrated the utility of ddPCR for the detection and quantification of various fungal pathogens, including Aspergillus, Candida, and Pneumocystis species, in clinical samples such as blood, respiratory specimens, and tissue biopsies. The high sensitivity, specificity, and reproducibility of ddPCR make it a promising tool for improving the diagnosis and management of fungal infections (Chen et al., 2021). The development of the polymerase chain reaction (PCR) in the 1980s revolutionized the field of molecular biology, enabling the exponential amplification of specific DNA sequences from minute amounts of starting material (Ishmael et al., 2008). This powerful technique has since become an indispensable tool in a wide range of applications, from genetic analysis and diagnostics to forensics and evolutionary studies. Over the past decades, PCR technology has continued to evolve, with each generation bringing significant advancements (Durmaz et al. 2015). Now, the third generation of PCR, known as droplet digital PCR (ddPCR), represents a transformative biotechnological refinement of conventional PCR methods (Singh et al., 2023). ddPCR offers unique capabilities that address some of the limitations of earlier PCR iterations, including the ability to directly quantify and clonally amplify target DNA molecules (Lund et al., 2016).

 DOI: 10.1201/9781003474036-5

ddPCR offers several important advantages that make it a valuable tool for rapid and accurate detection of fungal pathogens in today's clinical settings like ddPCR can detect and quantify fungal DNA with high sensitivity, often down to single-copy level, improving the ability to identify infections at early stages (Poh et al., 2020). The digital nature of ddPCR enhances specificity by reducing the impact of PCR inhibitors, leading to more reliable results. It provides absolute quantification of target DNA molecules without the need for standard curves, which are required in traditional qPCR. This allows for more accurate monitoring of fungal burden, which is crucial for guiding antifungal treatment decisions. Compared to culture-based methods that can take days to weeks, ddPCR can provide results within hours, enabling timely diagnosis and initiation of appropriate antifungal therapy (Fang et al., 2023). It requires only small sample volumes, making it particularly useful for testing clinical samples with limited availability, such as from pediatric or immune compromised patients. ddPCR platforms allow for the simultaneous detection of multiple fungal pathogens in a single test, improving efficiency and reducing costs. It can be used to monitor fungal burden during antifungal treatment, helping to assess the effectiveness of therapy and guide dose adjustments (Lei et al., 2021). These advantages of ddPCR make it a valuable tool for rapid and accurate detection of fungal pathogens in various clinical settings, from hospitals and diagnostic laboratories to resource-limited regions. By improving the speed and reliability of fungal pathogen identification, ddPCR can significantly impact patient outcomes by enabling earlier diagnosis and more targeted antifungal treatment (Fang et al., 2023).

5.2 EVOLUTION OF PCR TECHNOLOGIES IN PATHOGEN DETECTION

The evolution of PCR technologies in pathogen detection has been marked by significant advancements aimed at enhancing sensitivity, specificity, speed, and ease of use.

5.2.1 Real-time PCR (qPCR)

This technology enables the monitoring of DNA amplification in real-time, allowing for quantitative analysis of the starting target DNA concentration (Navarro et al., 2015). It eliminates the need for post-PCR processing, reducing the risk of contamination and speeding up the detection process. qPCR has greatly improved the accuracy and efficiency of pathogen detection in clinical diagnostics, environmental monitoring, and food safety (Chapela et al., 2015).

5.2.2 Multiplex PCR

Multiplex PCR allows for the simultaneous amplification of multiple targets DNA sequences in a single reaction (Markoulatos et al., 2002). By using multiple primer pairs specific to different target genes or pathogens, multiplex PCR enables the detection of multiple pathogens in a single sample (Caliendo et al., 2011). This approach improves throughput, conserves sample material, and reduces assay costs, making it particularly valuable in clinical diagnostics and epidemiological studies (De Paoli, 2005).

5.2.3 Digital PCR (dPCR)

Digital PCR partitions the PCR reaction mixture into thousands of individual microreactions, each containing either zero or one target DNA molecule (Cao et al., 2017). By counting the number of positive partitions, dPCR provides absolute quantification of the target DNA without the need for standard curves (Sanders et al., 2011). This technology offers enhanced precision and sensitivity, particularly useful for detecting low abundance pathogens and monitoring subtle changes in pathogen load over time (Sibley et al., 2012).

5.2.4 Isothermal Amplification Techniques

Isothermal amplification methods, such as loop mediated isothermal amplification (LAMP) and recombinase polymerase amplification (RPA), amplify DNA at a constant temperature without the need for thermal cycling (Naveen et al., 2020). These techniques offer rapid and sensitive detection of pathogens under simple conditions, making them suitable for point-of-care diagnostics, field based testing, and resource limited settings (Yager et al., 2008).

5.2.5 Next-Generation Sequencing (NGS)

While not strictly PCR based, NGS technologies have revolutionized pathogen detection by enabling the sequencing and identification of entire microbial genomes directly from clinical or environmental samples (Gu et al., 2019). PCR is often used as a preprocessing step in NGS workflows to amplify target DNA regions or enrich for specific microbial taxa (Hodkinson et al., 2015). NGS provides comprehensive genomic information, facilitating the identification of known and novel pathogens, characterization of microbial communities, and investigation of antimicrobial resistance (Angers et al., 2018). These advancements in PCR technologies have greatly expanded the capabilities of pathogen detection, allowing for faster, more accurate, and more comprehensive analyses in diverse applications ranging from clinical diagnostics to public health surveillance and outbreak investigation (Sibley et al., 2012).

5.3 PCR

Polymerase chain reaction (PCR) is a powerful molecular biology technique used to amplify specific DNA sequences from a small amount of starting material. Developed in the 1980s by Kary Mullis, PCR has revolutionized numerous fields, including genetics, microbiology, forensics, and diagnostics (Botes et al., 2011). The fundamental principle of PCR is the enzymatic replication of DNA. It involves three main steps:

1. *Denaturation:* The double-stranded DNA is heated to high temperatures (usually around 95°C) to separate the strands.
2. *Annealing:* Short DNA sequences called primers bind to complementary regions on the single-stranded DNA, serving as starting points for DNA synthesis.
3. *Extension:* A DNA polymerase enzyme uses the primers and available nucleotides to synthesize new complementary DNA strands (Kadri et al., 2019).

These three steps are repeated in cycles, typically 20-40 times, resulting in the exponential amplification of the target DNA sequence. Each cycle doubles the amount of DNA, leading to a million-fold or greater increase in the final product. PCR is highly sensitive, capable of detecting and amplifying even a single copy of a target DNA molecule (Bej et al., 1991). This sensitivity, combined with its specificity and speed, has made PCR a widely adopted technique in various applications, such asgenetic analysis and diagnostics, pathogen detection and identification, forensic DNA profiling, evolutionary studies and phylogenetic analysis, and quantitative gene expression analysis (Artika et al., 2022). Advancements in PCR technology, such as real-time PCR and digital PCR, have further expanded the capabilities of this versatile technique, allowing for accurate quantification, improved sensitivity, and greater multiplexing capabilities (Kreitmann et al., 2023).

5.4 DROPLET PCR

Droplet PCR, also known as ddPCR, is an advanced variation of the traditional PCR technique (Li et al., 2018). It offers precise and absolute quantification of nucleic acids, providing advantages over conventional PCR methods (Bustin et al., 2000). ddPCRpartitions the PCR reaction mixture into thousands of individual droplets, each containing a single DNA template molecule, or no molecule at all (Zhang et al., 2010). This partitioning process is typically achieved using microfluidic devices or emulsions. Each droplet serves as a separate reaction chamber, enabling absolute quantification of the target DNA without relying on standard curves (Hou et al., 2023).

5.4.1 Key Features

ddPCR offers several key features that distinguish it from traditional PCR methods and other nucleic acid quantification techniques. ddPCR enables absolute quantification of target DNA or RNA molecules without the need for standard curves (Dodd et al., 2013). By partitioning the PCR reaction mixture into thousands of individual droplets, each containing either one or no copies of the target molecule, ddPCR provides precise and absolute quantification of the target concentration (Zhang et al., 2010). It offers improved precision and accuracy compared to traditional qPCR methods. By partitioning the sample into numerous individual reactions, ddPCR reduces the effects of PCR inhibitors, amplification efficiency variations, and other sources of experimental variability, leading to more reliable and reproducible results (Wang et al., 2020). It enhances the sensitivity of nucleic acid detection, enabling the detection of rare targets present at low concentrations (Long et al., 2022). The ability to detect and quantify low abundance targets with high confidence makes ddPCR particularly well suited for applications such as detecting circulating tumor DNA (ctDNA) in liquid biopsy samples or monitoring minimal residual disease (MRD) in cancer patients (Tarazona et al., 2019). ddPCR offers a wide dynamic range of target quantification, spanning several orders of magnitude (Quan et al., 2018). This allows for accurate quantification of both high and low target concentrations within the same experiment, without the need for sample dilution or multiple assays (Trypsteen et al., 2015). ddPCR enables multiplexed detection and quantification of multiple targets in a single reaction. By using different fluorescent probes or dyes to label each target, ddPCR allows for simultaneous analysis of multiple targets, saving time, resources, and sample material (Dobnik et al., 2015). It is less susceptible to PCR inhibitors present in complex or challenging sample matrices compared to traditional qPCR methods. The partitioning of the sample into individual droplets helps mitigate the effects of inhibitors, improving assay performance and reliability (Sidstedt et al., 2023). It generates clear and straightforward data outputs, with positive and negative droplets easily distinguishable based on fluorescence intensity. This simplifies data interpretation and analysis, even for complex experiments involving multiple targets or samples (Fang et al., 2018). ddPCR is compatible with a wide range of sample types, including genomic DNA, cDNA, RNA, and cell free nucleic acids. This versatility allows for the analysis of diverse sample sources, including blood, tissue, plasma, urine, saliva, and environmental samples (Kojabad et al., 2021).

5.4.2 Experimental Protocols and Workflow

Sample preparation is a crucial step in the workflow of ddPCR for fungal pathogen detection. Proper sample preparation ensures the efficient extraction and purification of fungal DNA from clinical or environmental samples, maximizing the sensitivity and accuracy of ddPCR assays (Bope et al., 2018). Workflow for sample preparation in ddPCR fungal pathogen detection includes the following.

Sample Collection (Ensure proper labeling and documentation)

⇩

Sample Processing (Ensure proper homogenization of sample)

⇩

Fungal DNA Extraction (Choose suitable DNA extraction method)

⇩

DNA Quantification and Quality Assessment (Quantify the concentration by NanoDrop)

⇩

Quality Control Checks (to ensure reliability and integrity of extracted fungal DNA)

⇩

Storage of Extracted DNA (at -20°C or -80°C)

⇩

Documentation and Record Keeping (Maintain detailed records of sample)

5.4.3 Primer and Probe Design

Primer and probe design for ddPCR in fungal pathogen detection is crucial for ensuring specific and efficient amplification of target DNA sequences (Kokkoris et al., 2021). Some guidelines and considerations for primer and probe design include:

i. **Target Selection:** Choose target regions of the fungal pathogen's genome that are conserved among different strains and species. Select regions that are unique to the fungal pathogen of interest to minimize cross reactivity with nontarget DNA (Landlinger et al., 2009).
ii. **Primer Design:** Design primers with a length of 1825 nucleotides, with a melting temperature (Tm) of approximately 5565°C. Aim for a guanine cytocine content of 4060% to ensure stable primer annealing and efficient amplification. Avoid primer sequences with significant self complementarity or secondary structures that may interfere with primer annealing or PCR amplification (Borah, 2011).Use primer design software tools (e.g., Primer3, NCBI Primer BLAST) to design specific and efficient primer pairs targeting the selected genomic regions (Ye et al., 2012).
iii. **Probe Design:** Probes should be approximately 20 nucleotides in length, with a Tm of 6070°C. Design probes with a fluorescent reporter dye (e.g., FAM, VIC, HEX) at the 5' end and a quencher dye (e.g., TAMRA, BHQ) at the 3' end (Suseela et al., 2018). Position the probe annealing site between the forward and reverse primers to ensure efficient probe binding and cleavage by the DNA polymerase during PCR amplification (Apte and Daniel, 2009).
iv. **Considerations for Probe Placement:** Place the probe annealing site within the amplicon region targeted by the primers to maximize probebinding efficiency and signal generation. Avoid regions with known polymorphisms or sequence variations that may affect probe binding or detection specificity (Rodriguez et al., 2015). Position the probe annealing site away from primer binding sites to prevent interference with primer annealing and extension during PCR amplification (Apte and Daniel, 2009).
v. **Specificity Checking:** Perform in silico specificity checks using bioinformatics tools (e.g., BLAST) to ensure the specificity of primer and probe sequences against the target fungal genome (Davi et al., 2021). Avoid primer and probe sequences that exhibit significant homology to nontarget sequences, including host DNA, commensal organisms, or environmental contaminants (Green et al., 2014).
vi. **Experimental Validation:** Validate primer and probe sequences experimentally using standard PCR or quantitative PCR (qPCR) assays. Evaluate primer and probe performance for amplification efficiency, specificity, and sensitivity using synthetic DNA templates or known positive controls (Rodriguez et al., 2015).
vii. **Optimization and Iteration:** Optimize primer and probe concentrations, annealing temperatures, and other PCR conditions to maximize assay performance and sensitivity (Green et al., 2018). Iterate primer and probe design if necessary based on experimental results or feedback from initial validation experiments (Kalendar et al., 2009).

5.4.4 Reaction Setup for Droplet Generation

Setting up the reaction for droplet generation in ddPCR for fungal pathogen detection involves careful preparation of the PCR reaction mix and emulsification with droplet generation oil as shown in Figure 5.1.

Preparation of Master Mix

⇩

Prepare Droplet Generation Oil

⇩

Combine Master Mix and Droplet Generation Oil

⇩

Emulsification

⇩

Droplet Generation

⇩

Transfer to PCR Plate

⇩

Proceed to PCR Amplification

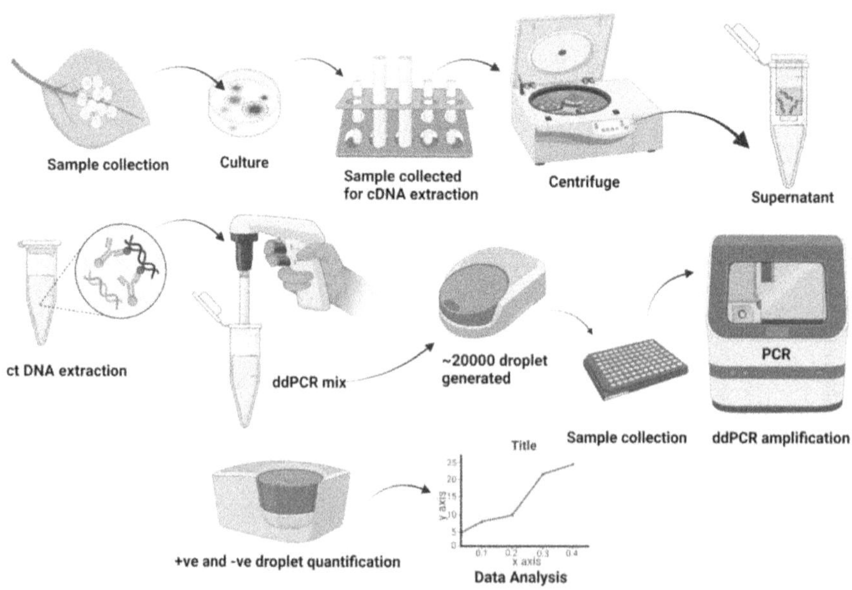

FIGURE 5.1 Workflow and Droplet Generation of ddPCR in Fungal Pathogen Detection.

5.4.5 Amplification and Detection in Individual Droplets of Droplet Digital PCR in Fungal Pathogen Detection

Amplification and detection in individual droplets of ddPCRfor fungal pathogen detection involve performing PCR amplification within each droplet followed by fluorescence detection to determine the presence or absence of the target DNA (Kokkoris et al., 2021).

a. **Amplification in Individual Droplets:** In ddPCR, the sample is first partitioned into thousands to millions of nanoliter-sized droplets. Each droplet acts as an individual reaction chamber, containing the PCR reagents, target DNA, and a single or few DNA molecules (Wainman et al., 2024). The droplets undergo the standard PCR thermal cycling steps, denaturation, annealing, and extension within the sealed droplets. The compartmentalization of the sample into discrete droplets allows for clonal amplification of the target DNA sequences within each droplet (Perkins et al., 2017).
b. **Detection in Individual Droplets:** After amplification, the droplets are passed through a detection system, typically a flow cytometer or a droplet reader (Zhu et al., 2020). Each droplet is individually interrogated for the presence of the amplified target DNA sequences (Zhang et al., 2010).
c. **Data Analysis and Interpretation:** Use dedicated software (e.g., BioRadQuantaSoft Analysis Pro) to analyze the fluorescence data and quantify the target DNA molecules. Calculate the absolute concentration of the target fungal pathogen in the original sample (Marusina, 2017).

5.4.6 Development of Point-of-Care Applications

The development of point-of-care (POC) applications of ddPCR in fungal pathogen detection holds immense promise for improving timely diagnosis, treatment, and management of fungal infections (Chen et al., 2022).While ddPCR is traditionally performed in laboratory settings, efforts are underway to adapt this technology for use at the point of care, such as in clinics, hospitals, and remote healthcare settings (Iliescu et al., 2021).Some key developments and considerations in the development of POC applications of ddPCR in fungal pathogen detection include:

i. **Miniaturization of Instrumentation:** The development of portable and miniaturized ddPCR platforms suitable for POC settings is underway. These compact instruments are designed to be lightweight, user friendly, and battery powered, enabling deployment in resource limited or remote environments (Messina et al., 2020). Integration of microfluidic technologies, microfabrication techniques, and handheld devices may further enhance the portability and accessibility of ddPCR for POC applications (Nazim et al., 2023).
ii. **Simplified Sample Preparation:** Streamlined and user-friendly sample preparation protocols compatible with POC settings are being developed to facilitate rapid and efficient extraction of fungal DNA from clinical specimens

(Nichols et al., 2021). Integration of sample toanswer platforms and automated sample processing systems may eliminate the need for complex laboratory equipment and technical expertise, making ddPCR more accessible to nonspecialized healthcare providers (Geng et al., 2020).

iii. **Rapid Turnaround Time:** Efforts are focused on reducing assay turnaround time to enable rapid onsite detection and diagnosis of fungal infections (Fang et al., 2023). Optimization of PCR cycling parameters, droplet generation, and fluorescence detection methods may accelerate the ddPCR workflow without compromising sensitivity or accuracy (Kojabad et al., 2021). POCddPCR assays with turnaround times of less than one hour are being explored for timely diagnosis and immediate clinical decision making in POC settings (Rajendran et al., 2021).

iv. **Single-Use Assays and Disposable Cartridges:** Disposable cartridges and single-use assay formats are being developed to minimize cross-contamination and simplify assay setup and disposal in POC settings (Gubala et al., 2012). These disposable cartridges may incorporate preloaded reagents, lyophilized PCR components, and integrated microfluidic channels for sample processing, droplet generation, and fluorescence detection, enabling rapid and convenient on-demand testing at the point of care (Liu et al., 2021).

v. **Integration with Portable Diagnostic Platforms:** Integration of ddPCR with other portable diagnostic platforms, such as smartphone-based devices, handheld spectrometers, and point-of-care molecular diagnostic systems, may further expand access to POC ddPCR testing (Liu et al., 2021). Smartphone-based applications for data analysis, result interpretation, and remote consultation may enhance user interface and connectivity, enabling real-time transmission of results to healthcare providers and public health authorities (Albahri et al., 2018).

vi. **Validation and Clinical Utility:** Rigorous validation studies and clinical trials are essential to evaluate the performance, reliability, and clinical utility of POC ddPCR assays for fungal pathogen detection (Hou et al., 2023). Collaborative efforts among researchers, clinicians, industry partners, and regulatory agencies are needed to establish standardized protocols, quality control measures, and regulatory guidelines for POC ddPCR testing in fungal infection management (Crowley et al., 2004).

5.5 CONCLUSION

Droplet digital PCR holds great promise for the detection and quantification of fungal pathogens, offering several advantages over traditional PCR methods. As evidenced by its growing adoption in research, clinical diagnostics, and environmental monitoring, ddPCR has demonstrated its utility in addressing the challenges associated with fungal pathogen detection. ddPCR enables precise and absolute quantification of fungal DNA, offering enhanced sensitivity and accuracy compared to traditional PCR methods. The ability to partition the PCR reaction into thousands of individual droplets allows for robust detection of low abundance fungal pathogens in complex samples, improving diagnostic sensitivity and specificity. Multiplexing capabilities enable simultaneous detection and quantification of multiple fungal species or genetic markers in a single reaction, facilitating comprehensive fungal profiling and differentiation of closely related species. ddPCR offers advantages in antifungal resistance testing, environmental monitoring, outbreak investigation, research, and epidemiological studies, contributing to improved disease management and public health outcomes.

5.6 FUTURE PERSPECTIVES

Further optimization and standardization of ddPCR protocols for fungal pathogen detection will enhance assay performance, reproducibility, and comparability across different laboratories and platforms. Continued development of novel primer and probe sets targeting conserved regions of fungal genomes, as well as emerging antifungal resistance markers, will expand the utility of ddPCR in clinical and environmental settings. Integration of ddPCR with other molecular diagnostic techniques, such as next-generation sequencing (NGS) and meta-genomic analysis, may provide comprehensive insights into fungal diversity, pathogenesis, and drug resistance mechanisms. Advancements in droplet generation technology, fluorescence detection methods, and analysis software will further improve the sensitivity, speed, and scalability of ddPCR assays for fungal pathogen detection. Exploration of ddPCRapplications in point-of-care testing (POCT) and resource-limited settings may facilitate rapid and accurate diagnosis of fungal infections in diverse clinical and environmental contexts. Collaborative efforts among researchers, clinicians, industry partners, and regulatory agencies are essential to address remaining challenges, validate ddPCR assays, and translate innovative technologies into clinical practice for improved patient care and public health outcomes.

REFERENCES

Albahri, O. S., Albahri, A. S., Mohammed, K. I., Zaidan, A. A., Zaidan, B. B., Hashim, M., & Salman, O. H. (2018). Systematic review of real-time remote health monitoring system in triage and priority-based sensor technology: Taxonomy, open challenges, motivation and recommendations. *Journal of Medical Systems*, *42*, 1–27.

Angers-Loustau, A., Petrillo, M., Bengtsson-Palme, J., Berendonk, T., Blais, B., Chan, K. G., ... & Van den Eede, G. (2018). The challenges of designing a benchmark strategy for bioinformatics pipelines in the identification of antimicrobial resistance determinants using next generation sequencing technologies. *F1000Research*, 7, 459.

Apte, A., & Daniel, S. (2009). PCR primer design. *Cold Spring Harbor Protocols*, *2009*(3), pdb–ip65.

Artika, I. M., Dewi, Y. P., Nainggolan, I. M., Siregar, J. E., & Antonjaya, U. (2022). Real-time polymerase chain reaction: Current techniques, applications, and role in COVID-19 diagnosis. *Genes*, *13*(12), 2387.

Bej, A. K., Mahbubani, M. H., & Atlas, R. M. (1991). Amplification of nucleic acids by polymerase chain reaction (PCR) and other methods and their applications. *Critical Reviews in Biochemistry and Molecular Biology*, *26*(3–4), 301–334.

Bope, A. M. (2018). *Using Digital PCR to Improve Quantitative Measurement of Bacterial and Fungal DNA from Carpet Dusts in the Built Environment* (Master's thesis, The Ohio State University).

Borah, P. (2011). Primer designing for PCR. *Science Vision*, *11*(3), 134–136.

Botes, M., de Kwaadsteniet, M., & Cloete, T. E. (2011). Application of quantitative PCR for the detection of microorganisms in water. *Food Microbiology*, *28*(5), 848–61.

Bustin, S. A. (2000). Absolute quantification of mRNA using real-time reverse transcription polymerase chain reaction assays. *Journal of Molecular Endocrinology*, *25*(2), 169–193.

Caliendo, A. M. (2011). Multiplex PCR and emerging technologies for the detection of respiratory pathogens. *Clinical Infectious Diseases*, *52*(suppl_4), S326–S330.

Cao, L., Cui, X., Hu, J., Li, Z., Choi, J. R., Yang, Q., ... & Xu, F. (2017). Advances in digital polymerase chain reaction (dPCR) and its emerging biomedical applications. *Biosensors and Bioelectronics*, *90*, 459–474.

Chapela, M. J., Garrido-Maestu, A., & Cabado, A. G. (2015). Detection of foodborne pathogens by qPCR: A practical approach for food industry applications. *Cogent Food & Agriculture*, *1*(1), 1013771.

Chen, B., Jiang, Y., Cao, X., Liu, C., Zhang, N., & Shi, D. (2021). Droplet digital PCR as an emerging tool in detecting pathogens nucleic acids in infectious diseases. *ClinicaChimica Acta*, *517*, 156–161.

Chen, S., Sun, Y., Fan, F., Chen, S., Zhang, Y., Zhang, Y., ... & Lin, J. M. (2022). Present status of microfluidic PCR chip in nucleic acid detection and future perspective. *TrAC Trends in Analytical Chemistry*, *157*, 116737.

Cheng, Y., Tang, X., Gao, C., Li, Z., Chen, J., Guo, L., et al. (2020). Molecular diagnostics and pathogenesis of fungal pathogens on bast fiber crops. *Pathogens*, *9*(3), 223.10.3390/pathogens9030223

Crowley Jr, W. F., Sherwood, L., Salber, P., Scheinberg, D., Slavkin, H., Tilson, H., ... & Rimoin, D. (2004). Clinical research in the United States at a crossroads: Proposal for a novel public-private partnership to establish a national clinical research enterprise. *Jama*, *291*(9), 1120–1126.

Davi, M. J. P., Jeronimo, S. M., Lima, J. P., & Lanza, D. C. (2021). Design and in silico validation of polymerase chain reaction primers to detect severe acute respiratory syndrome coronavirus 2 (SARS-CoV-2). *Scientific Reports*, *11*(1), 12565.

De Paoli, P. (2005). Biobanking in microbiology: From sample collection to epidemiology, diagnosis and research. *FEMS Microbiology Reviews*, *29*(5), 897–910.

Dobnik, D., Štebih, D., Blejec, A., Morisset, D., & Žel, J. (2016). Multiplex quantification of four DNA targets in one reaction with Bio-Rad droplet digital PCR system for GMO detection. *Scientific Reports*, *6*(1), 35451.

Dodd, D. W., Gagnon, K. T., & Corey, D. R. (2013). Digital quantitation of potential therapeutic target RNAs. *Nucleic Acid Therapeutics*, *23*(3), 188–194.

Durmaz, A. A., Karaca, E., Demkow, U., Toruner, G., Schoumans, J., & Cogulu, O. (2015). Evolution of genetic techniques: Past, present, and beyond. *BioMed Research International*, *2015*(1), 461524

Ellis, M. (2002). Invasive fungal infections: evolving challenges for diagnosis and therapeutics. *Molecular Immunology*, *38*(12-13), 947–957.

Fang, W., Wu, J., Cheng, M., Zhu, X., Du, M., Chen, C., ... & Pan, W. (2023). Diagnosis of invasive fungal infections: Challenges and recent developments. *Journal of Biomedical Science*, *30*(1), 42.

Fang, X., Zheng, Y., Duan, Y., Liu, Y., & Zhong, W. (2018). Recent advances in design of fluorescence-based assays for high-throughput screening. *Analytical Chemistry*, *91*(1), 482–504.

Geng, Z., Li, S., Zhu, L., Cheng, Z., Jin, M., Liu, B., ... & Liu, P. (2020). "Sample-to-answer" detection of rare ctDNA mutation from 2 mL plasma with a fully integrated DNA extraction and digital droplet PCR microdevice for liquid biopsy. *Analytical Chemistry*, *92*(10), 7240–7248.

Green, H. C., White, K. M., Kelty, C. A., & Shanks, O. C. (2014). Development of rapid canine fecal source identification PCR-based assays. *Environmental Science & Technology*, *48*(19), 11453–11461.

Green, M. R., & Sambrook, J. (2018). Optimizing primer and probe concentrations for use in real-time polymerase chain reaction (PCR) assays. *Cold Spring Harbor Protocols*, *2018*(10), pdb-prot095018.

Gu, W., Miller, S., & Chiu, C. Y. (2019). Clinical metagenomic next-generation sequencing for pathogen detection. *Annual Review of Pathology: Mechanisms of Disease*, *14*, 319–338.

Gubala, V., Harris, L. F., Ricco, A. J., Tan, M. X., & Williams, D. E. (2012). Point of care diagnostics: status and future. *Analytical Chemistry*, *84*(2), 487–515.

Hodkinson, B. P., & Grice, E. A. (2015). Next-generation sequencing: a review of technologies and tools for wound microbiome research. *Advances in Wound Care*, *4*(1), 50–58.

Hou, Y., Chen, S., Zheng, Y., Zheng, X., & Lin, J. M. (2023). Droplet-based digital PCR (ddPCR) and its applications. *TrAC Trends in Analytical Chemistry*, *158*, 116897.

Iliescu, F. S., Ionescu, A. M., Gogianu, L., Simion, M., Dediu, V., Chifiriuc, M. C., ... & Iliescu, C. (2021). Point-of-care testing—the key in the battle against SARS-CoV-2 pandemic. *Micromachines*, *12*(12), 1464.

Ishmael, F. T., & Stellato, C. (2008). Principles and applications of polymerase chain reaction: Basic science for the practicing physician. *Annals of Allergy, Asthma & Immunology*, *101*(4), 437–443.

Kadri, K. (2019). Polymerase chain reaction (PCR): Principle and applications. *Synthetic Biology-New Interdisciplinary Science*, *1,* 157–154 .

Kalendar, R., Lee, D., & Schulman, A. H. (2009). FastPCR software for PCR primer and probe design and repeat search. *Genes, Genomes and Genomics*, *3*(1), 1–14.

Kojabad, A. A., Farzanehpour, M., Galeh, H. E. G., Dorostkar, R., Jafarpour, A., Bolandian, M., & Nodooshan, M. M. (2021). Droplet digital PCR of viral DNA/RNA, currentprogress, challenges, andfutureperspectives. *Journal of Medical Virology*, *93*(7), 4182–4197.

Kokkoris, V., Vukicevich, E., Richards, A., Thomsen, C., & Hart, M. M. (2021). Challenges using droplet digital PCR for environmental samples. *Applied Microbiology*, *1*(1), 74–88.

Kreitmann, L., Miglietta, L., Xu, K., Malpartida-Cardenas, K., D'Souza, G., Kaforou, M., ... & Rodriguez-Manzano, J. (2023). Next-generation molecular diagnostics: Leveraging digital technologies to enhance multiplexing in real-time PCR. *TrAC Trends in Analytical Chemistry*, *160*, 116963.

Kumar, P., Gupta, V. K., Tiwari, A. K., & Kamle, M. (Eds.). (2016). *Current Trends in Plant Disease Diagnostics and Management Practices*. Springer International Publishing.

Landlinger, C., Preuner, S., Willinger, B., Haberpursch, B., Racil, Z., Mayer, J., & Lion, T. (2009). Species-specific identification of a wide range of clinically relevant fungal pathogens by use of Luminex xMAP technology. *Journal of Clinical Microbiology*, *47*(4), 1063–1073.

Lei, S., Chen, S., & Zhong, Q. (2021). Digital PCR for accurate quantification of pathogens: Principles, applications, challenges and future prospects. *International Journal of Biological Macromolecules*, *184*, 750–759.

Li, H., Bai, R., Zhao, Z., Tao, L., Ma, M., Ji, Z., ... & Liu, A. (2018). Application of droplet digital PCR to detect the pathogens of infectious diseases. *Bioscience Reports*, *38*(6), BSR20181170.

Liu, W., Yue, F., & Lee, L. P. (2021). Integrated point-of-care molecular diagnostic devices for infectious diseases. *Accounts of Chemical Research*, *54*(22), 4107–4119.

Long, S., & Berkemeier, B. (2022). Ultrasensitive detection and quantification of viral nucleic acids with Raindance droplet digital PCR (ddPCR). *Methods*, *201*, 49–64.

Lund, H. L., Hughesman, C. B., McNeil, K., Clemens, S., Hocken, K., Pettersson, R., ... & Haynes, C. (2016). Initial diagnosis of chronic myelogenous leukemia based on quantification of M-BCR status using droplet digital PCR. *Analytical and Bioanalytical Chemistry*, *408*, 1079–1094.

Markoulatos, P., Siafakas, N., & Moncany, M. (2002). Multiplex polymerase chain reaction: a practical approach. *Journal of Clinical Laboratory Analysis*, *16*(1), 47–51.

Marusina, K. (2017). Positioning digital PCR for sharper genomic views. *Genetic Engineering & Biotechnology News*, *37*(17), 1–8.

McDermott, G. P., Do, D., Litterst, C. M., Maar, D., Hindson, C. M., Steenblock, E. R., ... & Lowe, A. J. (2013). Multiplexed target detection using DNA-binding dye chemistry in droplet digital PCR. *Analytical Chemistry*, *85*(23), 11619–11627.

Messina, P., Luciano, B., & Placente, D. (2020). *Tomorrow's Healthcare by Nano-sized Approaches: A Bold Future for Medicine*. CRC Press.

Navarro, E., Serrano-Heras, G., Castaño, M. J., & Solera, J. J. C. C. A. (2015). Real-time PCR detection chemistry. *ClinicaChimicaActa*, *439*, 231–250.

Naveen, K. P., & Bhat, A. I. (2020). Development of reverse transcription loop-mediated isothermal amplification (RT-LAMP) and reverse transcription recombinase polymerase amplification (RT-RPA) assays for the detection of two novel viruses infecting ginger. *Journal of Virological Methods*, *282*, 113884.

Nazim, F., & Tabassum, S. (2023). Application of micro-fluidics in biological world. *Avicenna Journal of Medical Sciences*, *3*(1).

Nichols, Z. E., & Geddes, C. D. (2021). Sample preparation and diagnostic methods for a variety of settings: A comprehensive review. *Molecules*, *26*(18), 5666.

Perkins, G., Lu, H., Garlan, F., & Taly, V. (2017). Droplet-based digital PCR: application in cancer research. *Advances in Clinical Chemistry*, *79*, 43–91.

Poh, T. Y., Ali, N. A. T. B. M., Chan, L. L., Tiew, P. Y., & Chotirmall, S. H. (2020). Evaluation of droplet digital polymerase chain reaction (ddPCR) for the absolute quantification of Aspergillus species in the human airway. *International Journal of Molecular Sciences*, *21*(9), 3043.

Quan, P. L., Sauzade, M., & Brouzes, E. (2018). dPCR: Atechnology review. *Sensors*, *18*(4), 1271.

Rajendran, V. K., Bakthavathsalam, P., Bergquist, P. L., & Sunna, A. (2021). Smartphone technology facilitates point-of-care nucleic acid diagnosis: A beginner's guide. *Critical Reviews in Clinical Laboratory Sciences*, *58*(2), 77–100.

Rodríguez, A., Rodríguez, M., Córdoba, J. J., & Andrade, M. J. (2015). Design of primers and probes for quantitative real-time PCR methods. *PCR Primer Design*,31–56.

Sanders, R., Huggett, J. F., Bushell, C. A., Cowen, S., Scott, D. J., & Foy, C. A. (2011). Evaluation of digital PCR for absolute DNA quantification. *Analytical Chemistry*, *83*(17), 6474–6484.

Sibley, C. D., Peirano, G., & Church, D. L. (2012). Molecular methods for pathogen and microbial community detection and characterization: Current and potential application in diagnostic microbiology. *Infection, Genetics and Evolution*, *12*(3), 505–521.

Sidstedt, M., Rådström, P., & Hedman, J. (2020). PCR inhibition in qPCR, dPCR and MPS—mechanisms and solutions. *Analytical and Bioanalytical Chemistry*, *412*(9), 2009–2023.

Singh, Y., Anand, K. J., Singh, M., Ambawat, S., & Tiwari, G. (2023). Recent developments in PCR technology. *Biological Sciences*, *77*, 79.

Suseela, Y. V., Narayanaswamy, N., Pratihar, S., & Govindaraju, T. (2018). Far-red fluorescent probes for canonical and non-canonical nucleic acid structures: current progress and future implications. *Chemical Society Reviews*, *47*(3), 1098–1131.

Tarazona, N., Gimeno-Valiente, F., Gambardella, V., Zuñiga, S., Rentero-Garrido, P., Huerta, M., ... & Cervantes, A. (2019). Targeted next-generation sequencing of circulating-tumor DNA for tracking minimal residual disease in localized colon cancer. *Annals of Oncology*, *30*(11), 1804–1812.

Trypsteen, W., Vynck, M., De Neve, J., Bonczkowski, P., Kiselinova, M., Malatinkova, E., ... & De Spiegelaere, W. (2015). ddpcRquant: Threshold determination for single channel droplet digital PCR experiments. *Analytical and Bioanalytical Chemistry*, *407*, 5827–5834.

Wainman, L. M., Sathyanarayana, S. H., & Lefferts, J. A. (2024). Applications of digital polymerase chain reaction (dPCR) in molecular and clinical testing. *Journal of Applied Laboratory Medicine*, *9*(1), 124–137.

Wang, Y., Menon, N., Shen, S., Feschenko, M., & Bergelson, S. (2020). A qPCR method for AAV genome titer with ddPCR-level of accuracy and precision. *Molecular Therapy-Methods & Clinical Development*, *19*, 341–346.

Yager, P., Domingo, G. J., & Gerdes, J. (2008). Point-of-care diagnostics for global health. *Annual Review of Biomedical Engineering*, *10*, 107–144.

Ye, J., Coulouris, G., Zaretskaya, I., Cutcutache, I., Rozen, S., & Madden, T. L. (2012). Primer-BLAST: A tool to design target-specific primers for polymerase chain reaction. *BMC Bioinformatics*, *13*, 1–11.

Zhang, C., & Xing, D. (2010). Single-molecule DNA amplification and analysis using microfluidics. *Chemical Reviews*, *110*(8), 4910–4947.

Zhu, X., Liu, B., Su, S., Wang, B., Bai, Y., Huang, H., ... & Guo, Y. (2020). A "quasi" confocal droplet reader based on laser-induced fluorescence (LIF) cytometry for highly-sensitive and contamination-free detection. *Talanta*, *206*, 120200.

6 Molecular PLANT-Dx System for Point-of-Use Detection of Pathogens Associated with Plant Disease

Subaya Manzoor, Sajad Un Nabi, Tariq Rasool Rather, Farooq Ahmad Bhat, Zahoor Ahmad Mir, Shugufta Parveen, Uzma Rashid, and Sheikh Mansoor

6.1 INTRODUCTION

Agriculture represents a substantial economic value, estimated at $1,500 billion US dollars annually (Agrios, 2005). However, a significant portion of agricultural produce is lost every year due to various diseases, particularly in developing countries (Agrios, 2005; Strange, 2012). This emphasizes the urgent need for effective crop disease management approaches, mostly in economies heavily reliant on agriculture. In cases where resistant crop varieties are not available, early detection of pathogens in the field is essential to prevent disease outbreaks. Consequently, the success of integrated pest management strategies largely hinges on the availability of rapid, sensitive, and specific diagnostic methods. It is imperative to develop more efficient technologies for detecting crop pathogens and establish seamless connections with decision-makers to deploy timely responses and safeguard agricultural systems. Numerous methods have been devised for identifying plant pathogens (Punja *et al.*, 2007). Conventional methods historically relied on symptom observation, involving field inspections to recognize disease symptoms and subsequent laboratory tests like pathogen culture on selective media, followed by physiological, biochemical, and pathogenicity assessments (Horsfall and Cowling, 1977). Although conventional methods are reliable, they are time-consuming and necessitate skilled plant pathologists to identify the responsible pathogen. The surge in global trade and climate change has led to the introduction of foreign invasive species, emerging pests and pathogens worldwide. As a result, this demand has necessitated faster, accurate, and specific methods for identifying plant pathogens. There is a pressing need for cost-effective and efficient point-of-use DNA testing to deliver precise results from remote locations in real-time. Point-of-care (POC) diagnostic assays, which do not

DOI: 10.1201/9781003474036-6

rely on sophisticated equipment and can be swiftly and inexpensively conducted in the field, are in high demand (Yager *et al.*, 2006). Isothermal DNA amplification methods are ideally suited to address this challenge in agriculture (Gill and Ghaemi, 2008; Craw and Balachandran, 2012). For instance, isothermal amplification, when coupled with lateral flow strips and portable fluorometers, has proven successful in POC detection of plant pathogen nucleic acid (Rohrman and Richards-Kortum, 2012). Point-of-care tests (POCTs) are conducted at the location of sample collection and typically yield results within a short timeframe, often less than 2 hours. POCTs can range from handheld to benchtop size, and other rapid diagnostic tests performed in laboratories, providing results within 1-2 hours, may be referred to as near-POCTs. The agricultural sector stands to benefit significantly from the availability of convenient and cost-effective POC assays, as agricultural fields are often situated far from well-equipped analytical laboratories, and sample transport can present logistical challenges. Molecular PLANT-Dx (Plant Diagnostics) systems have emerged as a potent tool for point-of-use pathogen detection, with the potential to transform plant disease management strategies. This chapter delves into the pivotal features, applications, and potential implications of the molecular PLANT-DX system in modern agriculture.

6.2 OVERVIEW OF MOLECULAR PLANT-DX SYSTEM

The development of a molecular PLANT-Dx System for point-of-use detection of pathogens associated with plant diseases is a significant leap forward in agricultural technology. This innovative system combines the power of molecular biology with user-friendly design, offering opportunities for farmers and researchers alike. First and foremost, the molecular PLANT-Dx System provides rapid and highly accurate pathogen detection. It utilizes advanced molecular techniques such as polymerase chain reaction (PCR), isothermal amplification, DNA sequencing, and other genetic assays to identify the presence of specific pathogens in plant samples (Verosloff *et al.*, 2019). This level of precision is crucial in disease management, as it enables early diagnosis and targeted treatment, reducing the risk of crop loss and the unnecessary use of pesticides. One of the standout features of this system is its user-friendliness. Designed for point-of-use applications, it does not require specialized training or laboratory equipment. Farmers and agricultural extension workers can easily operate this device in the field, enabling on-the-spot disease diagnosis and immediate action. This accessibility is a game-changer for farmers in remote or resource-limited regions who may not have access to advanced laboratory facilities. Moreover, the molecular PLANT-Dx system offers portability and versatility. Its compact size allows for easy transport to various locations, and it can be used for the detection of a wide range of plant pathogens (Verosloff *et al.*, 2019). This flexibility is particularly valuable in scenarios where multiple diseases may threaten a crop, enabling quick assessments and appropriate treatment decisions. The benefits of this system extend beyond disease detection. By providing real-time information about the presence and prevalence of specific pathogens, it can contribute to the development of more effective disease management strategies. Farmers can make informed decisions about planting, crop rotation, and the selection of disease-resistant varieties, ultimately optimizing their agricultural practices for improved yields and reduced environmental impact. The technology can be used at border checkpoints to prevent the introduction of exotic plant pathogens, contributing to biosecurity efforts. Researchers and scientists can utilize the system for ongoing monitoring of pathogen populations, studying disease dynamics, and identifying potential novel threats to plant health. Researchers and scientists can utilize this system for ongoing monitoring of pathogen populations, studying disease dynamics, and identifying potential novel threats to plant health. The molecular PLANT-Dx system can also be employed in certification processes to ensure pathogen-free seeds and plant material, further improving the agricultural industry's health and productivity (Verosloff *et al.*, 2019)

6.3 OVERVIEW OF PLANT-DX SYSTEM

In response to the limitations of conventional detection methods, a point-of-use diagnostic system named PLANT-Dx (Point-of-use Laboratory for Nucleic acids in a Tube) was created. PLANT-Dx leverages the sensitivity of isothermal amplification techniques for amplifying specific plant pathogen genetic material while incorporating the versatility of synthetic gene regulatory systems and the resilience of cell-free gene expression reactions to generate visible color-based results discernible without the need for specialized equipment. The fundamental concept behind PLANT-Dx is to transform plant pathogen genetic material into constructs encoding synthetic RNA regulators. These regulators, when synthesized, activate an RNA-based genetic switch that controls the expression of an enzyme responsible for inducing a visible color change. This RNA genetic switch effectively acts as a filtering layer, discerning accurately amplified on-target viral sequences from the potentially noisy output of isothermal amplification products. Thus, PLANT-Dx operates by initially employing recombinase polymerase amplification (RPA) to amplify a specific region of a plant pathogen's genome, generating a double-stranded DNA construct that encodes the production of a synthetic RNA regulator known as a small transcription activating RNA (STAR). These DNA templates are then utilized to guide the transcription of STARs within a cell-free gene expression reaction. Once produced, these STARs activate the transcription of a construct regulated by STAR, encoding the enzyme catechol 2, 3-dioxygenase (CDO). Only in the presence of the pathogen is the RPA product formed, resulting in the expression of CDO, which subsequently transforms a colorless catechol compound into a visible yellow product. This design is capable of detecting Cucumber mosaic virus (CMV) in infected plant lysate with high sensitivity at the picomolar level. Furthermore, this design can be adapted to detect nucleic acids from various viral genomes without interference. Notably, this system only requires simple mixing and the application of body heat to trigger the color change, making it highly suitable for deployment in field settings (Verosloff *et al.*, 2019).

6.4 POINT-OF-CARE TESTING FOR PLANT PATHOGENS

Point-of-use plant pathogen detection involves the capability to identify and diagnose plant pathogens directly at the location where the plants are cultivated or where potential infections are suspected. This approach is of great value to farmers, agronomists, and researchers as it enables swift identification and response to plant diseases, which is crucial for preventing the spread of infections and minimizing crop damage. It eliminates the need to send samples to distant laboratories, thus reducing turnaround time and cost. Numerous portable devices and tools have been designed for point-of-use detection. These devices are typically handheld or field-deployable, making them suitable for use in remote or agricultural settings. Some point-of-use detection methods utilize sensor technologies that can detect specific pathogens by analyzing biological or chemical signals produced by the plant in response to infection. With technological advancements, data from point-of-use detection can be connected to databases and cloud systems, allowing for real-time monitoring, data sharing, and predictive analytics. Rapid pathogen detection can facilitate the implementation of timely control measures, such as isolating infected plants, applying appropriate pesticides, or adjusting irrigation and nutrient regimes to minimize disease spread. Continuous monitoring and early detection can contribute to more effective disease management strategies, helping farmers and agricultural experts make well-informed decisions about crop health (Attaluri and Dharavath, 2023).

6.5 POINT-OF-CARE NUCLEIC ACID EXTRACTION METHODS

Point-of-care (POC) nucleic acid extraction methods are critical for the development of fast and user-friendly molecular diagnostic tests, but when addressing nucleic acid-based diagnostic procedures, infield sampling is sometimes disregarded. This is a difficult process since it requires continuously producing high-quality nucleic acid samples that are devoid of possible inhibitors, especially when dealing with plant tissues (Ikeda *et al.*, 2008). In plant pathogen identification, a fast and effective lateral flow device (LFD) DNA extraction technique has been successfully applied. It entails employing metal ball bearings to break the sample in an extraction buffer, transferring the lysate onto an LFD nitrocellulose membrane, and using the extracted DNA directly in DNA amplification procedures like PCR or other isothermal amplification techniques (Tomlinson *et al.*, 2010a). Because the DNA remains stable on the membrane at ambient temperature, this approach allows for field-based extraction (Tomlinson *et al.*, 2010a). Solid Phase Reversible Immobilisation (SPRI) is another potential POC DNA extraction approach that employs ordinary filter pipette tips and SPRI technology to dependably extract DNA/RNA at a precise concentration for downstream isothermal amplification (Wee *et al.*, 2015). Because it simply takes a magnet and a micropipette, magnetic SPRI bead-based extraction is perfect for POC applications. Wee *et al.* (2015) utilised this technology to diagnose human and plant illnesses. An new cellulose-based dipstick that allows for quick plant sample processing in as little as 30 seconds has been created. This dipstick efficiently macerates plant tissues, can detect pathogen DNA and RNA, and works with various amplification methods, including PCR, LAMP, and RPA (Zou *et al.*, 2017). It has been successfully applied to multiple plant species, including rice, wheat, tomato, and sorghum (Zou *et al.*, 2017).

6.6 OTHER TECHNIQUES IN MOLECULAR PLANT-DX

6.6.1 CRISPR/Cas System

The CRISPR/Cas system, which encompasses CRISPR-associated proteins like Cas9, Cas12, and Cas13, is a powerful tool for plant pathogen detection through genome editing. In this technique, target sequences are pre-amplified and CAS proteins are employed for detection. When combined with various isothermal nucleic acid amplification methods, the CRISPR/CAS system facilitates a system for POC applications with exceptional specificity and sensitivity (Zhang *et al.*, 2022). In a recent study utilizing the CRISPR/Cas9 system, an isothermal amplification triggered by this system was developed for a visual assay to detect *Phytophthora infestans* genomic DNA. Gold nanoparticle (GNP) probes self-assemble and aggregate, leading to a noticeable change in hue within a short timeframe. This single-stranded RNA approach is user-friendly and straightforward in design. The technique can be adapted for detecting other DNA sequences and exhibits a linear relationship between the ratio of absorbance at 650 and 525 nm and DNA concentration in the range of 0.2 pM–20 nM, with an optical detection limit for DNA of 2 pM. When compared to earlier CRISPR-based amplification systems, it has much greater specificity for single-base mismatches and enables for visual interpretation (Chang *et al.*, 2019). A CRISPR-Cas12a-based dual recognition system was developed to identify Fusarium head blight (FHB), a wheat yield-reducing disease caused by *Fusarium graminearum*, with sensitivity as low as 1 fg/L of total DNA. The internal transcribed spacer (ITS) region and the transcription elongation factor were detected using guide RNA hybridization (Mu *et al.*, 2022). The researchers used a LAMP-coupled Cas12a approach to demonstrate point-of-use diagnostics for tomato yellow leaf curl virus (TYLCV) and tomato leaf curl New Delhi virus, with a diagnosis time of about 1 hour. Using a simple and affordable fluorescence visualizer, the results could be visually understood. In another study, CRISPR-Cas9 genome editing was employed for *Venturia inaequalis*, the pathogenic agent of apple scab, and a high-throughput screening method was presented using high-resolution melt curve analysis (Rocafort *et al.*, 2022).

6.6.2 Metallic Nanoparticles based Biosensors

Metallic nanoparticles based biosensors are frequently used to create highly sensitive electrochemical and optical biosensors due to their ability to enhance electrical conductivity and optical signals. One such application is surface plasm on resonance immune-osensors, utilizing metallic nanoparticles like AuNP. These sensors have been used for the detection of various plant pathogens, such as T. *indica* in wheat (Singh *et al.*, 2010). An AuNP-based immune-chromatographic assay for the detection of GLARaV-3, a pathogen causing significant grape yield loss and decreased quality, has been developed (Byzova *et al.*, 2018). This immune-chromatographic assay offers rapid visual detection within 10 minutes. The method involves the binding of gold nanoparticles (GNP) coupled with antibodies and a polyvalent antigen. Compared to standard lateral flow immuno assay (LFIA), pre-incubating the gold nanoparticle conjugates and samples for 30 seconds resulted in a lower limit of detection (LOD), reducing it from 330 ng/mL to 5.4 ng/mL. Another development is an electrochemical biosensor for *Pseudomonas syringae* detection that utilizes disposable screen-printed carbon electrodes and AuNPs. This sensor is based on a recombinase polymerase reaction after rapid amplification of the target pathogen's DNA. It demonstrates a 10,000-fold greater sensitivity compared to the traditional PCR gel electrophoresis technique (Lau *et al.*, 2017). For the identification of stone fruit and almond infective bacteria, an on-site nano-sensor based on LFIA was designed (Lopez-Soriano *et al.*, 2017).

6.6.3 Isothermal Nucleic Acid Amplification

In recent years, various isothermal nucleic acid amplification technologies have emerged and gained popularity (Table 6.1). Loop-mediated isothermal amplification (LAMP) is the most commonly used among them. LAMP offers advantages over PCR-based approaches, such as the ability to perform amplification at a constant temperature, short reaction times, high efficiency, sensitivity, and cost-effectiveness, as it requires simple equipment (Notomi *et al.*, 2000). Several portable devices for POC plant pathogen detection have been developed using LAMP technology. For instance, a battery-powered, user-friendly, handheld, and waterproof device has been used to identify different species of begomovirus infecting plants (Wilisiani *et al.*, 2019). Another example is a micropipette tip-based nucleic acid test (MTNT) capable of detecting DNA and RNA from various plant crude extracts (Lu *et al.*, 2016). Fast LAMP protocols for POC detection of plant pathogens have been developed for different plant species, combining rapid nucleic acid extraction methods with signal detection directly in the field. While LAMP holds great promise for POC analysis, it has limitations, such as the design of specific primers, which can sometimes be challenging (Notomi *et al.*, 2015). Nonetheless, optimized LAMP-based protocols have shown to be simpler, faster, cheaper, and more sensitive than traditional methods for the detection of various plant pathogens (Villari *et al.*, 2013). Multiplexing of samples is also possible with LAMP, similar to PCR-based approaches (Feng *et al.*, 2019).

6.6.3.1 Recombinase polymerase amplification (RPA)

Recombinase polymerase amplification (RPA) is an isothermal amplification technique introduced in 2006 by Piepenburg *et al.* Instead of relying on the temperature-dependent denaturation-annealing process used in PCR, RPA is based on recombinase-mediated primer hybridization with double-stranded DNA (dsDNA). In the first stage of RPA, primers form a complex with a phage recombinase called uvrX. This complex then binds with the target dsDNA and displaces one DNA strand at the appropriate site. An additional recombinase loading factor, facilitated by uvrY and ATP hydrolysis, is required for this process. The single strand released is secured by a single-strand binding (SSB) protein. Bsu I DNA polymerase is employed to synthesize DNA using these primers. To enable rapid amplicon detection after RPA, two-sided labeled amplicons are commonly used. There are two approaches to obtain these labeled amplicons: either by using forward and reverse primers with labels or by using one labeled primer and an oligonucleotide labeled probe. The oligonucleotide probe consists of two different-length oligonucleotides connected with

TABLE 6.1
Onsite detection methods for plant pathogens

Method of Detection	Detected Pathogen	Sample type and host	Minimum detectable	Time	Reference
RPA and SERS	*Botrytis cinerea, Pseudomonas syringae, and Fusarium oxysporum*	Field samples	2 genomic copies of pathogen DNA	40 min	(Lau *et al.*, 2017)
LAMP	*Pseudomo nae syringae pv.*	Lab and field samples	1.61 10 fg/µL	60 min	(Chen *et al.*, 2020)
	Penicillium oxalicum	artificially/naturally contaminated	100 pg genomic DNA	60 min	(Vogt *et al.*, 2017)
	Phytophthora agathidicida	Field samples	1 fg DNA	60 min	(Winkworth *et al.*, 2020)
	Neofabraea perennans	Artificially inoculated apples	0.001 ng/µL	10 min	(Enicks *et al.*, 2020)
	Sclerotinia sclerotiorum	Artificially inoculated soil samples	10 pg of DNA	Less than 60 min	(Grabicoski *et al.*, 2020)
RT-LAMP	Prunus necrotic ringspot virus (PNRSV)	RNA extracted from tomato Torrado virus-infected plants	10^6 of cDNA	–	(Zong *et al.*, 2014)
LFA	Potato virus Y	Field samples	330 ng/mL to 5.4 ng/ mL	15 min	(Razo *et al.*, 2018)
	Stone fruit and almond infective bacteria	naturally infected plants	10^4 CFU/mL	15 min	(Lopez-Soriano *et al.*, 2017)
Amperometric immune-sensor	Capsicum chlorosis virus (CaCV)	Field samples	800 to 1000 times more sensitive	50 min	(Sharma *et al.*, 2017)
LAMP	*Clavibacter michiganensis* subsp. *sepedonicus* 16 S rDNA intergenic spacer region AF001266.1	Potato tuber	Pos/neg test (validation of LAMP)	65 min	(Sagcan *et al.*, 2019)
LAMP	*Leifsonia* xyli subsp. *Xyli* ** *ISLxx5* transposase gene NC_006087.1	Sugarcane xylem sap and leaves	Pos/neg test, 1:5 diluted infected plant extract	<20 min	(Naidoo *et al.*, 2017)
HDA	*Pseudomonas*	Cherry			(Illicic *et al.*, 2022)
RCA	Begomoviruses	Pulses			(Pandey *et al.*, 2022)
RPA	*Xanthomonas oryzae*	Rice	1×10^3 copies/µL	40 min	(Buddhachat *et al.*, 2022)

tetrahydrofuran (THF). The longer oligonucleotide (50) is modified with fluorescein (FAM), while the shorter oligonucleotide (30) is blocked. An annealing site for the probe must be chosen within the region between the primer annealing sites. With the 30 oligonucleotide blocked, the probe cannot function as a primer. However, the probe can intercalate into the dsDNA of the amplified product in a recombinase-dependent manner. DNA-endonuclease Nfo or exonuclease ExoIII cleaves the THF abasic sites, resulting in the release of short oligonucleotides from the duplex. The cleaved labeled probe has a normal 30 end and can be utilized as a primer. The new product contains FAM and biotin at the terminals of the DNA molecules. The use of this probe-based approach is intended to enhance the assay's specificity by forming labeled amplicons only at double recognition sites (oligonucleotide probe and primer). Despite its advantages, RPA has certain limitations. It has a product length limitation of 1500 bp (Piepenburg *et al.*, 2006), with the manufacturer recommending designs within 800 bp. The optimal RPA product length for lateral flow assays (LFAs) is 150 bp. RPA is also tolerant to mismatches in primers, which can lead to false-positive results when up to nine mismatches are present in both primers. The impact of mismatches is unpredictable and depends on factors like primer length, position, and distribution, further reducing the number of target sequences that can be detected and hindering discrimination of closely related organisms. Additionally, RPA reactions tend to have high viscosity, which can hinder diffusion and affect local amplification and amplification volume, particularly when amplifying low-titre target DNA. The manufacturer recommends intensive shaking after 4 minutes of the reaction. Currently, RPA is the only commercially available isothermal method for detecting plant pathogens. Since 2012, RPA-based tests for various pathogens have been developed, with TwistDx's RPA kit becoming available for researchers. There are also specialized commercial kits designed for the detection of specific plant pathogens, such as plum pox virus, *Candidatus Liberibacter asiaticus,* grapevine red blotch-associated virus, *Fusarium oxysporum* f. sp. *vasinfectum* Race 4, banana bunchy top virus, little cherry virus 2, and tomato chlorotic dwarf viroid (AmplifyRP® by Agdia). In 2014, the first reverse transcription RPA (RT–RPA) test was created for little cherry virus 2, offering rapid in-field diagnostics with slightly lower sensitivity compared to RT–qPCR. In recent years, there has been a surge in original research on the development of RPA tests for various plant pathogens, which will be discussed in detail concerning their detection methods (Ivanov *et al.*, 2021).

6.6.3.2 Loop-mediated amplification (LAMP)

LAMP is an isothermal amplification method based on the precise design of four primers (Notomi *et al.*, 2000). In contrast to RPA, LAMP employs only the Bst polymerase enzyme, which functions optimally at a temperature of 60–65 °C. The reaction typically takes about 1 hour to complete. The Bst polymerase possesses strand displacement activity, eliminating the need for high-temperature denaturation as seen in PCR. A simplified LAMP method requires four primers: two for generating loops and two for displacing newly synthesized strands. After the initial amplification stage, the newly created DNA molecule forms a dumbbell-like structure with two terminal loop regions connected by a stem-loop region. This structure serves as the initial unit for amplification, which is then extended to generate longer products. LAMP can be modified by adding two more primers that anneal to these loops, thereby enhancing the reaction and accelerating it by up to 30 minutes. Additionally, LAMP for RNA, coupled with reverse transcriptase, has been developed. One of the key advantages of LAMP is its cost-effectiveness, as it requires minimal equipment and reagents. However, the precise design of primers and the abundance of primer options pose challenges to this method. Suboptimal primers and temperature conditions can lead to nonspecific amplification and the formation of primer-dimer products (Rolando *et al.*, 2020). Another limitation of LAMP is the complexity of multiplexing due to the intricate design of two or more sets of primers. Despite these challenges, several multiplexed LAMP systems have been developed, primarily for non-plant pathogens. Various methods are available for detecting LAMP products, including dye staining for coloration, fluorescent labeling, lateral flow assays, electrochemical detection, surface plasma on resonance, lab-on-chip technology, gold nanoparticle aggregation induced by DNA binding, and pyrosequencing (Becherer *et al.*, 2020).

6.6.3.3 Rolling circle amplification (RCA)

Rolling circle amplification (RCA) was the first isothermal amplification method developed. It involves Phi29 DNA polymerase binding to a circular single-stranded DNA (ssDNA) template and a primer, and then synthesizing long ssDNA strands through its strand displacement activity. This results in the generation of multiple copies of circular DNA in the form of repeated linear ssDNA. The optimal temperature for RCA is 37 °C, and the reaction typically takes about 1 hour (Fire and Xu, 1995). However, it is worth noting that only certain viruses have DNA suitable for RCA-based detection. To address this limitation, padlock-ligated probes (PLP) have been proposed. These probes consist of short ssDNA sequences with terminal regions complementary to the target DNA sequence. When hybridized with the target dsDNA, the complex resembles a padlock, and the probe ends are ligated using a ligase. Non-ligated probes are removed by exonuclease, and the circular ssDNA is then amplified through RCA. One of the primary challenges of RCA is the efficiency of the ligation stage, which can be influenced by substrate concentration, reaction conditions, and potential inhibition by compounds present in the sample. Nevertheless, RCA offers flexibility and can be combined with various enhancement methods, including hyper-branching, multiple aptamer formation encoded in a padlock probe, and labeled probe annealing. RCA with padlock ligation probes and its modifications have found applications in the detection of plant pathogens. Some plant viruses, such as geminiviruses or nanoviruses, feature circular ssDNA, making them ideal targets for RCA-based detection. However, most of the tests developed for these viruses are more suitable for research purposes than for field diagnostics, involving techniques like gel electrophoresis, restriction endonuclease digestion assays, and deep sequencing of RCA products to detect and distinguish different geminiviruses (e.g., Abutilon mosaic virus, tomato golden mosaic virus, beet curly top virus, sida golden mosaic, costa rica virus, African cassava mosaic virus, and tomato golden mosaic virus-yellow vein) (Jeske, 2018). Another research application for RCA-PLP is the identification of geminiviruses (e.g., maize streak virus, digitaria streak virus, chloris striate mosaic virus, miscanthus streak virus, wheat dwarf virus, barley dwarf virus, or oat dwarf virus) through sequencing (Schubert *et al.*, 2007).

6.6.3.4 Nucleic acid sequence-based amplification

It is frequently referred to as NASBA and is based on transcription and reverse transcription processes (Compton, 1991). NASBA may amplify either RNA or DNA templates, with RNA being the more common target. The linear steps of RNA-based NASBA: At 65 °C, a DNA oligonucleotide having a T7 promoter sequence on the 5' end anneals with the target RNA. At 41 °C, AMV reverse transcriptase produces cDNA. RNAseH is an enzyme that degrades RNA inside the complex cDNA-RNA structure. AMV expands the cDNA when a second primer attaches to it. The reaction produces double-stranded DNA with the T7 promoter, which starts the amplification cycle via repeated RNA transcriptions by T7 DNA polymerase, followed by AMV-RNAseH reaction processes with the RNA. For the synthesis of cDNA from single-stranded DNA using DNA NASBA, an initial denaturation stage is necessary. Because NASBA produces RNA molecules, detection possibilities are restricted to hybridization of labeled probes. The majority of products are identified using fluorescently labeled probes or immunodetection. NASBA may potentially be utilized to identify numerous diseases at the same time. It is, however, not an extensively used approach for identifying plant diseases. Only one research team has developed a relatively quick NASBA-based assay for *Xanthomonas citri* identification (Scuderi *et al.*, 2010). They used the commercial Nuclisens basic kit (bioMérieux, Marcy l'Etoile, France) for pathogen amplification (120 minutes at 41 °C), followed by amplicon hybridization on a nylon membrane. The amplified DNA was cross-linked with UV light before going through a variety of hybridization operations with the Hybrimax device (Hybribio, Chaozhou, China), which uses flow-through hybridization. Prehybridization, hybridization with dioxygenin-labeled DNA oligonucleotides, washing, incubation with anti-dioxygenin antibodies, additional washing, and color development steps were all part of the total procedure. Even with the equipment, hybridization and detection took 30 minutes. The test showed that it could detect 1 fg of the target RNA. Other NASBA-based plant pathogen tests employ real-time fluorescence detection for viruses and pathogens such as sugarcane yellow leaf virus, apple scar skin virus, strawberry vein banding virus, Arabis mosaic virus, apple stem pitting virus, potato virus Y, *Fusarium poae*, and *Ralstonia solanacearum* (Szemes *et al.*, 2003), and apple stem pitting virus.

6.6.3.5 Helicase-dependent amplification (HDA)

Instead of thermal denaturation of double-stranded DNA, HDA uses DNA helicase. The original HDA employs the E. coli uvrD helicase, which unwinds double-stranded DNA held by single-stranded DNA-binding proteins (SSB) and permits primer binding and extension by DNA polymerase (Klenow fragment). At 37 °C, this mesophilic HDA (mHDA) occurs, with the repair protein MutL activating the uvrD helicase (Vincent *et al.*, 2004). Another HDA modification, thermophilic HDA (tHDA), uses a thermostable uvrD helicase and Bst DNA polymerase that operates at 60-65 °C.

This method does not necessitate the presence of SSB or MutL (An *et al.*, 2005). These changes make tHDA more appropriate for commercial kits. HDA is resistant to inhibition and can multiply DNA in the presence of human blood. However, one disadvantage of HDA is its relatively slow reaction time, which is advised to be 1 hour (An *et al.*, 2005). HDA is frequently utilised in human pathogen diagnostics, and it can be detected utilising methods such as gel electrophoresis, staining, and fluorescence detection, which are similar to PCR. On-site visualisation approaches such as lateral flow tests, chip devices, and gold nanoparticle aggregation are also used. While there are a few HDA-based tests for plant pathogen detection, the majority of research papers have focused on human pathogen diagnosis. One HDA-based test for *Phytophthora kernoviae* in particular stood out (Schwenkbier *et al.*, 2015). The innuPREP MP Basic kit was used to extract total DNA from *Rhododendron ponticum*. Asymmetric HDA was created by using too many biotinylated primers. For HDA, a commercial kit was used. On a dot chip, biotinylated single-stranded DNA was introduced to a particular probe. Following washing, enzymatic-dependent silver NP growth was carried out using a commercial kit, resulting in grey spots containing amplified DNA. The assay was sensitive to 1 pg/L pure genomic DNA. However, due to the lengthy amplification (90 minutes) and visualisation (60 minutes), this approach was not especially ideal for field detection. Furthermore, the washing stages rendered the test less suitable for field diagnosis. Portable heat blocks with precise temperature control are also suited for LAMP, HDA, RCA, and NASBA.

6.7 VISUAL DETECTION AFTER AMPLIFICATION

6.7.1 The Simplest Methods-coloration

The easiest method for visual equipment-free or low-equipment detection after amplification is coloration. There are methods for color development for DNA-containing solutions. One method is the direct binding of a dye with DNA molecules. dsDNA molecules can interact with dyes in three different ways: The dye can intercalate between strands of the DNA, it can bind with a major groove of the DNA, and it can bind with a minor DNA groove. Various dyes can interact with DNA in different ways, with the most common mechanism being intercalation. Dyes like ethidium bromide, SYBR Green I (SGI), and propidium iodide are frequently used for this purpose. SGI, in particular, is popular due to its low cost, low toxicity, and increased fluorescence upon binding to dsDNA. However, it has some limitations, including moderate stability and potential interference with certain amplification methods (Ivanov *et al.*, 2021). Some techniques rely on by-products from the amplification process to produce color changes, such as hydroxy naphthol blue (HNB) for LAMP detection, which changes from violet to sky blue as magnesium pyrophosphate precipitates (Goto *et al.*, 2009). Other pH-sensitive dyes, like phenol red, cresol red, and neutral red, have been developed for visual LAMP product detection, offering a straightforward differentiation between positive and negative samples based on color. Fluorescence detection, on the other hand, is widely used with portable equipment both in and outside the laboratory. These devices are employed for on-site pathogen detection, particularly in the context of real-time amplification. Various portable tools are available for techniques like RPA and LAMP, allowing for more reproducible diagnostics and the quantification of disease severity directly in the field. This enables more accurate and efficient on-site assessments of pathogen presence and concentration (Almassian *et al.*, 2013). A colorimetric assay, utilizing SYBR Green and a portable device, was developed for the detection of *Bursaphelenchus xylophilus* in pinewood chips using Recombinase Polymerase Amplification (RPA). In this approach, a 129 bp product generated from the ribosomal intergenic spacer was visualized with SYBR Green after a 20-minute amplification period. Remarkably, this method could detect pure genomic DNA at an astonishingly low concentration of 1.6 femtograms. Notably, this colorimetric RPA exhibited a tenfold higher sensitivity compared to real-time LAMP for spiked samples. However, it is important to note that the researchers found it more appropriate to measure adsorption as they could not visually differentiate color changes, although the portable device used may have some limitations for in-field applications. Therefore, UV visualization might be a more practical choice for on-site diagnostics, but it should be used with caution, as it can occasionally yield false-positive results (Cha *et al.*, 2020).

One of the simplest and most cost-effective methods for detecting LAMP products is through colorimetric assays. When designing LAMP analyses, colorization, often using SYBR Green I (SGI), is frequently employed for additional detection, making it compatible with visual inspection, gel electrophoresis, and quantitative analysis. Numerous studies have utilized SGI, which can be detected with either UV light or the naked eye, to detect various groups of plant pathogens, including bacteria, oomycetes, fungi, and viroids. Notably, most pathogens detected using SYBR Green are viruses. Some LAMP assays even propose SGI detection without the need for UV light, making the tests more cost-effective and user-friendly (Sui *et al.*, 2018). Another observable colorimetric detection method involves pH changes. A color change, typically from yellow to pink, in a positive LAMP reaction is readily distinguishable from a negative sample. The reverse color change is also a viable option. Commercial premixes containing pH indicators have been developed for detection, eliminating the need to open the tube after the LAMP mix, thus reducing the risk of contamination. Additionally, the HNB dye undergoes a color change from purple to sky blue during LAMP reactions. HNB-based detection methods have been established for viruses, bacteria, and fungi. There is even a prototype for direct detection of plant pathogens through the precipitation of magnesium pyrophosphate. These colorimetric methods offer simplicity and convenience for on-site diagnostics (Ivanov *et al.*, 2021). Colorimetric detection of RCA amplicons has been applied to the development of tests for plant pathogens. In one instance, a straightforward visualization method was devised for the detection of four Neofabraea fungi species (N. *perennans*, N. *kienholzii*, N. *vagabunda*, and N. *malicorticis*) through PLP-RCA. The aim of this test was to discriminate between the four species, with a focus on specificity rather than sensitivity. The DNA from these fungal species was extracted from apple roots, and SYBR Green I (SGI) was used for visualizing the DNA products, with the assay taking approximately 110-140 minutes (Lin *et al.*, 2018).

6.7.1.1 RCA-PLP Test

It was developed for the rapid differential diagnosis of fungal pathogens within the *Fusarium graminearum* species complex, including F. *oxysporum*, F. *incarnatum-equiseti*, and F. *tricinctum*. Fungal DNA was extracted from plant tissues using the CTAB method, and the target genes were selected as elongation factor 1a. Amplified products were visualized using SGI under UV light and validated through gel electrophoresis (Davari et al., 2012). In another interesting study, Cas9-triggered strand displacement amplification was followed by RCA. In this method, the amplification process commenced when the Cas9-sgRNA complex recognized the model DNA of *Phytophthora infestans*, leading to cleavage in one strand. Bst polymerase, with strand displacement activity, then synthesized ssDNA. These ssDNA copies served as a template for PLP-RCA. The extended ssDNA product could bind to GNP conjugated with short complementary DNA oligos, resulting in DNA aggregation and a color change. This approach exhibited high sensitivity, with a limit of detection at 0.2 pM for the model fragment. However, the substantial enzymatic reaction involved in this test might limit its practicality for field applications (Chang *et al.*, 2019).

6.7.1.2 Recombinase Polymerase Amplification (RPA)-lateral Flow Assay (LFA)

In the realm of plant pathogen detection, the combination of isothermal amplification, specifically RPA, with LFA is gaining popularity for in-field diagnostics. RPA, often using commercial kits, has been described in approximately 50 original papers, with 16 of these incorporating LFA for visualization. The choice of target gene for RPA was based on maximal specificity for the pathogen of interest, often using highly conserved viral coat protein sequences for viruses. Most RPA-LFA tests employed labeled primers and a DNA probe for cleavage, while some adopted simpler strategies with FAM- and biotin-labeled primers. Extraction of DNA or RNA from samples was generally conducted with commercial kits, enabling rapid and simplified analysis. RPA-LFA tests typically used GNP conjugates for visual detection, although some experiments employed anti-digoxygenin conjugates with carbon nanoparticles. The average analysis time for these tests was within 2 hours, with the potential for acceleration through disc extraction of DNA from crude homogenates. While RPA-LFA is generally less sensitive than quantitative PCR, it remains a valuable method for in-field diagnostics, offering the highest sensitivity among field-based diagnostic approaches for plant pathogens (Ivanov *et al.*, 2021).

6.7.1.3 LAMP-Based Tests

LFA detection of LAMP products involves various combinations of labeled primers, including inner, loop, and hybridization probes. Over 100 LAMP-based tests have been reported, with a focus on plant pathogenic species. These tests are primarily developed for POC applications, showing sensitivity comparable to PCR. Various lateral flow strip manufacturers have been utilized, such as Millenia, Foresite Diagnostics, Biohelix, and AMODIA Bioservice, along with homemade dipsticks.

The overall workflow includes DNA extraction, LAMP amplification, optional hybridization, and LFA, with LAMP taking at least 30 minutes, LFA 5 minutes, and hybridization 5 minutes. LAMP–LFA offers a promising field diagnostic approach, with visual detection comparable in sensitivity to PCR or canonical LAMP (Ivanov *et al.*, 2021).

6.7.2 Lab-on-a-chip Technology

It offers a versatile approach to DNA detection, with compatibility for various isothermal amplification methods, including fluorescence, colorimetry, and nanoparticle (NP) detection. These compact and portable devices can provide accurate and rapid pathogen detection, making them valuable for on-site applications (Chiriaco *et al.*, 2018). While some lab-on-a-chip devices require specialized equipment for manufacturing and signal interpretation, we focus on the development of portable and autonomous biosensors. Only a few such tests, meeting these criteria and based on a combination of lab-on-a-chip and LAMP, have been proposed (Li *et al.*, 2019). Notably, paper-based lab-on-a-chip approaches have been utilized, enabling the detection of plant pathogens as well as genetically modified organisms. One example involves a microfluidic device for virus detection from orchid leaves, albeit requiring complex equipment unsuitable for field diagnostics. Another approach utilizes a paper chip device for diagnosing multiple Aspergillus species, offering a more straightforward and field-friendly option. This chip facilitates LAMP reactions and product detection through a clever use of paper, GNP, and hybridization, providing an innovative solution for pathogen detection (Ivanov *et al.*, 2021).

6.7.3 Lateral Flow Assays (LFAs)

LFAs are a quick and inexpensive way to identify amplicons. LFAs have been widely used in clinical and veterinary diagnostics, agriculture, food chemistry, environmental monitoring, and immunoassays since the 1980s (Parolo *et al.*, 2020) (Table 6.2). LFAs work by directing sample molecules across test strip membranes and identifying target molecules in a specific test zone using colored labels. The pore structure of the membrane separates bigger molecules, cells, and fragments from smaller components by diffusion. While this method is comparable to NP-based methods, nanoparticles are often used as detection markers in LFAs. Colloidal gold, colored latex, fluorescent particles (e.g., quantum dots), and carbon particles are some of the NPs that may be used for conjugation (Bahadr *et al.*, 2016). In contrast to NP aggregation methods discussed earlier, LFAs leverage specific affinity recognition. Receptors, often antibodies, are immobilized on the surface of NPs. These NPs become concentrated in the binding zones (test and control) due to the formation of specific complexes, resulting in visible color changes in these regions. LFAs are known for their rapidity, typically providing results within 10–20 minutes, making them highly convenient for in-field diagnostics. Two primary approaches are used for detecting nucleic acid targets with LFAs:

TABLE 6.2
Commercially available point-of-use kits for detection of plant pathogens

Device/ Product	Target application	Underlying principle	Company
AgriStrip	Pathogens of ornamental crops	Lateral flow immunoassay	BIOREBA Switzerland
Rapid plant disease tests	*Erwinia amylovora, Potato virus Y, Phytophthora, Ralstonia solanacearum*	Lateral fiow assay	Pocket Diagnostic, UK
Alert Test Kits	*Pythium, Phytophthora, Rhizoctonia*	Lateral fiow assay	Neogen Corporation
ImmunoStrip	Plant virus	Lateral fiow immunochromatography	Agdia, Elkhart
TwistAmp® Exo RT kit	Yam mosaic virus (YMMV)	Recombinase Polymerase Amplification (RPA)	Cepheid

1. Direct Detection of DNA Sequences: In this method, single-stranded nucleic acid targets hybridize with complementary nucleotide fragments immobilized in the test zone (Jahanpeyma et al., 2019).
2. NALFIA: This is a more popular method that includes labeling DNA molecules with low-molecular-weight markers like as biotin, fluorescein, or digoxigenin. Target DNA or RNA is commonly amplified with 5'-end labeled primers, one of which is biotinylated and the other with fluorescein (FAM). After amplification of the target RNA/DNA, the amplified sample is applied to the LFA. This technique produces a complex composed of streptavidin, biotin-labeled dsDNA, FAM, and anti-FAM antibodies conjugated with NPs in the test zone. Some researchers have proposed a hybridization-based technique combined with NALFIA, in which target DNA is amplified using a single labeled primer. Following that, high-temperature hybridization is conducted with a labeled oligonucleotide probe (Rigano *et al.*, 2014). The availability of commercially accessible LFT (lateral flow test) strips intended to detect biotin- or FAM-labeled amplicons is one of the key advantages of LFAs. Millenia (Gießen, Germany), BioUSTAR (Hangzhou, China), Agdia (Elkhart, IN, USA), and Abingdon Health's PCRD nucleic acid detector (York, UK) manufacture these strips. Furthermore, as observed in certain research (Hsieh *et al.*, 2017), components of LFT strips may be customized and optimized to improve signal intensity in the test zone. Another significant benefit of LFAs is the availability of portable instruments for registering color changes in the test zone. These portable gadgets are ideal for capturing results during on-site testing. These portable devices are highly convenient for recording results during in-field diagnostics. With the use of portable devices, labels such as fluorescent dyes or quantum dots can further reduce detection limits (Urusov *et al.*, 2019).

6.8 CONCLUSION

For the on-site detection of plant pathogens, it is essential to have a user-friendly and efficient process for plant tissue processing, nucleic acid extraction, amplification, and detection. Various portable equipment, kits, and protocols are available for on-site preparation, and the nature of the target plant tissue should guide the choice of suitable homogenization methods. Processing soft tissues or exudates can be accomplished within 5 minutes using basic homogenization equipment. For more thorough homogenization and when employing inhibition-tolerant amplification methods, crude extraction without DNA purification may be feasible. Isothermal amplifications are particularly well-suited for on-site diagnostics due to their features. To summarize the findings from different approaches to on-site plant pathogen detection, this chapter highlights the key attributes of an ideal field-deployable kits, such as duration and sensitivity. Each approach possesses its unique advantages and drawbacks for field applications. However, none of the reviewed tests fully met all the proposed requirements. In essence, on-site plant pathogen detection involves a series of stages, from sample collection to result visualization, and various techniques and tools can be applied to make this process more efficient and accessible, but each has its limitations and trade-offs.

REFERENCES

Aglietti, C., Luchi, N., Pepori, A. L., Bartolini, P., Pecori, F., Raio, A., ... & Santini, A. (2019). Real-time loop-mediated isothermal amplification: an early-warning tool for quarantine plant pathogen detection. *AMB Express*, *9*, 1–14.

Agrios, G. N. (2005). Plant Pathology, 5th Edn. Amsterdam: Elsevier Academic Press.

Almassian, D. R., Cockrell, L. M., & Nelson, W. M. (2013). Portable nucleic acid thermocyclers. *Chemical Society Reviews*, *42*(22), 8769–8798.

Ammour, S. I., M., Bilodeau, G. J., Tremblay, D. M., Van der Heyden, H., Yaseen, T., Varvaro, L., & Carisse, O. (2017). Development of real-time isothermal amplification assays for on-site detection of *Phytophthora infestans* in potato leaves. *Plant Disease*, *101*(7), 1269–1277.

An, L., Tang, W., Ranalli, T. A., Kim, H. J., Wytiaz, J., & Kong, H. (2005). Characterization of a thermostable UvrD helicase and its participation in helicase-dependent amplification. *Journal of Biological Chemistry*, *280*(32), 28952–28958.

Attaluri, S., & Dharavath, R. (2023). Novel plant disease detection techniques-a brief review. *Molecular Biology Reports*, 2023 Nov; 50(11), 9677–9690..

Bahadır, E. B., & Sezgintürk, M. K. (2016). Lateral flow assays: Principles, designs and labels. *TrAC Trends in Analytical Chemistry*, *82*, 286–306.

Becherer, L., Borst, N., Bakheit, M., Frischmann, S., Zengerle, R., & von Stetten, F. (2020). Loop-mediated isothermal amplification (LAMP)–review and classification of methods for sequence-specific detection. *Analytical Methods*, *12*(6), 717–746.

Boyle, D. S., Lehman, D. A., Lillis, L., Peterson, D., Singhal, M., Armes, N., ... & Overbaugh, J. (2013). Rapid detection of HIV-1 proviral DNA for early infant diagnosis using recombinase polymerase amplification. *MBio*, *4*(2), 10–1128

Buddhachat, K., Sripairoj, N., Ritbamrung, O., Inthima, P., Ratanasut, K., Boonsrangsom, T., ... & Sujipuli, K. (2022). RPA-assisted cas12a system for detecting pathogenic Xanthomonas oryzae, a causative agent for bacterial leaf blight disease in rice. *Rice Science*, *29*(4), 340–352

Byzova, N. A., Vinogradova, S. V., Porotikova, E. V., Terekhova, U. D., Zherdev, A. V., & Dzantiev, B. B. (2018). Lateral flow immunoassay for rapid detection of grapevine leafroll-associated virus. *Biosensors*, *8*(4), 111.

Cha, D., Kim, D., Choi, W., Park, S., & Han, H. (2020). Point-of-care diagnostic (POCD) method for detecting *Bursaphelenchus xylophilus* in pinewood using recombinase polymerase amplification (RPA) with the portable optical isothermal device (POID). *PLoS One*, *15*(1), e0227476.

Chang, W., Liu, W., Liu, Y., Zhan, F., Chen, H., Lei, H., & Liu, Y. (2019). Colorimetric detection of nucleic acid sequences in plant pathogens based on CRISPR/Cas9 triggered signal amplification. *Microchimica Acta*, *186*, 1–8.

Chen, Z. D., Kang, H. J., Chai, A. L., Shi, Y. X., Xie, X. W., Li, L., & Li, B. J. (2020). Development of a loop-mediated isothermal amplification (LAMP) assay for rapid detection of *Pseudomonas syringae* pv. *tomato* in planta. *European Journal of Plant Pathology*, *156*, 739–750.

Chiriacò, M. S., Luvisi, A., Primiceri, E., Sabella, E., De Bellis, L., & Maruccio, G. (2018). Development of a lab-on-a-chip method for rapid assay of *Xylella fastidiosa* subsp. pauca strain CoDiRO. *Scientific Reports*, *8*(1), 7376.

Compton, J. (1991). Nucleic acid sequence-based amplification. *Nature*, *350*(6313), 91–92.

Craw, P., & Balachandran, W. (2012). Isothermal nucleic acid amplification technologies for point-of-care diagnostics: a critical review. *Lab on a Chip*, *12*(14), 2469–2486.

Daher, R. K., Stewart, G., Boissinot, M., Boudreau, D. K., & Bergeron, M. G. (2015). Influence of sequence mismatches on the specificity of recombinase polymerase amplification technology. *Molecular and Cellular Probes*, *29*(2), 116–121.

Davari, M., van Diepeningen, A. D., Babai-Ahari, A., Arzanlou, M., Najafzadeh, M. J., van der Lee, T. A., & de Hoog, G. S. (2012). Rapid identification of Fusarium graminearum species complex using Rolling Circle Amplification (RCA). *Journal of Microbiological Methods*, *89*(1), 63–70.

De Jonghe, K., De Roo, I., & Maes, M. (2017). Fast and sensitive on-site isothermal assay (LAMP) for diagnosis and detection of three fruit tree phytoplasmas. *European Journal of Plant Pathology*, *147*, 749–759.

DeAngelis, M. M., Wang, D. G., & Hawkins, T. L. (1995). Solid-phase reversible immobilization for the isolation of PCR products. *Nucleic Acids Research*, *23*(22), 4742.

Du, X. J., Zhou, T. J., Li, P., & Wang, S. (2017). A rapid Salmonella detection method involving thermophilic helicase-dependent amplification and a lateral flow assay. *Molecular and Cellular Probes*, *34*, 37–44.

Enicks, D. A., Bomberger, R. A., & Amiri, A. (2020). Development of a portable LAMP assay for detection of *Neofabraea perennans* in commercial apple fruit. *Plant Disease*, *104*(9), 2346–2353.

Feng, W., Hieno, A., Kusunoki, M., Suga, H., & Kageyama, K. (2019). LAMP detection of four plant-pathogenic oomycetes and its application in lettuce fields. *Plant Disease*, *103*(2), 298–307.

Fire, A., & Xu, S. Q. (1995). Rolling replication of short DNA circles. *Proceedings of the National Academy of Sciences*, *92*(10), 4641–4645.

Gétaz, M., Bühlmann, A., Schneeberger, P. H., Van Malderghem, C., Duffy, B., Maes, M., ... & Cottyn, B. (2017). A diagnostic tool for improved detection of *Xanthomonas fragariae* using a rapid and highly specific LAMP assay designed with comparative genomics. *Plant Pathology*, *66*(7), 1094–1102.

Gill, P., & Ghaemi, A. (2008). Nucleic acid isothermal amplification technologies—a review. *Nucleosides, Nucleotides and Nucleic Acids*, *27*(3), 224–243.

Goto, M., Honda, E., Ogura, A., Nomoto, A., & Hanaki, K. I. (2009). Colorimetric detection of loop-mediated isothermal amplification reaction by using hydroxy naphthol blue. *Biotechniques*, *46*(3), 167–172.

Grabicoski, E. M. G., Jaccoud-Filho, D. D. S., Lee, D., Henneberg, L., & Pileggi, M. (2020). Real-time quantitative and ion-metal indicator LAMP-based assays for rapid detection of *Sclerotinia sclerotiorum*. *Plant Disease*, *104*(5), 1514–1526.

Heo, S., Kim, H. R., & Lee, H. J. (2019). Development of a quantitative real-time nucleic acid sequence based amplification (NASBA) assay for early detection of apple scar skin viroid. *The Plant Pathology Journal*, *35*(2), 164.

Horsfall, J. G., and Cowling, E. B. (1977). *Plant Disease: An Advanced Treatise*. New York, NY: Academic Press.

Hsieh, H. V., Dantzler, J. L., & Weigl, B. H. (2017). Analytical tools to improve optimization procedures for lateral flow assays. *Diagnostics*, *7*(2), 29.

Ikeda, S., Tsurumaru, H., Wakai, S., Noritake, C., Fujishiro, K., Akasaka, M., & Ando, K. (2008). Evaluation of the effects of different additives in improving the DNA extraction yield and quality from Andosol. *Microbes and Environments*, *23*(2), 159–166.

Iličić, R., Jelušić, A., Marković, S., Barać, G., Bagi, F., & Popović, T. (2022). *Pseudomonas cerasi*, the new wild cherry pathogen in Serbia and the potential use of recG helicase in bacterial identification. *Annals of Applied Biology*, *180*(1), 140–150.

Vincent, M., Xu, Y., & Kong, H. (2004). Helicase-dependent isothermal DNA amplification. *EMBO Reports*, *5*(8), 795–800.

Ivanov, A. V., Safenkova, I. V., Zherdev, A. V., & Dzantiev, B. B. (2021). The potential use of isothermal amplification assays for in-field diagnostics of plant pathogens. *Plants*, *10*(11), 2424.

Jahanpeyma, F., Forouzandeh, M., Rasaee, M. J., & Shoaie, N. (2019). An enzymatic paper-based biosensor for ultrasensitive detection of DNA. *Frontiers in Bioscience-Scholar*, *11*(1), 122–135.

Jeske, H. (2018). Barcoding of plant viruses with circular single-stranded DNA based on rolling circle amplification. *Viruses*, *10*(9), 469.

Lai, D., Zhang, Y., Huang, Q., Yin, G., Pennerman, K. K., Liu, Z., & Guo, A. (2018). Reverse transcription loop-mediated isothermal amplification to rapidly detect Rice ragged stunt virus. *Saudi Journal of Biological Sciences*, *25*(8), 1577–1584.

Lau, H. Y., Wu, H., Wee, E. J., Trau, M., Wang, Y., & Botella, J. R. (2017). Specific and sensitive isothermal electrochemical biosensor for plant pathogen DNA detection with colloidal gold nanoparticles as probes. *Scientific Reports*, *7*(1), 38896.

Leal, I., Allen, E., Foord, B., Anema, J., Reisle, C., Uzunovic, A., ... & James, D. (2015). Detection of living *Bursaphelenchus xylophilus* in wood, using reverse transcriptase loop-mediated isothermal amplification (RT-LAMP). *Forest Pathology*, *45*(2), 134–148.

Leung, C. H., Zhong, H. J., Lu, L., Chan, D. S. H., & Ma, D. L. (2013). Luminescent and colorimetric strategies for the label-free DNA-based detection of enzyme activity. *Briefings in Functional Genomics*, *12*(6), 525–535.

Li, R., Chen, J., Zhang, X., Cui, J., Tao, S., & Yang, L. (2019). Mini-disk capillary array coupling with LAMP for visual detection of multiple nucleic acids using genetically modified organism analysis as an example. *Journal of Agricultural and Food Chemistry*, *68*(3), 899–906.

Li, X., Nie, J., Ward, L. J., Nickerson, J., & De Boer, S. H. (2011). Development and evaluation of a loop-mediated isothermal amplification assay for rapid detection and identification of *Pectobacterium atrosepticum*. *Canadian Journal of Plant Pathology*, *33*(4), 447–457.

Lillis, L., Siverson, J., Lee, A., Cantera, J., Parker, M., Piepenburg, O., ... & Boyle, D. S. (2016). Factors influencing recombinase polymerase amplification (RPA) assay outcomes at point of care. *Molecular and Cellular Probes*, *30*(2), 74–78.

Lin, H., Jiang, X., Yi, J., Wang, X., Zuo, R., Jiang, Z., ... & Zhou, E. (2018). Molecular identification of *Neofabraea* species associated with bull's-eye rot on apple using rolling-circle amplification of partial EF-1α sequence. *Canadian Journal of Microbiology*, *64*(1), 57–68.

Lopez-Soriano, P., Noguera, P., Gorris, M. T., Puchades, R., Maquieira, A., Marco-Noales, E., & López, M. M. (2017). Lateral flow immunoassay for on-site detection of *Xanthomonas arboricola* pv. *pruni* in symptomatic field samples. *PLoS One*, *12*(4), e0176201.

Lu, W., Wang, J., Wu, Q., Sun, J., Chen, Y., Zhang, L., ... & Jiang, X. (2016). High-throughput sample-to-answer detection of DNA/RNA in crude samples within functionalized micro-pipette tips. *Biosensors and Bioelectronics*, *75*, 28–33.

Mahas, A., Hassan, N., Aman, R., Marsic, T., Wang, Q., Ali, Z., & Mahfouz, M. M. (2021). LAMP-coupled CRISPR–Cas12a module for rapid and sensitive detection of plant DNA viruses. *Viruses*, *13*(3), 466.

Mehle, N., Kogovšek, P., Rački, N., Jakomin, T., Gutiérrez-Aguirre, I., Kramberger, P., & Ravnikar, M. (2017). Filling the gaps in diagnostics of Pepino mosaic virus and Potato spindle tuber viroid in water and tomato seeds and leaves. *Plant Pathology*, *66*(7), 1191–1201.

Mekuria, T. A., Zhang, S., & Eastwell, K. C. (2014). Rapid and sensitive detection of Little cherry virus 2 using isothermal reverse transcription-recombinase polymerase amplification. *Journal of Virological Methods*, *205*, 24–30.

Mu, K., Ren, X., Yang, H., Zhang, T., Yan, W., Yuan, F., ... & Zeng, Q. (2022). CRISPR-Cas12a-based diagnostics of wheat fungal diseases. *Journal of Agricultural and Food Chemistry*, *70*(23), 7240–7247.

Naidoo, N., Ghai, M., Moodley, K., Mkize, L., Martin, L., McFarlane, S., & Rutherford, S. (2017). Modified RS-LAMP assay and use of lateral flow devices for rapid detection of *Leifsonia xyli* subsp. *xyli*. *Letters in Applied Microbiology*, *65*(6), 496–503

Notomi, T., Mori, Y., Tomita, N., & Kanda, H. (2015). Loop-mediated isothermal amplification (LAMP): principle, features, and future prospects. *Journal of Microbiology*, *53*(1), 1–5.

Notomi, T., Okayama, H., Masubuchi, H., Yonekawa, T., Watanabe, K., Amino, N., & Hase, T. (2000). Loop-mediated isothermal amplification of DNA. *Nucleic Acids Research*, *28*(12), e63–e63

Pandey, S., Girish, T. R., Basavaraj, S., Padmaja, A. S., & Nagaraju, N. (2022). RCA-based detection of begomoviruses in weed genera associated with legumes in Southern Karnataka. *Legume Research-An International Journal*, *45*(2), 237–245

Parolo, C., Sena-Torralba, A., Bergua, J. F., Calucho, E., Fuentes-Chust, C., Hu, L., ... & Merkoçi, A. (2020). Tutorial: design and fabrication of nanoparticle-based lateral-flow immunoassays. *Nature protocols*, *15*(12), 3788–3816.

Piepenburg, O., Williams, C. H., Stemple, D. L., & Armes, N. A. (2006). DNA detection using recombination proteins. *PLoS Biology*, *4*(7), e204.

Punja, Z. K., Boer, S. D., & Sanfaçon, H. (Eds.). (2007). *Biotechnology and Plant Disease Management*. CAB International.

Razo, S. C., Panferov, V. G., Safenkova, I. V., Varitsev, Y. A., Zherdev, A. V., Pakina, E. N., & Dzantiev, B. B. (2018). How to improve sensitivity of sandwich lateral flow immunoassay for corpuscular antigens on the example of potato virus Y?. *Sensors*, *18*(11), 3975.

Rigano, L. A., Malamud, F., Orce, I. G., Filippone, M. P., Marano, M. R., Do Amaral, A. M., ... & Vojnov, A. A. (2014). Rapid and sensitive detection of *Candidatus Liberibacter asiaticus* by loop mediated isothermal amplification combined with a lateral flow dipstick. *BMC Microbiology*, *14*(1), 1–9.

Rocafort, M., Arshed, S., Hudson, D., Sidhu, J. S., Bowen, J. K., Plummer, K. M., ... & Mesarich, C. H. (2022). CRISPR-Cas9 gene editing and rapid detection of gene-edited mutants using high-resolution melting in the apple scab fungus, Venturia inaequalis. *Fungal Biology*, *126*(1), 35–46.

Rohrman, B. A., & Richards-Kortum, R. R. (2012). A paper and plastic device for performing recombinase polymerase amplification of HIV DNA. *Lab on a Chip*, *12*(17), 3082–3088.

Rolando, J. C., Jue, E., Barlow, J. T., & Ismagilov, R. F. (2020). Real-time kinetics and high-resolution melt curves in single-molecule digital LAMP to differentiate and study specific and non-specific amplification. *Nucleic Acids Research*, *48*(7), e42–e42.

Safenkova, I. V., Ivanov, A. V., Slutskaya, E. S., Samokhvalov, A. V., Zherdev, A. V., & Dzantiev, B. B. (2020). Key significance of DNA-target size in lateral flow assay coupled with recombinase polymerase amplification. *Analytica Chimica Acta*, *1102*, 109–118.

Sagcan, H., & Turgut Kara, N. (2019). Detection of Potato ring rot Pathogen Clavibacter michiganensis subsp. s epedonicus by Loop-mediated isothermal amplification (LAMP) assay. *Scientific Reports*, *9*(1), 20393.

Schubert, J., Habekuß, A., Kazmaier, K., & Jeske, H. (2007). Surveying cereal-infecting geminiviruses in Germany—diagnostics and direct sequencing using rolling circle amplification. *Virus Research*, *127*(1), 61–70.

Schwenkbier, L., Pollok, S., Rudloff, A., Sailer, S., Cialla-May, D., Weber, K., & Popp, J. (2015). Non-instrumented DNA isolation, amplification and microarray-based hybridization for a rapid on-site detection of devastating *Phytophthora kernoviae*. *Analyst*, *140*(19), 6610–6618.

Scuderi, G., Golmohammadi, M., Cubero, J., López, M. M., Cirvilleri, G., & Llop, P. (2010). Development of a simplified NASBA protocol for detecting viable cells of the citrus pathogen *Xanthomonas citri* subsp. *citri* under different treatments. *Plant Pathology*, *59*(4), 764–772.

Sharma, A., Kaushal, A., & Kulshrestha, S. (2017). A Nano-Au/C-MWCNT based label free amperometric immunosensor for the detection of capsicum chlorosis virus in bell pepper. *Archives of Virology*, *162*, 2047–2052.

Sillo, F., Giordano, L., & Gonthier, P. (2018). Fast and specific detection of the invasive forest pathogen Heterobasidion irregulare through a Loop-mediated isothermal AMP lification (LAMP) assay. *Forest Pathology*, *48*(2), e12396.

Singh, S., Singh, M., Agrawal, V. V., & Kumar, A. (2010). An attempt to develop surface plasmon resonance based immunosensor for Karnal bunt (Tilletia indica) diagnosis based on the experience of nano-gold based lateral flow immuno-dipstick test. *Thin Solid Films*, *519*(3), 1156–1159.

Strange, R. (2012). Almost 40 Per Cent of Worldwide Crops Lost to Diseases. The Crop Site. Available at: www.thecropsite.com/

Sui, X., Zhang, S., Wu, Z., & Ling, K. S. (2018). Reverse transcription loop-mediated isothermal amplification for species-specific detection of tomato chlorotic spot orthotospovirus. *Journal of Virological Methods*, *253*, 56–60.

Szemes, M., & Schoen, C. D. (2003). Design of molecular beacons for AmpliDet RNA assay—characterization of binding stability and probe specificity. *Analytical Biochemistry*, *315*(2), 189–201.

Tomlinson, J. A., Boonham, N., & Dickinson, M. (2010). Development and evaluation of a one-hour DNA extraction and loop-mediated isothermal amplification assay for rapid detection of phytoplasmas. *Plant Pathology*, *59*(3), 465–471.

Tomlinson, J. A., Dickinson, M., Hobden, E., Robinson, S., Giltrap, P. M., & Boonham, N. (2010). A five-minute DNA extraction method for expedited detection of Phytophthora ramorum following prescreening using Phytophthora spp. lateral flow devices. *Journal of Microbiological Methods*, *81*(2), 116–120

Urusov, A. E., Zherdev, A. V., & Dzantiev, B. B. (2019). Towards lateral flow quantitative assays: detection approaches. *Biosensors*, *9*(3), 89.

Verosloff, M., Chappell, J., Perry, K. L., Thompson, J. R., & Lucks, J. B. (2019). PLANT-Dx: a molecular diagnostic for point-of-use detection of plant pathogens. *ACS Synthetic Biology*, *8*(4), 902–905.

Villari, C., Tomlinson, J. A., Battisti, A., Boonham, N., Capretti, P., & Faccoli, M. (2013). Use of loop-mediated isothermal amplification for detection of Ophiostoma clavatum, the primary blue stain fungus associated with Ips acuminatus. *Applied and Environmental Microbiology*, *79*(8), 2527–2533.

Vogt, E. I., Kupfer, V. M., Bechtner, J. D., Frisch, L. M., Vogel, R. F., & Niessen, L. (2017). Detection of penicllium oxalicum in grapes with a loop-mediated isothermal amplification (LAMP) assay. *Journal of Microbiology, Biotechnology and Food Sciences*, *7*(3), 265.

Wee, E. J. H., Lau, H. Y., Botella, J. R., & Trau, M. (2015). Re-purposing bridging flocculation for on-site, rapid, qualitative DNA detection in resource-poor settings. *Chemical Communications*, *51*(27), 5828–5831.

Wilisiani, F., Tomiyama, A., Katoh, H., Hartono, S., Neriya, Y., Nishigawa, H., & Natsuaki, T. (2019). Development of a LAMP assay with a portable device for real-time detection of begomoviruses under field conditions. *Journal of Virological Methods*, *265*, 71–76.

Winkworth, R. C., Nelson, B. C., Bellgard, S. E., Probst, C. M., McLenachan, P. A., & Lockhart, P. J. (2020). A LAMP at the end of the tunnel: A rapid, field deployable assay for the kauri dieback pathogen, *Phytophthora agathidicida*. *PLoS One*, *15*(1), e0224007.

Yager, P., Edwards, T., Fu, E., Helton, K., Nelson, K., Tam, M. R., & Weigl, B. H. (2006). Microfluidic diagnostic technologies for global public health. *Nature*, *442*(7101), 412–418.

Zhang, L., Jiang, H., Zhu, Z., Liu, J., & Li, B. (2022). Integrating CRISPR/Cas within isothermal amplification for point-of-Care Assay of nucleic acid. *Talanta*, *243*, 123388.

Zong, X., Wang, W., Wei, H., Wang, J., Chen, X., Xu, L., ... & Liu, Q. (2014). Rapid detection of Prunus necrotic ringspot virus using magnetic nanoparticle-assisted reverse transcription loop-mediated isothermal amplification. *Journal of Virological Methods*, *208*, 85–89.

Zou, Y., Mason, M. G., Wang, Y., Wee, E., Turni, C., Blackall, P. J., ... & Botella, J. R. (2017). Nucleic acid purification from plants, animals and microbes in under 30 seconds. *PLoS Biology*, *15*(11), e2003916.

7 DNA Microarrays as a Versatile Tool for the Detection of Plant Pathogens

Falak Fayaz and Mohammad Waris Haider

7.1 INTRODUCTION

Agriculture has been a cornerstone of human civilisation for millennia, providing sustenance and economic stability. However, one of the most significant challenges that agriculture faces is the relentless threat posed by plant diseases. Crop losses due to the pest invasion is estimated to be $220 billion USD annually, out of which 14% of the crop loss share is due to plant diseases (FAO,2022). These diseases are caused by various pathogens like viroids fungi, viruses, bacteria, nematodes, and phytoplasma and have the potential to devastate crops, leading to significant economic losses. Thus, plant disease detection plays a pivotal role in ensuring global food security and sustainable agriculture. Early and accurate detection becomes paramount for timely intervention. While morphological identification of some of the plant pathogens is possible, such practices can be augmented with the utilisation of serological and nucleic-acid based detection, which not only helps in accurate but also efficient and specific detection of pathogen invading the plant system as well as for the pathogens that lack morphological significance. Leveraging conventional detection methods such as biochemical assays and dilution plate technique with innovative technologies and practices such as DNA microarrays and sequencing methods has proven revolutionary in the early and tangible detection of pathogens not only under *in vitro* conditions but also in their real-time ecological niches, along with accurate species and sub-species level characterisation of bacteria that usually remains a challenge for most the classical detection techniques. Further, DNA microarray complements conventional gene-by-gene approach and acts as more wholesome platform to study cellular physiology (Wildsmith and Elcock 2001). DNA microarray is a viable technique to determine and measure the expression levels of numerous genes all at once, which allows to monitor the transcriptome profile of the various distinct cell types via production of complementary DNA (cDNA) of the respective RNA transcripts through reverse transcription, where each DNA copy can be spotted on the chip via a particular colour coding system, which is in tandem with the gene expression algorithm (Aslam *et al.*, 2022).

DOI: 10.1201/9781003474036-7

The technique introduced first by Schena *et al.* (1995) has become paramount in genome mining programs as it allows to study the expression profile of thousands of genes parallelly by subjecting the nucleotides to a solid surface and allowing simultaneous interrogation of all the sequences (Hadidi *et al.,* 2008). Precision medicine, drug development, biomedical research, and microbial detection are some examples of the fields where this tool is proving to be substantial and versatile in its application. DNA microarray has become one of the most powerful profiling tools with broad range applicability for the detection of plant pathogens owing to the bulk availability of genomic information from continuous sequencing of plant pathogens (Mitra 2021).

7.2 DECODING DNA MICROARRAYS: AN OVERVIEW

The underlying concept of microarray strategy is not novel and finds its origin in the 1970s with advent expression profiling techniques like northern blotting (Alwine *et al.,* 1977), which involved scoring of only one gene at a time. Over time innovations led to emergence of new techniques that although involved analysis of multiple genes like subtractive hybridisation (Zimmermann *et al.,*1980), differential hybridisation (Sargent,1987), and differential display (Liang and Pardee,1992), they were found to be labour and technical expertise intensive and highly complex to gauge besides required bulk of starter RNA. Hence, the shortcomings of all these techniques led to the evolution of microarrays, wherein glass slides were used as solid support owing to their less background fluorescence, and such surfaces allow much higher probe density than the porous flexible membranes (Schena *et al.,* 1995, 1996).

Microarray is an analytical tool that houses an assortment of molecules such as oligonucleotides, cDNA, antibodies, PCR products, clones, polypeptides, and tissue sections. These molecules are precisely immobilised in an ordered manner at micrometre to millimetre sized locations on a solid surface and function by capitalising on the selective binding that occurs between complementary single stranded nucleic acids. (Barret 2003). Analogous to biological space, this device houses molecules like DNA or proteins referred to as probes, which are systemically arrayed on a microscopic scale. This precise configuration is established on a solid surface either being a membrane or a glass slide, whilst typically being a glass slide or an analogous material, wherein DNA molecules are affixed at predetermined positions (spots), individual microarray can encompass a multitude of these designated sites, each hosting numerous identical DNA molecules or fragments, exhibiting lengths that range from twenty to several hundred nucleotides

DNA microarray is presented as high-throughput, quantitative method of scoring expression profile of thousands of genes parallelly via exploiting DNA complementarity through base pairing discovered by Watson and Crick and immobilisation of DNA probes on the membrane that binds to complementary DNA or RNA via specific hybridisation (Gillispie and Spiegelman,1965). Sample DNA or RNA binds to a specific complementary region of the immobilised DNA probe, which paves way for the hybridisation. Immobilised probe consists of thousands of cDNA or oligonucleotides, each specified for a particular gene or sequence of DNA or RNA. A few nanolitres of DNA probe is deposited over the solid surface to develop an array model containing a well-ordered arrangement of samples, which acts as a medium to facilitate the hybridisation between nucleic acids based on rules of complementarity. Automated printing leads to consistent spotting (Missoum 2018). The core feature of microarray technology involves simultaneously hybridising a labelled mixture of nucleic acids, referred to as "target," with the multitude of individual nucleic acid sequences known as "probes." These probes can be uniquely identified based on their spatial placement in a single experimental setup. The specific position of the given probe on the array is often called as "spot" or "feature." The probes are securely attached to a solid membrane, and the targets, after fluorescent labelling, are introduced over the array as a solution (Bonant *et al.,* 2002). Figure 7.1 gives an overview DNA microarray.

The microarray chip is designed with thousands of spots fabricated with probes that hybridise with the target consisting of cDNA prepared by reverse transcription of RNA isolated from the samples that helps to determine the pathological samples from the background noise.

The colour coding of the chip helps in determining the expression level in one condition (treatment), which is usually red (i.e., pathological cell) and expression level in another condition (control), which is usually green (i.e., healthy cell). The colour code in DNA microarrays is used to represent the relative expression levels of genes in different samples. By comparing the intensity of fluorescent signals,

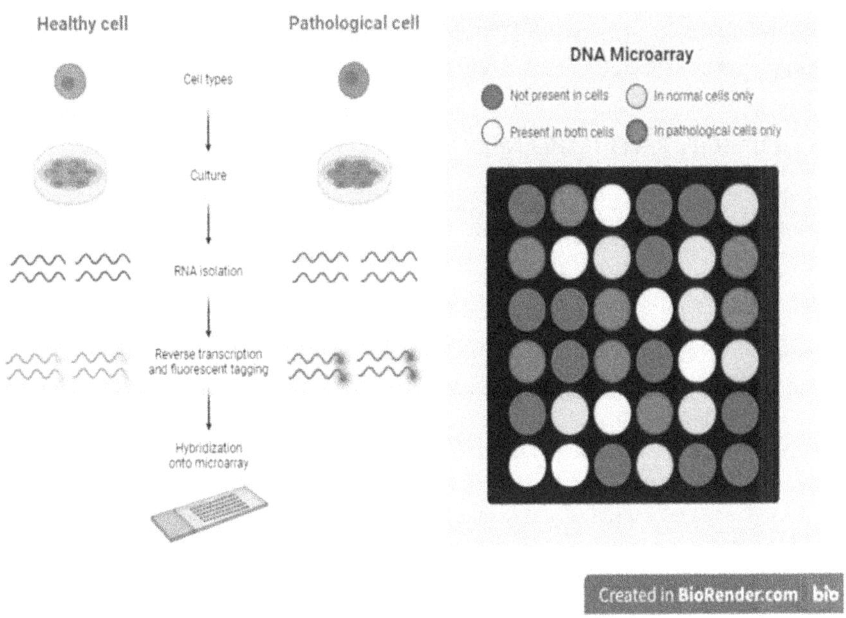

FIGURE 7.1 Flow chart that depicts the basic principle of DNA Microarray.

Template retrieved from https://app.biorender.com/biorender-templates

we can identify genes that are differentially expressed under various experimental conditions, providing valuable insights into biological processes and disease mechanisms.

7.2.1 Array Types

Depending on the size of sample areas, arrays can be called either microarray or macro arrays, with microarray sample spots being typically less than 200 microns in diameter and 300 microns or greater size for macroarray spots.

Microarray can be referred as 'low density' when the spot density follows the order of $100cm^2$. Commercial low-density microarray chips consist of glass slides or nylon coated glass slides duly treated with chemicals or UV light to facilitate irreversible immobilisation of DNA probe over the surface. To enhance nucleic acid binding to glass, Poly-L-Lysine or saline is used. Poly-L-Lysine relies on ionic interactions, while saline creates covalent bonds to secure probe DNA during washing. For detecting radioactively labelled nucleic acids, glass slides are coated with nylon membrane and DNA probes are cross-linked using UV light. DNA samples on the array may vary in size and composition, including plasmids (500-5000 base pairs), complementary DNA strands (hundreds of bases), PCR products (100-500 base pairs), or synthetic oligonucleotides with modified 5' ends containing either amino or thiol group (Hadidi *et al.,* 2004).

High-density microarrays, generally referred to as DNA chip, have a spot density of 1000-10000 per cm^2, containing glass or silicon as a mobilisation substrate.

Simultaneous confrontation of thousands of nucleotides all at once makes this technique an advanced, updated, and superior method of detection.

7.3 DESIGN AND IMPLEMENTATION

A cascade of steps needs to be followed in DNA microarray experiment implementations including the following.

7.3.1 Target-specific Probes

A variety of probe types, such as cDNAs, short or long oligonucleotides, chromosomes, whole organisms, or defined entities are employed. These probes can be generated using established methods like cloning, restriction enzymes, and PCR, or they can be separately obtained or attached to surfaces like chips, glass slides, or nylon supports. Gene expression for libraries, containing cDNA or nucleotides probes, can be purchased or synthesized. Each probe contains multiple cDNAs or oligonucleotides representing known genes in baker's yeast *Saccharomyces cerevisiae*. The arrays had an excess of 65,000 sites forsynthesising DNA. Each 50x50 μm element was produced using a 25-mer oligonucleotide with over 10^7 duplicates. On average, there were 400 synthesis elements for each of the 6321 recognised genes in yeast. A similar technique was utilised for each of the 6321 recognised genes in yeast (Lipshutz *et al.,* 1999). A similar technique was also utilised to generate oligonucleotides having known sequences. cDNA can be advantageous due to unsequenced genes, including those in viral genomes.

7.3.2 Array Fabrication

Numerous methods are available for attaching DNA to array surfaces, including techniques like micro-contact printing, contact-dip deposition printing, microfluid networks, wet printing, and electrocapture piezoelectric printing. Photolithography is used exclusively for in situ DNA synthesis. Arrays are created by depositing a small amount of purified DNA onto a solid surface with robotic assistance to ensure precise and identical spotting. Macroarrays can be relatively large (10-15cm^2) with fewer DNA pieces per cm^2, while microarrays are smaller (2-3cm^2) but can accommodate thousands of genes per cm^2 (Bertucci *et al.*,2001). Affymetrix developed oligo-chips (Genechips) that involve synthesizing short DNA segments directly on a solid surface using photolithography and solid-phase DNA synthesis. This approach allows the construction of arrays containing hundreds or thousands of DNA probes per cm^2, with some experimental versions reaching up to a million probes per cm^2 (Lipshutz *et al.*,1999).

7.3.3 Sample Preparation

Initial single-stranded cDNA is produced via reverse transcription of purified RNA or poly (A) RNA in totality. During reverse transcription process, fluorescently labelled target dCTP labelled with Cy3 or Cy5 are integrated into the cDNA on a glass support material (Lockhart *et al.*, 1996). A different strategy includes employing reverse transcription with a chemically reactive nucleotide analog referred to as amino allyl-dUTP. After this step, the cDNA is subsequently labelled through "post-labelling" with Cy3 or Cy-NHS esters, which attach specifically to the altered nucleotides. These techniques yield uniform and strong signal intensities during the process of microarray hybridisation. In the context of nylon membrane support, the labelling of cDNA is accomplished by using radiolabelled nucleotides like a32 P-dCTP during the reverse transcription process. Detection of the hybridisation can be performed through either chemiluminescence or autoradiography (Bertucci *et al.*, 1999a,1999b). Alternatively, cDNA can be marked by integrating biotin-16-dUTPor digoxigenin-11-dUTP. After the hybridisation process, detection is carried out using either b-galactosidase-conjugated streptavidin or alkaline-phosphatase-conjugated anti-digoxigenin antibodies, and the resulting hybrid complexes are detected through colorimetric methods (Chen *et al.*, 1998). It is crucial to emphasize that RNA purity is of paramount importance at this stage to prevent any interference in the detection of both radioactive and fluorescent signals.

7.3.4 Hybridisation

The process begins with the denaturation of probe DNA through heating, followed by the application of labelled target solution. The quantity of original total RNA required for hybridisation varies depending on the type of support used. It ranges from 300ng (Bertucci *et al., 199a)* to 5μg of total RNA when employing nylon membranes and up to 20μg of total RNA on glass supports. In cases where the array is used for hybridisation with the single type of cDNA, single-dye or radioactive labelling during hybridisation is adequate. However, in the context of exploring variations in gene expression using the array, control cDNA can be tagged with Cy3-dCTP for control cDNA and Cy5-dCTP for experimental cDNA. In such cases, it is crucial to ensure the equal use of both cDNA and dye. Subsequently, the slide or chip is placed in a humid environment at 65°C for several hours during the hybridisation process. Finally, the slide or chip undergoes a washing step to eliminate any unbound nucleotides in preparation for detection phase.

7.3.5 Scanning and Imaging

The detection process offers two methods: a scanning system and a direct imaging system. In scanning fluorescently labelled DNA is detected using a laser excitation source. This method processes each spot individually, with a detector measuring the emitted light. On the other hand, direct imaging depends on an excitation source that illuminates the entire array simultaneously, and a camera captures the resultant images. Direct imaging is preferred when working with radioactive hybridisation or colorimetry. Scanning typically provides superior resolution and sensitivity in comparison to direct imaging. Subsequent to the detection step, the results are used to generate graphic computer files in various formats, including TIFF and others.

7.3.6 Data Analysis

Effectively deciphering DNA array data necessitates employing advanced bioinformatics techniques to present findings in an informative manner. When working with dyes, specialised software is used for the analysis of image files, quantifying light intensity across different wavelengths. In the cases where hybridisations involve a mix of cDNAs labelled with, for instance, Cy3/Cy5 ratio is computed. Subsequently, a false-colour overlay is created, where spots representing varying levels of hybridisation of one cDNA population are visualised as red or green, while equal hybridisation of both populations is represented as yellow. Thus, the red-to-green ratio at each gene spot reflects the relative abundance of each cDNA (experimental vs. control). The software combines these measurements to create a visual depiction of the comparative expression levels between the experimental and control samples. The results of this analysis can be conveyed in various formats, including a two-dimensional grid featuring red, green, and yellow spots of varying intensities corresponding to the array positions. Alternatively, the output may be presented as a scatterplot, effectively highlighting variations in gene expression levels. Managing complex datasets often necessitates the utilisation of software solutions such as GeneSpring and Spotfire (Amaratunga *et al.*,2014).

7.4 DNA MICROARRAYS: A VERSATILE TOOL FOR MICROBIAL DETECTION

Microarray technology emphasises its application in a multiplex format for similar or highly diverse pathogens, capitalising on the capacity to utilise numerous probes on a single chip (Schoen *et al.*, 2002; Bonants *et al.*, 2002; and 2003; *Franke Whittle et al.*, 2005; Fessehaie *et al.*,2003; Bonants *et al.*, 2005; Boonham *et al.*, 2007; Pasquini *et al.*, 2008; van Doorn *et al.*, 2007).

This potential of microarray technology in detecting and diagnosing plant diseases is significant, and is mainly attributed to its multiplex capabilities. Additionally, it can be integrated with other systems, such as performing nucleic acid extraction on the chip (Liu *et al.*, 2007), conducting PCR reactions, and detecting them on the same device (Van Doorn *et al.*, 2007). There's also the option to combine all these systems into one (Lee *et al.*, 2006), offering automation possibilities that hold considerable importance and utility. This potential for automation, combined with integration into preceding analysis steps (extraction, PCR, detection), suggests broader applications in future protocols (Lee *et al.*, 2006; Liu *et al.*, 2007; Bonants *et al.*, 2005).

7.4.1 Plant Pathogen Interactions

The levels at which plants interact with the pathogens have been examined comprehensively from both ends. Specific molecular communication follows in each interaction between the plant and invading pathogen (Dodds *et al.*, 2010; Hammond-Kosack *et al.*, 2000). Plants experience the wrath of different types of microbes like bacteria, viruses, oomycetes, pathogenic fungi, nematodes, and insects (Schenk *et al.*,2000). Pathogens successfully breach the physiological integrity of plants and mount a series of morbidities following compatible interaction (Schenk *et al.*,2000); in case of incompatible interactions, pathogen invasion triggers a cascade of downstream defence signalling pathways and activation of local and systemic defence responses (Schenk *et al.*, 2009). Local response involves hypersensitive reaction and build-up of active oxygen species and localised cell death, expression, and amplification of PR proteins (Hammond-Kosack *et al.*, 2000). Systemic response involves production of nitric oxide (NO), ethylene, salicylic acid (SA), and jasmonic acid and induction of SAR (Vallad *et al.*,2004 and Torres *et al.*, 2006).

Several discoveries related to the genes involved in plant defence occurred after the advent of this technology and led to deciphering of an overall picture of the genetic profile of plants upon invasion by pathogens and have also helped in understanding the intricate complexities of the interactions that occur between plant and pathogen and unravelling the molecular signatures that take part in interaction and induce a response either in the form of defence or disease. Transcriptome profiling helps in understanding the plant pathogen interactions; the key factors that get activated or induced and determine the host or non-host resistance, besides understanding the mechanism of necrotrophy and biotrophy and thus microarray, helps to frame a rationale between various host-pathogen interactions (Wise *et al.*,2007). This profiling also helps in deciphering the novel physiological processes involved in defence response. cDNA microarray analysis of Arabidopsis leaves infected with biotrophic pathogen *Pseudomonas syringae*, comprising 13,000 unique ESTs, a robust alteration in the expression profile of genes occurred from housekeeping genes to defence pathway with an accentuated demand in the physiology of the plant to fend off the pathogen (Scheideler *et al.*,2002). Microarray technology not only unveils the signal transduction pathways initiated by pathogens but also detects the genes specifically activated to facilitate the infection process. This includes modulation of host factors by the viruses. From cDNA library of potato An EST microarray was created to investigate the response of *Nicotiana benthamiana* to biotic stress induced by *Impatiens necrotic spot virus* (INSV) – a phyto-virus that reproduces in the host cytoplasm. The analysis of microarray revealed a notable increase in the expression levels of small heat-shock protein genes, specifically HSP18, HSP20, and HSP70, in reaction to INSV (Senthil *et al.,* 2005). Carr *et al.* (2006) demonstrated that the HSP70 and HSP101 expression remains unaffected by the JA and SA in *Arabidopsis thaliana* defence signalling pathways when facing *Oilseed Rape Mosaic virus* (ORMV) infection. This affirms that ORMV triggers the expression of HSP independently. Additionally, in cotyledons of pea, it was found that in pea early browning virus (PEBV) RNA1 alone induces the HSP70 expression (Witham *et al.*,2003). Numerous noteworthy microarray experiments have been conducted, offering insights into the modulation of plant genes by viruses (Chen *et al.*,2002).

7.4.2 Detection of Virus

Several factors impact the evolution and utilisation of assessment techniques. Phyto viruses, being a diverse group without a common nucleotide sequence, pose challenges in detection due to global trade, especially in the case of ornamental plants and the impact of climatic variability on movement of virus. Diagnostic techniques for identifying specific viruses in seeds, propagation materials, and field samples include electron microscopy, biological indexing antibody-based methods like ELISA, microarray detection, and PCR. Microarray-aided detection stands out for its parallel testing ability, offering sensitivity comparable to ELISA. While PCR is highly sensitive, its simultaneous detection capability is limited even in "multiplexed" use. Array techniques, due to their inherently parallel nature, are more apt for screening or diagnostic applications rather than the precise detection of individual targets essential for survey tasks (Boonham *et al.,* 2007). Moreover, as arrays enable the concurrent identification of various viruses, this approach is poised to be applied to address diseases resulting from combination of viruses. The previously documented virus detection arrays have primarily focused on narrow domains, concentrating on limited virus quantities as prototypes for assay advancement. Initial techniques employed diverse strategies to detect and identify various potato viruses (PVY & PVX) (Bystricka *et al.,* 2003; Boonham *et al.,* 2003) and tobamoviruses infecting cucurbits (Lee *et al.*,2003). In all instances, these approaches utilised cDNA microarrays, where PCR amplified products specific to distinct virus types were amplified and deposited onto a glass slide. In high-density microarray designed for the detection of pathogens, control probes play a crucial role in ensuring the accuracy of the assay, preventing false positives, enhancing specificity. The results obtained demonstrate clear differentiation between fluorescence signals from pathogen and control probes in all research conducted. A comparative analysis of the performance of microarray's was made against a traditional PCR-based method for pathogen detection, and the assay's versatility was also assessed through in silico testing (Krzysztof *et al.,* 2017). A novel 70-mer oligonucleotide microarray, with 570 unique probes, efficiently identifies a broad spectrum of grapevine viruses simultaneously. The microarray, validated against RT-PCR for 10 viruses, proves accurate for high-throughput plant certification and virus discovery, even for novel sequences of genome with partial similarity. Utilising random primed RT-PCR on grapevine samples, the microarray detects viral agents in single and mixed infections (Engel *et al.,*2010)

7.4.3 Detection of Bacteria

Arabidopsis was among the initial plant species investigated to understand the impact of bacteria on gene expression. The analysis of cDNA microarrays revealed an upregulation of transcripts associated with defence in Arabidopsis shoots infected with *Pseudomonas thivervalensis* (Calvo *et al*., 2019). Bacterial pathogens of maize including *Pantoea aagglomerans*, *Pantoea ananatis,* and *Enterobacter cloaceae* subsp. *Dissolvens* were simultaneously detected by targeting the entire genome of pathogens through DNA microarray analysis, and tested alongside the probes that align with genome of organisms that are closely related. The results obtained suggest that, in all conducted experiments, there is a clear differentiation between the fluorescence signals emitted by the pathogen and control probes (Kryzsztof *et al.,* 2017).

7.4.4 Detection of Fungi

The progress in DNA microarray technology has enabled efficient highly-throughput and simultaneous interrogation of diverse phytopathogens, including fungi (Tiberini and Barbara 2012). Fluorescent probes and PCR primers designed for the ITS region of *Rhizoctonia solani, Spongospora subterranea,* the Alt-a1gene of *Alternaria solani* and *A. alternata,* TEF-1a of *Fusarium sp*., and TUB2 of *Colletotrichum coccodes* were employed in microarray qPCR technology, utilising 48-spot silicon microarrays (Nikitin *et al.,* 2018). Liebe *et al.* (2016) introduced a pioneering microarray assay, "Array tube," which utilised effective probe targeting marker genes 16S rDNA, TEF-1a, and ITS. This approach facilitated the detection of diverse sugarbeet root rot inciting pathogens, including *Botrytis cinerea, Pencillium expansum,* and *Aphanomyces cochlioides.* It is an innovative microfluidic approach that streamlines probe creation by eliminating the need for pin-spotting. Rapid hybridisation on the same microarray assembly enabled identification of following green-house pathogens (*Didymella bryoniae, Botrytis squamosa,* and *Botrytis cinerea*). The study outlines optimisation details for probe immobilisation and sample hybridisation. The method successfully distinguished between two PCR products of 260 bp with a one-base-pair difference from closely related fungal pathogens (*B. cinerea and B. squamosa*) in greenhouse plants (Wang *et al.,* 2007). Custom DNA microarrays were created to detect specific Fusarium species (*F. graminareum, F. culmorum,* and *F. poae*) by employing oligonucleotide probes and utilising flourescent-labeled PCR products originating from a portion of the translation elongation factor 1 alpha. The hybridisation conditions were optimised with fungal template DNA. The validation of this DNA microarray approach included testing on artificially infected wheat samples and evaluating unknown infected wheat samples. The comprehensive verification involved simultaneous analyses using traditional mycological methods and specific PCR techniques (Pavlatova *et al.*, 2011)

7.4.5 Detection of Nematodes

Meloidogyne javanica, known as the root-knot nematode, operates and sustains a prolonged root-restricting feeding site, leading to the conversion of roots into multinucleate feeding cells, often termed "giant cells," as often reported in tomato infected with root knot nematode (Bar-Or *et al.,* 2005). In their research, scientists observed variations in transcript levels across diverse functional categories, including hormone-related transcripts, PR genes, and development-related transcription factors (TFs). The sole concern of the investigation was to comprehend how Arabidopsis plants respond to root-knot nematodes, utilising root tissues enriched with large cells (Jammes *et al.*,2005). A significant discrepancy in expression level of genes was noted among 3373 out of the 22,089 genes scrutinised. The reduction in the activity of specific genes was recognised as crucial for the proper development of galls. As per Jammes *et al.* (2005) the establishment of nematode feeding sites was made possible by the suppression of plant defences.

7.5 CONCLUSION AND FUTURE DIRECTIONS

DNA microarrays a revolutionary diagnostic tool have found wider application in the realm of plant pathology via detection and quantification of plant pathogens. Not only detection but also unravelled wider gene regulatory pathways and the expression profile of both the pathogen and plants in question and have helped to decipher at the molecular level the microbial signatures that actually take part in the mauling of plant tissues. DNA microarrays have also helped in finding a clear distinction between closely related strains of microbes, which usually remains a concern with classical detection techniques. Besides pathogens, from the host side, they provide insights into the complex and transient physiological responses that plants exhibit when affected by pathogens. DNA microarrays will continue find relevance as the bulk of genome sequencing information is now available, which will help in exploring the diversity of microbial communities. In the future, a synergistic collaboration between microarray technology and high-throughput sequencing will open avenues to explore novel inquiries within the realm of plant pathogen interaction. Attributes of large-scale multiplexing and gauging the relative differences among samples makes this diagnostic tool more advantageous over other detection tools. The immense discriminatory power, which is often lacking in other molecular techniques like sequencing, may act as a thrust for DNA microarrays to continue being the stand-alone diagnostic tool for all humans, plants, and animals, due to their commercialisation and user friendliness.

Microarray holds immense potential in the specification of unnamed genes involved in complex processes via gene expression analysis (Tan *et al.*, 2003). The future of microarray technology, despite potential challenges, appears promising. The rapid advancements in this field are unlikely to slow down and will continue to grow, making significant contributions to our understanding of life's fundamental processes. There is a quite possibility that biosensors bearing the biomolecular signatures of the microbes may be able to overpower gene chips. Theoretically, the binding of target DNA probe could induce the attraction of micron-sized magnetic beads, leveraging the electric charge of DNA. Detection of this interaction could be accomplished using highly sensitive thin-film magnetic detectors. Magnetic biosensors have the potential to rapidly identify the presence of a single pathogenic DNA molecule. In recent developments, a transistor network has been developed to electronically detect point mutations in DNA (Pouthas *et al.*, 2004). Further advances in array technology such as higher probe densities, automated sample preparation, faster hybridisation, label-free detection, and efficient algorithms will enhance their use in pathogen detection. These improvements aim to enable mobile, point-of-care devices resulting within an hour, with metagenomic sequence analysis enhancing array designs for non-culturable species ensuring better specificity (Kevin S. McLoughlin, 2011).

REFERENCES

Alwine, J.C., Kemp, D.J., Stark, G.R. (1977). Method for detection of specifi c RNAs in agarose gels by transfer to diazobenzyloxymethyl-paper and hybridization with DNA probes. *Proceedings of the National Academy Science USA* 74:5350–5354.

Amaratunga, D., Cabrera, J ., Shkedy, Z. (2014). *Exploration and Analysis of DNA Microarray and Other High-Dimensional Data*. Wiley-Blackwell.

Aslam, H.M.U., Riaz, H., Killiny, N. Zhou, X.G., Thomashow, L.S., Peters, N.T., Chanda, A.K. (2022). Microarray Technology for Detection of Plant Diseases. In *Trends in Plant Disease Assessment* (pp. 203–223). Springer Nature. https://doi.org/10.1007/978-981-19-5896-0_11

Bar-Or, C., Kapulnik, Y ., Koltai, H. (2005). A broad characterization of the transcriptional profile of the compatible tomato response to the plant parasitic root knot nematode *Meloidogyne javanica. European Journal of Plant Pathology* 111(2):181–192.

Barrett, J.C., Kawasaki, E.S. (2003). Microarrays: the use of oligonucleotides and cDNA for the analysis of gene expression. *Drug Discovery Today* 8:134–141.

Bertucci, F., Bernard, K., Loroid, B., Chang, Y., Granjeaud, S., Bimbaum, D., Nguyen, C., Peck, K., Jordan, B. (1999a). Sensitivities issues in DNA array-based expression measurements and performance of nylon microarrays for small samples. *Human Molecular Genetics* 8: 1715–1722.

Bertucci F., Loriod B., Tagett R., Granjeaud S., Bimbaum D., Nguyen C., Houlgattte R., 2001. DNA arrays: technological aspects and applications. *Cancer Bulletin* 88: 243–255.

Bertucci, F., Van Hulst, S., Bernard, K., Loroid, B., Granjeaud S., Tagette, R.,Starkey, M., Nguyen, C., Jordan, B., Brimbaum, D. (1999b). Expression scanning of an array of growth control genes in human cell lines. *Oncogene* 18:3905–3912.

Bisen. P.S., Debnath, M., Prasad, G.B. (2010). Microarray. In: Molecular Diagnostics: Promises and Possibilities. Springer, Dordrecht. https://doi.org/10.1007/978-90-481-3261-4_13.

Bonants, P., De Weerdt, M., Van Beckhoven, J., Hilhorst, R., Chan, A., Boender, P., Zijlstra, C., Schoen, C. (2002). Multiplex detection of plant pathogens by microarrays: an innovative tool for plant health management. In J.E. Frey and F. Pasquer. Abstracts Agricultural Biomarkers for Array Technology, Management Committee Meeting, Wadenswil, 20.

Bonants, P.J.M., Schoen, C.D., Szemes, M., Speksnijder, A., Klerks, M.M., van den Boogert, P.H.J.F., Waalwijk, C., van der Wolf, J.M., Zijlstra, C. (2005). From single to multiplex detection of plant pathogens: pUMA, a new concept of multiplex detection using microarrays. *Phytopathologia Polonica* 35:29–47.

Boonham, N., Walsh, K., Smith, P., Madagan, K., Graham, I., Barker, I. (2003). Detection of potato viruses using microarray technology: towards a generic method for plant viral disease diagnosis. *Journal of Virological Methods* 108:181–87.

Boonham, N., Tomlinson, J., Munford, R. (2007). Microoarrays for rapid identification of plant viruses. *Annual Review of Phytopathology* 45:307–328.

Bystricka, D., Lenz, O., Mraz, I., Dedic, P., Sip, M. (2003). DNA microarray: parallel detection of potato viruses. *Acta Virology* 47:41–44.

Calvo, P., Zebelo, S., Mcnear, D., Kloepper, J., Fadamiro, H. (2019). Plant growth promoting rhizobacteria induce changes in *Arabidopsis thaliana* gene expression of nitrate and ammonium uptake genes. *Journal of Plant Interactions* 14(1):224–231.

Carr, T., Wang, Y., Huang, Z., Yeakley, J.M., Fan, J.B., Whitham, S. (2006). Tobamovirus infection is independent of HSP101 mRNA induction and protein expression. *Virus Research* 121:33–41.

Chen, W., Provart, N., Glazebrook, J., Katagiri, F., Chang, H.S., Eulgem, T., et al., (2002). Expression profile matrix of Arabidopsis transcription factor genes implies their putative functions in response to environmental stresses. *Plant Cell* 14:559–574.

Chen J., Wu R., Yang P., Huang J., Sher Y., Han M., Kao W., Lee P., Chiu T., Chang F., Chu Y., Wu C., Peck K., 1998.Profiling expression patterns and isolating differentially expressed genes by cDNA microarray system with colorimetry detection. *Genomics* 51:313–324.

Debnath, M., Prasad, G.B.K.S., Bisen, P.S. (2010). Microarray. In Molecular Diagnostics: Promises and Possibilities (pp. 193–208). Dordrech Heidelberg London: Springer.

Dodds, P.N., Rathjen, J.P. (2010). Plant immunity: Towards an integrated view of plant-pathogen interactions. *Nature Reviews Genetics* 11:539–548.

Engel, E.A., Escobar, P.F., Rojas, L.A., Rivera, P.A., Fiore, N., Valenzuela, P.D.T. (2010). A diagnostic oligonucleotide microarray for simultaneous detection of grapevine viruses. *Journal of Virological Methods* 163(2):445–451.

FAO. (2022). World Food and Agriculture – Statistical Yearbook 2022. Rome. https://doi.org/10.4060/cc2211en

Fessehaie, A., De Boer, S.H., Le´vesque, C.A. (2003). An oligonucleotide array for the identification and differentiation of bacteria pathogenic on potato. *Phytopathology* 93(3):262–269.

Franke-Whittle, I.H., Klammer, S.H., Mayrhofer, S., Insam, H. (2006). Comparison of different labeling methods for the production of labeled target DNA for microarray hybridization. *Journal of microbiological methods, 65* 1, 117-26.

Franke-Whittle I.H., Klammer S.H., Insam H., Design and application of an oligonucleotide microarray for the investigation of compost microbial communities J. Microbiol. Methods 62 2005 375615823393.

Gillespie, D., Spiegelman, S.A. (1965). Quantitative assay for DNA-RNA hybrids with DNA immobilized on a membrane. *Journal of Molecular Biology* 12:829–842.

Hadidi, A., Barba, H.M. (2008). DNA microarrays 21st century pathogen detection. *Acta Horticulturae*:48–781.

Hammond-Kosack, K.E., Jones, J.D.G. (2000). Responses to plant pathogens. In *Biochemistry and molecular biology of plants* (pp. 1102–1156). Rockville: American Society of Plant Physiology.

Huang, Z., Yeakley, J.M., Garcia, E.W., Holdridge, J.D., Fan, J.B., Whitham, S.A. (2005). Salicylic acid-dependent expression of host genes in compatible Arabidopsis-virus interactions. *Plant Physiology* 137:1147–1159.

Jammes, F., Lecomte, P., DeAlmeida Engler, J., Bitton, F., Martin Magniette, M.L., Renou, J.P., Abad, P., Favery, B. (2005). Genome-wide expression profiling of the host response to root-knot nematode infection in Arabidopsis. *The Plant Journal* 44(3):447–458.

Krzysztof, K., Barbara, U., Alicja, M., Natasza, B. (2017). DNA microarray-based detection and identification of bacterial and viral pathogens of maize. *Journal of Plant Diseases and Protection* 124:577–583.

Lee, G.P., Min, B.E., Kim, C.S., Choi, S.H., Harn, C.H., et al. (2003). Plant virus cDNA chip hybridization for detection and differentiation of four cucurbit-infecting tobamoviruses. labeling methods for the production of labeled target DNA for microarray hybridization. *Journal of Microbiological Methods* 65:117–26.

Lee, J.G., Cheong, K.H., Huh, N., Kim, S., Choi, J.W., Ko, C. (2006). Microchip-based one step DNA extraction and real-time PCR in one chamber for rapid pathogen identification. *Lab on a Chip* 6:886–895.

Liang, P., Pardee, A.B. (1992). Differential display of eukaryotic messenger RNA by means of the polymerase chain reaction. *Science* 257:967–971.

Liebe, S., Daniela, S., Christ, D.S., Ehricht, R., Varrelmann, M. (2016). Development of a DNA microarray-based assay for the detection of sugar beet root rot pathogens. *Phytopathology* 106(1):76–86. 10.1094/PHYTO-07-15-0171-R.

Lipshutz, R., Fodor, S., Gingeras, T., Lockhart, D. (1999). High density synthetic oligonucleotide arrays. *Nature Genetics* 21(Suppl. 1):20–24.

Liu, Y., Cady, N.C., Batt, C.A. (2007). A plastic microchip for nucleic acid purification. *Biomedical Microdevices* 9:769–776.

Lockhart, D., Dong, H., Byme, M., Follettie, M., Gallo, M., Chee, M., Mittmann, M., Wang, C., Kobayashi, M., Horton, H., Brown, E. (1996). Expression monitoring by hybridization to high-density oligonucleotide arrays. *Nature Biotechnology* 14:1675–1680.

McLoughlin, K.S. (2011). Microarrays for pathogen detection and analysis. *Briefings in Functional Genomics* 10(6): 342–353.

Missoum, A. (2018). DNA microarray and bioinformatics technologies: a mini review. *Proceedings of the Nature Research Society* 2(1):02010.

Mitra, D. (2021). Emerging plant diseases: research status and challenges. *Emerging Trends in Plant Pathology*:1–17.

Nikitin, M., Deych, K., Grevtseva, I., Girsova, N., Kuznetsova, M., Pridannikov, M., et al. (2018). Preserved microarrays for simultaneous detection and identification of six fungal potato pathogens with the use of real-time PCR in matrix format. *Biosensors* 8:129. 10.3390/bios8040129.

Pasquini G., Barba M. Hadidi A., Faggioli F., Negri R., Sobol I., Tiberini A., Caglayan K., Mazyad H., Anfoka G., Ghanim M (2008). Oligonucleotide microarray-based detection and genotyping of Plum pox virus. *J Virol Methods* 147(1):118–126.

Pavlátová, L., Novotný, D., Hodek, J., Chrpová, J., Ovesná, J. (2011). Utilization of DNA microarrays for detection and identification of selected Fusarium species from the Czech Republic. *Czech Journal of Food Sciences* 29(Special Issue):S93–S101.

Pouthas, F., Gentil, C., Cote D., Bockelmann, U. (2004). DNA detection on transistor arrays following mutation-specific enzymatic amplification. *Applied Physics Letters* 84:1594–1596.

Sargent, T.D. (1987). Isolation of differentially expressed genes. *Methods in Enzymology* 152:423–432.

Scheideler, M., Schlaich, N.L., Fellenberg, K., Beissbarth, T., Hauser, N.C., Vingron, M., et al. (2002). Monitoring the switch from housekeeping to pathogen defence metabolism in *Arabidopsis thaliana* using cDNA Arrays. *Journal of Biological Chemistry* 277:10555–10561.

Schena, M., Shalon, D., Davis, R.W., Brown, P.O. (1995). Quantitative monitoring of gene expression patterns with a complementary DNA microarray. *Science* 270(5235):467–470.

Schena, M., Shalon, D., Heller, R., Chai, A., Brown, P. O., and Davis, R. W. (1996). Parallel human genome analysis: Microarray-based expression monitoring of 1000 genes. *Proc. Natl. Acad. Sci.* U.S.A. 93, 10614-10619

Schenk, P.M., Choo, J.H., Wong, C.L. (2009). Microarray analyses to study plant defence and rhizosphere microbe interaction. *CAB: Perspective in Agriculture, Veterinary, Science, Nutrition, and Natural Resources* 4:45–46.

Schenk, P.M., Kazan, K., Wilson, L., Anderson, J.P., Richmond, T., Somerville, S.C., et al. (2000). Coordinated plant defence responses in Arabidopsis revealed by microarray analysis. *Proceedings of the National Academy of Science* 97:11655–11660.

Schoen C., De Weerdt M., Hillhorst R., Chan A., Boender P.,Zijlstra C., Bonants P., (2002). Use of novel 3D microarray flow through system for plant pathogen multiplex detection. In: Abstracts Agricultural Biomarkers for Array Technology, Management Committee Meeting, Wadenswil 2002, 11.

Senthil, G., Liu, H., Puram, V.G., Clark, A., Stromberg, A., Goodin, M.M. (2005). Specific and common changes in *Nicotiana benthamiana* gene expression in response to infection by enveloped viruses. *Journal of General Virology* 86:2615–2625.

Tan, P.K., Downey, T.J., Spitznagel, E.L, Jr, Xu, P., Fu, D., Dimitrov, D.S., Lempicki, R.A., Raaka, B.M., Cam, M.C. (2003). Evaluation of gene expression measurements from commercial microarray platforms. *Nucleic Acids Research* 31:5676–5684.

Tiberini A., Barbara. (2012). Optimisation and improvement of oligonucleotide microarray-based detection of tomato viruses and pospiviroids. *Journal of Virological Methods* 185:43–51.10.10.1016/j.jviromet.2012.05.028.

Torres, M.A., Jonathan, D.G., Dangl, J.L. (2006). Reactive oxygen species signalling in response to pathogen. *Plant Physiology* 141:373–378.

Vallad, G.E., Goodman, R.M. (2004). Systemic acquired resistance and induced systemic resistance in conventional agriculture. *Crop Science* 44:1920–1934.

van Doorn R., Szemes M., Bonants P., Kowalchuk G.A., Salles J.F., Ortenberg E., Schoen CD (2007) Quantitative multiplex detection of plant pathogens using a novel ligation probe-based system coupled with universal, high-throughput realtime PCR on open arrays. *BMC Genomics* 8:276.

Wang, L., Paul C.H. Li. (2007). Flexible microarray construction and fast DNA hybridization conducted on a microfluidic chip for greenhouse plant fungal pathogen detection. *Journal of Agricultural and Food Chemistry* 55(26):10509–10516.

Whitham, S.A., Quan, S., Chang, H.S., Cooper, B., Estes, B., Zhu, T., et al. (2003). Diverse RNA viruses elicit the expression of common sets of genes in susceptible *Arabidopsis thaliana* plants. *The Plant Journal* 33:271–283.

Wildsmith, S.E., Elcock, F.J. (2001). Microarrays under the microscope. *Molecular Pathology* 54(1):8–16

Wise, R.P., Moscou, M.J., Bogdanove, A.J., Whitham, S.A. (2007). Transcript profiling in host–pathogen interactions. *Annual Review of Phytopathology* 45:329–369.

Zimmermann, C.R., Orr, W.C., Leclerc, R.F., Barnard, E.C., Timberlake, W.E. (1980). Molecular cloning and selection of genes regulated in Aspergillus development. *Cell* 21:709–715.

8 Applications of Robust Isothermal Amplification Assays in the Detection of Fungal and Viral Plant Pathogens

Mailem Yazing Shimray, Abnish Kumar, Utkarsh Singh Chauhan, Pooja Bhardwaj, Nishant, Nishant Srivastava, Rakesh Kumar, Susheel Kumar Sharma, Virendra Kumar Baranwal, Deeksha Joshi, and Nitika Gupta

8.1 INTRODUCTION

Plant diseases occur when some causative agent consistently harms the plant, leading to improper physiological processes disrupting its regular development, structure, function, or other vital processes. These diseases can be broadly classified based on whether the primary causative agent is infectious, originating from pathogenic organisms, such as viruses, worms, bacteria, fungi, and phytoplasma, or non-infectious like arising from genetic, environmental, and nutritional factors. Infectious crop diseases have the potential to significantly impact agricultural yields by severely affecting the quality and produce of the crop. Thus, early and accurate detection of plant diseases is essential for effective disease managementto minimize the impact of plant diseases and ensure timely interventions to protect the crops. The use of molecular diagnosis involving nucleic acid amplification allows early detection and enables highly accurate and rapid identification of disease-causing pathogens, often before visible symptoms appear or in case of latent infections. This proactive approach reduces the risk of extensive damage and crop loss.

In the past few decades, there has been extensive advancements in the molecular techniques used for the detection of different pathogens (e.g., bacteria, viruses, fungi, etc.) that have made them much more specific and sensitive. However, most of these techniques still have some limitations such as higher cost, requirement of power, expensive instruments, and longer processing times.

Recently, the technique of isothermal amplification has provided a viable alternative to the thermocycling-based methods. Isothermal amplification is a molecular method employed to rapidly and effectively amplify the specific DNA or RNA

DOI: 10.1201/9781003474036-8

sequences, while maintaining a constant temperature. Unlike polymerase chain reaction (PCR), which necessitates a series of thermal cycling steps, the isothermal amplification method works at a stable and constant temperature during the entire process. Alternatively, even when utilizing two or more different temperatures, the process still remains classified as isothermal as long as it avoids thermal cycling between different temperature levels.

The basic principle of isothermal amplification is to employ special enzymes and reaction mechanisms that can efficiently copy specific nucleic acid sequences at a constant temperature. This means there is no need for cycling through different temperature phases, making the process simpler and much more cost-effective. It is a valuable tool for rapidly generating many copies of specific genes for various applications in research and diagnostics.

Isothermal assays present several advantages, making them well-suited for disease diagnosis, exceeding conventional techniques like PCR. By operating at a constant temperature, these assays eliminate the need for thermal cycler, making the process simpler and more user-friendly, requiring minimal training. Additionally, some isothermal techniques offer rapid results within 30-60 minutes, thereby making them suitable for timely diagnosis of the disease with higher sensitivity and specificity. Moreover, they often require minimal equipment, reducing cost and contamination risks. In addition, some isothermal assays allow real-time monitoring of the results, and their flexibility and adaptability make them valuable tool for on-site diagnosis of various plant diseases.

The primary objective of these innovative and advanced isothermal amplification techniques is to simplify diagnostic procedures, enhance their cost-effectiveness, and maintain the sensitivity of the reaction. This chapter presents some of the important isothermal nucleic acid amplification techniques.

8.2 ISOTHERMAL AMPLIFICATION ASSAYS

8.2.1 Loop-Mediated Isothermal Amplification (LAMP)

A relatively new molecular method that has gained popularity during the past few years for plant pathogen detection is the loop-mediated isothermal amplification (LAMP) assay. The technique usually amplifies DNA at a constant temperature between 60-70°C. Due to its higher specificity, sensitivity, efficiency, and ability to be used in field situations under isothermal conditions, this method was rapidly adopted for the detection of plant diseases (Notomi et al., 2000; Iwamoto et al., 2003).

8.2.2 Mechanism

In LAMP assay, enzyme with strand displacement activity (*Bst* DNA polymerase from *Geobacillus stearothermophilus*) is usually used in conjunction with a set of six primers that are complementary to eight regions in the target DNA (Notomi et al., 2000; Nagamine et al., 2002). During amplification, the initial steps are primed using the inner primers and for self-priming during the later stages of amplification. Six different primers are used during LAMP assay. The primers used are as follows: Two inner primers [i.e.,Forward inner primer (FIP) in the 5' to 3' direction designated as F1c and F2 and Backward inner primer (BIP) as B2 and B1c (Notomi et al., 2000). The two outer primers [Forward outer primer (F3) and Backward outer primer (B3)] plays an important role during the noncyclic step of strand displacement (Parida et al., 2008). While two optional forward (LF) and reverse (LB) loop primers bind to additional DNA sequences that the four internal primers do not target (Nagamine et al., 2002). These primers are known to speed up and enhance amplification, and work together to produce DNA products with stem loops and multiple loops that resemble cauliflower structure, produced by auto-cycling and strand displacement DNA synthesis (Iwamoto et al., 2003; Parida et al., 2008).

FIP initiates the reaction, wherein the target DNA is denatured by heating at 95°C and immediately cooled on ice (Figure 8.1). This allows the FIP to invade and anneal to the target sequence (Notomi et al., 2000). A complementary strand synthesis initiates when the F2 region of the FIP anneals to the F2c section in the target DNA using strand displacement polymerase enzyme at a constant temperature of 65 °C. The strand displacement DNA synthesis is also initiated when the outer primer F3 hybridizes to its complementary F3c sequence, liberating a FIP-linked complementary strand, thereby forming a loop structure at one end when the F1c half of the primer anneals to its F1 complementary sequence in the DNA strand. The single-stranded DNA (ssDNA) serves as a template for the synthesis of new DNA, which is started by the B2 sequence at the 3′end of the BIP and finished by B3-primed strand displacement DNA synthesis, which releases a complementary strand that is linked to

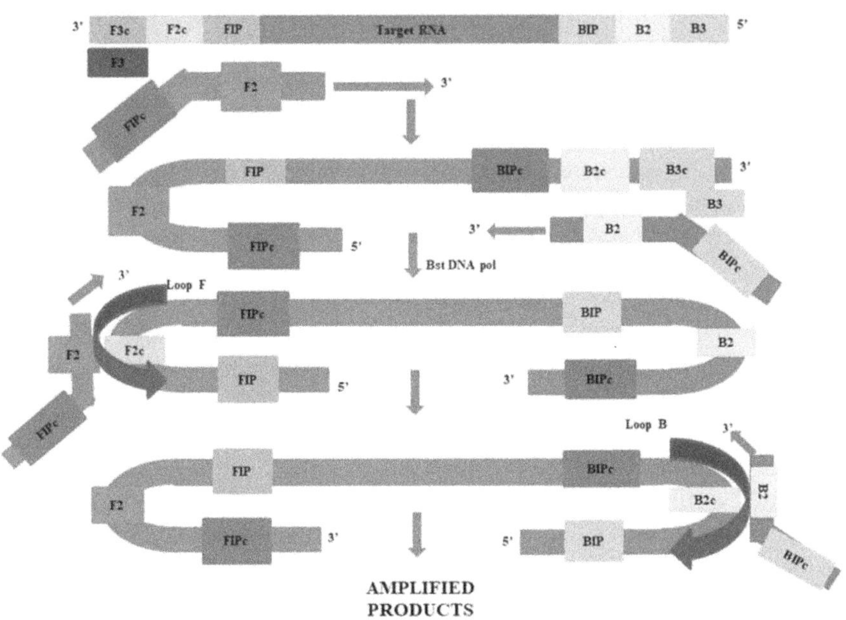

FIGURE 8.1 Loop-mediated isothermal amplification (LAMP) assay: It employs 4-6 primers that identify 6-8 distinct regions of target DNA. A strand-displacing DNA polymerase initiates synthesis, and two primers form loop structures to facilitate subsequent rounds of amplification.

the BIP. This ssDNA creates a dumbbell structure, which ultimately transforms into a stem-loop DNA.

The stem-loop DNA generated during the noncyclic process serves as a precursor for LAMP cycling, which is initiated by the FIP primer followed by strand displacement DNA synthesis and annealing to the F2c sequence in the loop of the stem-loop DNA. Further, when the LB primer is added to the reaction it binds to the stem-loop between the B1 and B2 regions leading to more stem-loop structures to which LB and LF primers can anneal, thereby promoting exponential amplification. Simultaneously, strand displacement occurs from the 3′end of the F1 region, opening the 5′end of the loop followed by a second strand displacement which takes place from the 3′end of the loop at B1 region. Thus, two intermediate structures are generated; one is complementary double stem-loop DNA while the other is new stem-loop DNA with comparatively much longer stem.

Further, the products generated serve as a template for a BIP-primed strand displacement reaction in the subsequent elongation and recycling steps. The BIP initiates subsequent DNA synthesis for the complementary double stem-loop DNA produced in further steps as it did for the original stem loop. If present, the LF primer binds to the loop between the F1 and F2 regions of this stem-loop structure and acts in a similar way as LB primer so as to speed up LAMP. In the subsequent elongation and recycling cycles, the products generated will serve as a template for a FIP-primed strand displacement reaction thereby leading to more elongated structures. Thus, the DNA target sequence is amplified almost three times at every half cycle (Notomi et al., 2000; Bruce et al., 2015; Li and Macdonald, 2015).

The results of LAMP assay can be detected visually by observing the turbidity of the solution, or determined quantitatively by using a photometer (Parida et al., 2005; Goto et al., 2009; Zhang et al., 2013). Further, the results can also be analyzed using 2% agarose gel electrophoresis wherein a ladder-like pattern of amplified products is observed.

SomeLAMP modification enables target discrimination in multisample analyses and is helpful in reducing common mistakes that arise from post-amplification detection in conventional LAMP experiments.

8.2.2.1 Reverse Transcription-LAMP (RT-LAMP)

RT-LAMP amplifies RNA target material using reverse transcriptase from Avian Myeloblastosis Virus (AMV) and the reaction can be completed in 60 minutes at a constant temperature, just as with LAMP (Tomita et al., 2008).

8.2.2.2 Multiplex LAMP

Multiplex LAMP (mLAMP) is achieved by incorporating an endonuclease recognition site into the LAMP primer (FIP). This results in the production of endonuclease-digested amplicons that have a specific length for the target species. This allows for the separation of target-specific amplicons from a pool of LAMP products. Additionally, sequence-based barcodes are used in combination with nicking endonuclease-mediated pyrosequencing.

In summary, this technique utilizes a short sequence as a target-specific barcode located within the FIP. The FIP includes a recognition site for a nicking endonuclease (NEase), followed by a LAMP reaction. The recognition site allows barcodes to be decoded using pyrosequencing, a sequence-by-synthesis method that does not require primer annealing (Liang et al., 2012).

8.2.2.3 Quantitative or Real Time-LAMP

The development of quantitative or real-time LAMP minimizes potential errors during post-amplification detection. It also enables quantitative detection of amplified products, making it easier to conduct multisample analyses. The most common methods for real-time LAMP include measuring solution turbidity or fluorescence emission with dyes such as ethidium bromide or SYBR Green I. A real-time turbidimeter is used to measure the amount of magnesium pyrophosphate produced by the LAMP reaction. Mori et al. (2004) designed a commercial real-time turbidimeter (Eiken Chemical Co., Ltd., Japan) for quantitative LAMP. This device operates at an optimal temperature of 60-65°C and can measure the turbidity of multiple samples at once (Mori et al., 2004; Panno et al., 2020).

8.2.2.4 Micro LAMP (μLAMP)

This technique combines LAMP and microfluidic assays, with LAMP conducted on a microfluidic chip. The results can be easily analyzed with the naked eye or via an optical sensor. This technique is simple, rapid, sensitive, and cost-effective. It uses a small volume (0.4μl) of reaction, making it highly portable for on-site analyses (Panno et al., 2020).

8.2.2.5 Digital LAMP (dLAMP)

This technique introduces the first on-chip loop-mediated DNA amplification in a digital format utilizing a sample self-digitization (SD) chip (Gansen et al., 2012). The SD chip creates nanoliter compartments, in which the total sample is distributed, with each compartment containing less than one template molecule. DNA amplification occurs in each compartment, and the number of compartments exhibiting amplification corresponds to the number of initial templates in the original sample. The SD chip is a cost-effective and user-friendly device that enables the digitization of a sample into nanoliter wells without any loss of volume. This technique enables accurate quantification of DNA or RNA in a target sample (Gansen et al., 2012).

8.2.2.6 Electric LAMP (eLAMP)

The electric LAMP (eLAMP) is a device that runs a Practical Extraction and Report Language (PERL) script and has a Tk graphical interface. It electronically stimulates the LAMP method and effectively tests potential LAMP primers against target sequences to determine compatibility. This technique has been used to test a pre-existing set of primers for detecting newly discovered sequence variants (Wong et al., 2018).

8.2.2.7 Lyophilized LAMP (iD-LAMP)

The purpose of lyophilized LAMP is to streamline the reaction process. This is accomplished by providing a lyophilized LAMP mix that includes all required reagents (Le and Vu, 2017). Once water and the sample or DNA-RNA template have been added to the lyophilized mixture, incubation can begin at the desired temperature. Certain lyophilized LAMP kits available on the market work with portable thermal cyclers and other devices. As a result, they can be used effectively in field settings (Panno et al., 2020).

8.2.3 Advantages of LAMP

LAMP is a promising method for detecting plant pathogens. It provides the benefits of increased sensitivity, specificity, and accuracy. One significant advantage is that it can operate at a constant temperature of 60-65 °C, which is easily maintained with a water bath even in field conditions. Furthermore, LAMP allows for the direct use of crude template DNA or tissue samples, with the results visible with the naked eye.

8.2.4 Disadvantages of LAMP

The LAMP method is known for its high sensitivity. However, this sensitivity makes it more prone to contamination. This contamination can occur during electrophoresis, a technique used to visualize results. Furthermore, using post-amplification dyes in LAMP necessitates opening the tubes, which raises the risk of contamination. This risk is heightened by the large amount of DNA generated during the reaction. Contamination can cause false positives or nonspecific detection of amplicons (Karthik et al., 2014; Fischbach et al., 2015). Another drawback of LAMP is the difficulty in designing appropriate primers. Because of the method's limited multiplexing ability, it requires 4-6 primers. Furthermore, LAMP produces a mixture of stem-loop DNAs and a cauliflower-like structure (Wong et al., 2018).

8.2.5 Applications of LAMP in Plant Disease Detection

Wheat dwarf virus (Hao et al., 2021), *Pyricularia oryzae* triticum lineage (Yasuhara-Bell et al., 2018), *Tilletia* species (Pieczul et al., 2018), *F. graminearum* (Gupta et al., 2020), and a trichothecene mycotoxin produced by *Fusarium* species (Denschlag et al., 2014) have all been detected using real-time LAMP. Similarly, a highly specific one-step RT-LAMP assay was developed for the rapid detection of various viruses and viroids, including 35 isolates of Indian citrus ringspot virus (ICRSV), columnea latent viroid (CLVd), pepper chat fruit viroid (PCFVd), potato spindle tuber viroid (PSTVd), tomato apical stunt viroid (TASVd), tomato chlorotic dwarf viroid (TCDVd), and tomato planta macho viroid (TPMVd) (Kokane et al., 2021; Tseng et al., 2021). In addition, RT-LAMP was used to detect potato virus A (PVA) in potato leaves, tuber samples, and insect vectors (aphids). It also detected potato virus X (PVX) and groundnut bud necrosis virus (GBNV) in potato samples (Raigondet al., 2019; Raigondet al., 2020; Raigondet al., 2023).

8.3 RECOMBINASE POLYMERASE AMPLIFICATION (RPA)

Recombinase polymerase amplification (RPA) is an isothermal amplification technique that has been recently developed (Piepenburg et al., 2006). It uses recombinase activity to denature DNA and synthesize new strands, thereby amplifying the target DNA. One of the advantages of RPA is that it can amplify as few as 1-10 DNA copies in less than 20 minutes, using minimally prepared crude sample extracts. The amplification can be done in solution, solid, or bridge format. The RPA assay can be completed within a short period of time, typically between 10 to 30 minutes, using two primers at a temperature of 37 to 42°C. By repeating this cycle, the target gene can be exponentially amplified (Erek et al., 2018). Furthermore, the RPA amplification technique is highly resistant to inhibitors and known for its durability.

8.3.1 Mechanism

The RPA enzyme mixtures are used to exponentially amplify specific DNA regions within a template. Three main enzymes are required for this amplification process: a recombinase, a ssDNA-binding protein, and a strand displacing polymerase. The amplification of the target DNA sequences can be completed in less than 20 minutes at a temperature range of 37-42°C (Li et al., 2019). The recombinase used, called UvsX, is derived from the T4-like bacteriophage. UvsX forms a nucleoprotein filament and searches for a single homologous target sequence within the DNA template, along with two oligonucleotide primers. This search is facilitated by the presence of ATP and a crowding agent (Piepenburg et al., 2006). Once a homologous sequence is found, the complex examines the double-stranded DNA (dsDNA)and allows strand invasion by the primer at the corresponding site. The displaced DNA strand is stabilized by single-stranded binding proteins, which control the release of the inserted primer through branch migration. Finally, three main enzymes are required for RPA amplification of any target DNA, among which "a strand displacing DNA polymerase" is one which binds to the 3' end of the primer and amplifies DNA template during amplification (Daher et al., 2016; James and Macdonald, 2015).

8.3.2 Advantages of RPA Assay

RPA, or recombinase polymerase amplification, is a method for amplifying DNA at a constant temperature of 37-42°C. RPA, unlike PCR, does not require thermal cyclers. Furthermore, RPA uses only one pair of oligonucleotide primers to synthesize nucleic acids of varying sizes. RPA can use a crude plant tissue extract as a template, in addition to DNA or cDNA. Agarose gel electrophoresis can be used to visualize the amplified products and analyze the RPA results. Alternatively, real-time monitoring can be accomplished using probe-based fluorescence or lateral flow dipsticks. These methods allow for quick evaluation of test results. In summary, RPA is a highly sensitive, simple, quick, and user-friendly technique for DNA amplification.

8.3.3 Applications in Plant Disease Detection

RPA assays have been used to detect various plant viruses and viroids, including plum pox virus, tomato yellow leaf curl virus, apple stem grooving virus, banana bunchy top virus (BBTV), cucumber mosaic virus (CMV), banana streak Mysore virus (BSMYV), onion yellow dwarf virus (OYDV), citrus tristeza virus (CTV), chilli veinal mottle virus (ChiVMV), grapevine geminivirus A (GGVA), grapevine leafroll-associated virus-3 (GLRaV-3), and peach latent mosaic viroid (Kapoor et al., 2017; Babu et al., 2018;Srivastava et al., 2019; Kapoor et al., 2020; Kumar et al., 2021; Sharma et al., 2022;Devi et al., 2023; Kishanet al., 2023a; Kishanet al., 2023b). It has also been designed and used to detect specific fungal pathogens. For instance, it can identify *Bipolaris sorokiniana*, which causes root rot and spot blotch on wheat based on calmodulin gene sequence (Zhao et al., 2021). It can also detect *Phytophthora infestans* on potatoes at 25-45°C by targeting the Ras-related protein (Ypt1) gene (Lu et al., 2020). The uncharacterized protein Pst134EA_003354 of the airborne pathogen *Puccinia striiformis* f. sp. *tritici*, which causes wheat stripe rust, was detected using visual detection within 10 minutes at 39°C (Liu et al., 2023). Similarly, the assay has been used to detect other important fungal pathogens, such as *Fusarium oxysporium* causing crown rot, wilt, and root rot on strawberry plants, by targeting the CYP51C gene unique to *Fusarium* spp. (Hu et al., 2023), as well as *Phytophthora hibernalis* (Dai et al., 2019a) and *Phytophthora sojae* (Dai et al., 2019b).

8.4 ROLLING CIRCLE AMPLIFICATION (RCA)

The Rolling Circle Amplification (RCA) assay was first described in 1998. This assay uses circular templates and DNA polymerases to generate long linear ssDNA. It can be combined with a padlock probe to detect nucleic acids with high sensitivity (Baner et al., 1998). The RCA method requires only DNA polymerases to complete the reaction. Furthermore, an additional ligase is required for the specific circularization of a padlock probe, which uses the target DNA sequence as a template during the ligation process (Ali et al., 2014).

8.4.1 Mechanism

RCA is a DNA synthesis reaction that can quickly generate multiple copies of circular DNA molecules (Johne et al., 2009). Proofreading can be performed at a constant temperature (room temperature to 37 °C) using phi29 (ϕ 29) DNA polymerase, which has polymerization activity from 5' to 3' and exonucleolytic activity from 3' to 5'. According to Dean et al. (2001), phi29 DNA polymerase is capable of displacing the complementary strand during replication, resulting in a new template for amplification. The use of random hexamers in RCA is useful for amplifying circular templates without prior knowledge of the virus sequence, which aids in the diagnosis of new plant diseases. In RCA, the polymerase continuously adds nucleotides (nt) to a primer, which is then annealed to a circular template, producing long ssDNA with tens to hundreds of tandem repeats. The ability of RCA to form a long DNA chain on a solid support or inside a cell from a single binding event allows for the detection of individual molecules (Lau et al., 2017). As a result, RCA is a promising technique for detecting a variety of plant diseases (Inoue-Nagata et al.,2004; Wambulwaet al.,2012).

A characteristic RCA reaction consists of four components: a reaction buffer with DNA polymerase (e.g., Phi29 DNA polymerase);a short DNA or RNA primer;a circular DNA template; and deoxynucleotide triphosphates (dNTPs) (monomers or building blocks of the RCA product).

Phi29 DNA polymerase is the most widely used polymerase due to its high processivity and ability to displace strands. It can handle complex circular DNA templates with multiple cross-overs, four-way junctions, and topological restrictions. In fact, the target DNA and RNA can be used to circularize a padlock probe, which will serve as the circular template for the RCA reaction. The circular DNA template, which ranges in length from 15 to 200 nt, can be synthesized enzymatically or chemically by ligating phosphate and hydroxyl end groups intramolecularly. The most common method is template-mediated enzymatic ligation with T4 DNA ligase or template-free ligation with specialized DNA ligases like Circ Ligase.

8.4.2 Advantages and Disadvantages

RCA has the following advantages:

(1) Increased sensitivity: RCA can detect single copies and has a strong amplification ability; exponential RCA's efficiency can reach 10^9 fold.
(2) Excellent sequence specificity, as it can distinguish between single nucleotide polymorphisms.
(3) Amplified products can be directly used for sequencing.
(4) High throughput: In-situ and slide amplification can be accomplished using RCA, which forms a closed circular sequence on the target, concentrating the signal it generates at a single point.

Nevertheless, the RCA method still has disadvantages. For example, the padlock probe is frequently close to 100 bp, and background interference causes problems during signal detection, so the synthesis cost is high.

8.4.3 Applications of RCA in Plant Disease Detection

The isothermal nature of the RCA assay opens up new possibilities for point-of-care diagnostics when compared to other techniques, like PCR, which require sophisticated and expensive instruments such as a thermal cycler. Because of these appealing characteristics, RCA-based platforms have been developed to detect a variety of targets, including DNA, RNA, SNPs, proteins, pathogens, and diseased cells (Schubert et al., 2007). Over 40 strains of *Fusarium* spp. were identified with the naked eye using a fluorescent dye to visualize RCA products (Davari et al., 2012). A technique for identifying fungal pathogens has also been developed, which involves the ligation of padlock probes followed by RCA (Najafzadeh et al., 2011). This technique can detect plant viruses with smaller genomes (<3 kb; families *Gemniviridae* and *Nanoviridae*), as well as viruses with larger genomes, such as badnaviruses, sugarcane bacilliform virus (SCBV), banana streak virus (BSV), and caulimovirus-like cauliflower mosaic virus (CaMV) (Inoue-Nagata et al., 2004; Haible et al., 2006; Grigoras et al., 2009; James et al., 2011; Sukal et al., 2019).

Combining RCA with various readout technologies, such as microarrays, DNA biosensors, and immune assays, has significantly increased the sensitivity of this technique compared to gel electrophoresis (Ji et al., 2012; Russell et al., 2014). These readout methods are ideal substitutes for RCA-based assays because they are more expensive and involve additional steps than a simple visual observation of a color change.

8.5 NUCLEIC ACID SEQUENCE-BASED AMPLIFICATION (NASBA)

Nucleic Acid Sequence-Based Amplification (NASBA), an isothermal amplification technique, was first introduced in 1991. It is a primer-dependent technology that allows for continuous amplification of nucleic acids at a fixed temperature in a single reaction mixture (Compton et al., 1991).

NASBA has a significant advantage in that it operates under isothermal conditions(i.e., at a constant temperature of 41°C or at two different temperatures depending on the primers and enzymes used). Importantly, even though this method uses two different temperatures, NASBA is still classified as isothermal because it avoids cycling between the two.

8.5.1 Mechanism

To generate multiple copies of single-stranded RNA, the NASBA process goes through several steps. At 65°C, single-stranded RNA binds to the primer that initiates transcription. Then, amplification occurs at 41°C. This successful amplification necessitates the use of an enzyme cocktail that includes Avian Myeloblastosis Reverse Transcriptase (AMV-RT), RNase H, and RNA polymerase, usually T7 DNA dependent RNA polymerase (DdRp) (Malek et al. 1994).

DNA amplification does occur in some NASBA systems, but it is inefficient. This type of amplification occurs when there is no RNA target present or when there is more target DNA (more than 1000-fold) than RNA (Deiman et al., 2002). As a result, the most common starting material and end product for this method is single-stranded RNA.

Primers are carefully designed to target a specific region of interest in order to achieve successful amplification. It is important to note that one of these primers includes the T7 RNA polymerase promoter sequence at the 5' end. This distinct feature enables the synthesis of single-stranded RNA molecules. These single-stranded RNAs are then reverse-transcribed by a reverse transcriptase enzyme, producing complementary DNA (cDNA) and resulting in the formation of DNA-RNA hybrids. RNase H enzyme activity removes the RNA portion of these hybrids (Figure 8.2). The enzyme reverse transcriptase then converts the single-stranded cDNA into dsDNA.

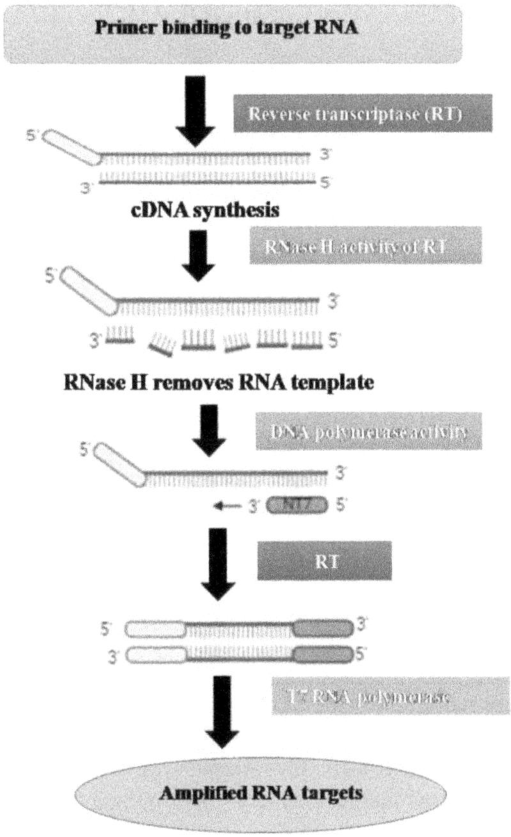

FIGURE 8.2 Schematic diagram illustrating the NASBA assay: Targeted primers, one containing a T7 RNA polymerase promoter, which binds to the RNA region of interest. The RNA is then reverse transcribed into cDNA by reverse transcriptase (RT), which degrades the RNA in the DNA-RNA hybrid through its RNase H activity. The resulting dsDNA template is then transcribed into RNA by T7 RNA polymerase, leading to exponential RNA target amplification.

This dsDNA is then used as a template for T7 RNA polymerase, resulting in exponential amplification of the target RNA sequences (Figure 8.2).

The "Boom" technique is the preferred method for nucleic acid isolation in NASBA, as it relies on the denaturing properties of guanidine isothiocyanate and the binding of nucleic acids to silica particles. After amplification, the obtained amplicons are detected using electro-chemiluminescence (ECL) (Deima et al., 2002).

8.5.2 Applications in Plant Disease Detection

NASBA can be a superior and more sensitive alternative to PCR in medical diagnostics. For rapid diagnosis of plant pathogenic viruses and viroids having single-stranded RNA genomes this technique has proven to be one of the quickest methods.

8.6 STRAND DISPLACEMENT AMPLIFICATION (SDA)

Strand Displacement Amplification (SDA) is an isothermal method of amplifying DNA in vitro. It employs a strand-displacing DNA polymerase and a specific DNA nicking event that is precisely targeted by primer design and the action of a nicking endonuclease. In less than 15 minutes, SDA can amplify a target DNA exponentially by a factor of 10^{10} amplification (Hellyer et al., 2004). This process has significant advantages, including lower costs and simpler instrumentation requirements. Furthermore, a universal detection format enables the use of identical fluorescent detector probes for multiple analytes.

8.6.1 Mechanism of SDA

SDA can be carried out in three steps as follows:

1. The initial step involves the heat denaturation of target DNA sample in the presence of primers and various reagents.
2. Following denaturation, the strand displacement takes place after the addition of nicking endonuclease and strand-displacing polymerase.
3. Subsequently, the sample is incubated at a constant temperature (i.e., at 37°C).

Four primers are required for the SDA reaction. These primers comprise of two outer "bump" primers that initiate the displacement of the target sequence, typically covering a short region of less than 100 base pairs. Additionally, there are two inner primers with 5' tail regions that include a recognition site for a nicking enzyme like as Nt, Bst, and NBI (Walker et al., 1992).

The SDA reaction begins with the generation of the target, which is accomplished through a series of steps that include primer extension and displacement. To begin, a primer containing restriction endonuclease identification site is combined with the single-stranded target molecule, yielding a dsDNA fragment. This fragment contains a replica of the target sequence, as well as restriction endonuclease recognition sites on both sides. These recognition sites are derived from the 5' tails of the SDA primer. One of the strands within the restriction site is modified by the addition of a phosphorothioate-containing nucleotide, making it resistant to cleavage by the endonucleases. In contrast, the unmodified strand is still susceptible to cleavage and is subsequently cleaved by the endonuclease, resulting in a nicked restriction site. DNA polymerase then constructs a new strand, beginning at the 3' end of the nick. The original strand carrying the target is displaced during the extension process, resulting in a fully dsDNA fragment. This new fragment contains the target sequence and is surrounded by a nick able restriction site. As a result, with each polymerase displacement step, the nicking site is continuously renewed. The cycling step consists of nicking, extension, and displacement, with the number of displaced target strands gradually increasing over time (Hellyer et al., 2004) (Figure 8.3).

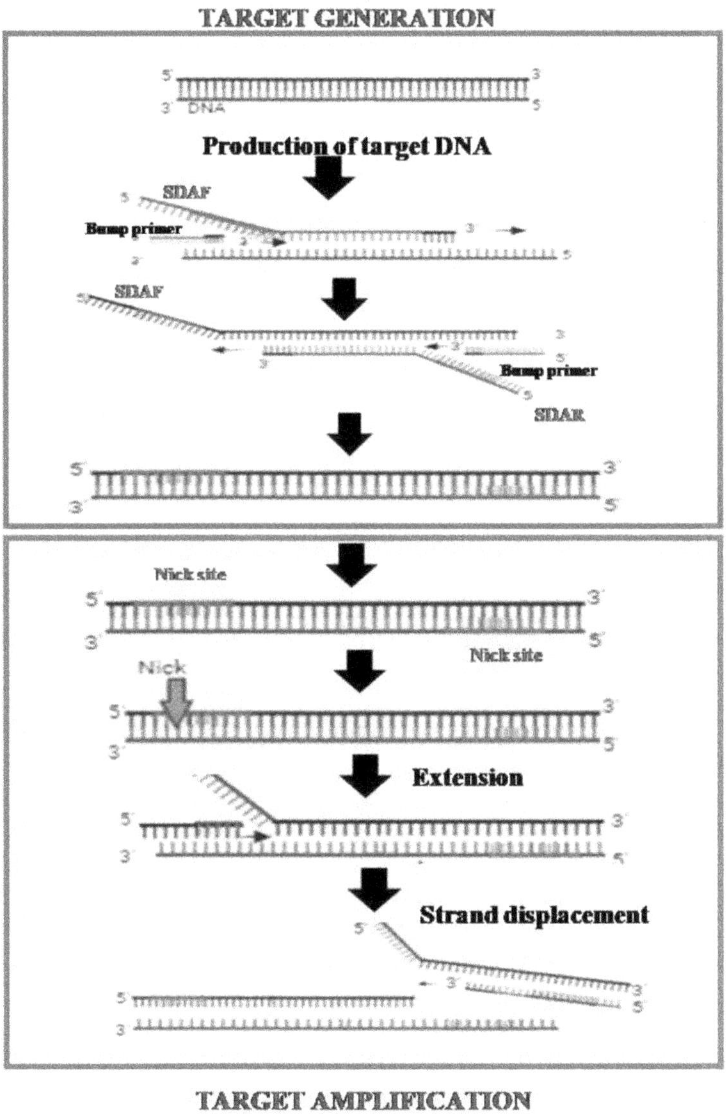

FIGURE 8.3 Schematic illustration of strand displacement isothermal amplification (SDA): The SDA primers contain a nicking endonuclease recognition sequence, a linker, and a sequence that is complementary to the target. Bumper primers are located upstream of SDA primers and complement the target sequence. These primers bind to the target sequence and are extended by specific DNA polymerases which have strand displacement activity. The nicking enzyme nicks at the nicking site, and the DNA polymerase initiates a new round of replication. Exponential amplification of the target sequence begins with a cycle of single and double nicking, extension and displacement.

8.6.2 Applications in Plant Disease Detection

Strand Displacement Amplification (SDA) is the first nucleic acid amplification technique to be integrated with real-time fluorescence detection for widespread use in clinical laboratories (Hellyer et al., 2004). This technique is very promising for genetic analysis, particularly for determining plant disease susceptibility and evaluating therapeutic efficacy based on the assessment of multiple nucleic acid markers.

8.7 WHOLE GENOME AMPLIFICATIONS (WGA)

Whole Genome Amplification (WGA) is a technique for robustly amplifying an entire genome. Typically, it begins with small amounts of DNA, ranging from picograms to nanograms, and ends with amplified products in micrograms. WGA is required when working with a small number of cell samples and limited content of DNA. Though PCR-based WGA approaches have been evolving for more than a decade, their limitations have led many researchers to switch to non-PCR-based amplification techniques.

Some notable isothermal whole genome amplification approaches include multiple displacement amplification (MDA), T7-based linear amplification of DNA (TLAD), and restriction and circularization aided-rolling circle amplification (RCA-RCA).

8.7.1 Multiple Displacement Amplification (MDA)

MDA is the most widely used approach for WGA because it is an isothermal amplification method that does not require PCR and eliminates the need for thermal cycling. MDA generates highly branched copies by utilizing the strand-displacement activity of DNA polymerases like phi29 or Bst. This process is commonly referred to as hyperbranched strand displacement amplification (HSDA) or simply strand displacement amplification (SDA) (Ahrabi et al., 2010).

8.7.1.1 Mechanism

The process typically begins with a short denaturation phase at 94°C for 2-3 minutes, followed by 6-18 hours of enzyme-specific incubation (Phi29: 30°C and Bst: 50°C), and finally inactivation of enzyme for 15 minutes at 65°C.

This method involves mixing high concentrations (10-50 µM) of random hexamer primers with template and phi29. The mixture is then incubated for the desired level of amplification. Random hexamer primers initially bind to the single-stranded target molecule. The upstream DNA strands are displaced as the primer is extended by DNA polymerase. These displaced DNA strands then serve as templates for the subsequent priming events, resulting in reverse elongation. This process continues, resulting in the generation of new DNA strands that displace further to form a hyper-branched framework, eventually yielding many copies of the initial molecule of DNA (Ahrabi et al., 2010). This process continues, resulting in the formation of new DNA strands that displace further to form

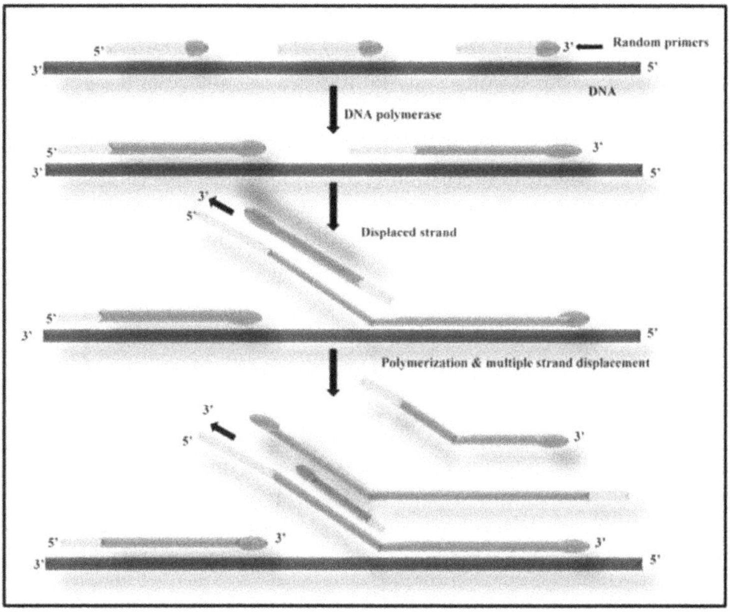

FIGURE 8.4 Multiple Displacement Amplification (MDA) assay: The initiation of the MDA process entails the DNA template binding with random hexamer primers. Subsequently, amplification is triggered using a strand-displacing polymerase, typically the phi29 DNA polymerase, resulting in the generation of complex, branched DNA network. To prepare the DNA products for subsequent steps, these branches can be disentangled by employing T7 Endonuclease.

a hyperbranched framework, eventually yielding many copies of the initial molecule of DNA (Ahrabiet al., 2010).

The Phi29 editing exonuclease activity ensures that the template DNA replicates with high fidelity. It has been discovered that the inclusion of phosphorothioate linkages in hexamer primers enhances efficiency. Pyrophosphatase can be added to high-yield reactions to prevent polymerase inhibition caused by high concentrations of pyrophosphate byproducts. The reaction's products are remarkably lengthy, exceeding 30 kb, and exhibit high branching as a result of the multiple displacement mechanism (Figure 8.4).

8.7.1.2 Applications in Plant Disease Detection

MDA has multiple applications, such as detecting disease causing agents such as bacteria, viruses, and fungus in plants and animals. It also enables the sequencing of genomic DNA from single cells. Moreover, MDA has proven to be a valuable technique for working with limited samples of valuable source material. Some examples of plant fungal pathogens that MDA can detect include fungi causing powdery mildew (Fernandez-Ortuno et al., 2007) *Penicillium paxilli* and *Epichloe festucae* (Foster and Monahan, 2005).

8.7.1.3 Disadvantages

One significant limitation of the MDA method is that the effectiveness of MDA diminishes quickly as the starting material reduces its molecular weight, rendering it unsuitable for amplifying typical templates, like DNA from formalin-fixed paraffin-embedded (FFPE) samples, corrupted DNA, DNA that are cross-linked, and short fragments of DNA. In addition, its time-consuming nature also limits its usability.

To tackle these issues, alternative WGA methods have been proposed. These include T7-based linear amplification of DNA (TLAD), balanced-PCR amplification, and restriction and circularization-assisted rolling circle amplification (RCA-RCA). These methods are intended to address the difficulties associated with amplifying these types of DNA templates.

8.7.2 T7-Based Linear Amplification of DNA (TLAD)

T7-Based Linear Amplification (TLAD) is a linear amplification technique that generates a significant quantity of genomic DNA. It can preserve variations observed in fragment sizes in the starting material with as little as 2.5 ng of input DNA (Ahrabi et al., 2010).

8.7.2.1 Mechanism

The first step involves the digestion of DNA with the restriction endonuclease AluI. The 3' ends of the digested DNA strands are then capped with a poly-T tail via terminal transferase. The genomic fragments are bound by an oligonucleotide primer with a T7 promoter at the 5' end and a poly A tract at the 3' end. The Klenow enzyme facilitates the synthesis of the second strand, and the resulting templates are used in an *in vitro* transcription reaction.

8.7.2.2 Applications and Limitations

The TLAD assay is an effective tool for amplifying DNA extracted from formalin-fixed paraffin-embedded (FFPE) tissue specimens, which frequently contain low-quality DNA. Nonetheless, it is worth noting that one disadvantage of this method is the requirement for sample purification following every stage. This requirement prolongs the duration of the assay, adds level of complexity, and may result in loss of sample. These challenges have slowed the widespread use of TLAD in practical applications.

8.7.3 Restriction and Circularization Aided-Rolling Circle Amplification (RCA-RCA)

RCA-RCA is a novel amplification technique designed to overcome issues associated with degradation of nucleic acids. This technique simultaneously preserves variations in alleles among genomes that are amplified attaining the coverage of nearly complete genome.

8.7.3.1 Mechanism

The RCA-RCA method is based on fragmenting down the genome with a suitable restriction enzyme, such as NlaIII. In formalin-fixed and paraffin-embedded (FFPE) samples, this enzyme cuts the DNA at least two times between adjacent DNA damage sites. As a result, intact DNA fragments are formed that can be circularized. Following circularization, any noncircular DNA is removed using exonuclease activity. The circular molecules are then denatured to allow for the start of exponential and highly branched rolling circle isothermal amplification. The approach utilizes random primers and Phi29 DNA polymerase (Ahrabi et al., 2010).

8.7.3.2 Applications in Plant Disease Detection

The RCA-RCA protocol's flexibility allows for the substitution of NlaIII with any single enzyme or combination of enzymes, without the need for additional modifications. This means that replacing the restriction enzyme used in the initial RCA-RCA step may allow for the recovery of samples with significant degradation. This increased utility and potential applications in plant pathology enhances the protocol's value.

8.8 CROSS-PRIMING AMPLIFICATION (CPA)

Cross-Priming Amplification (CPA) is an isothermal nucleic acid amplification technique that employs several primers and probes, at least one of which is a cross-primer. CPA differs from traditional amplification techniques in that it employs a strand displacement DNA polymerase. This eliminates the need for a preliminary denaturation step or the use of a nicking enzyme. As a result, CPA achieves exponential amplification of target DNA and is known for its exceptionally high specificity and sensitivity.

8.8.1 Mechanism

This technique employs multiple linked primers, typically ranging from 3 to 8, and amplifies a DNA target at a fix temperature, utilizing a simple heating apparatus like a water bath or dry bath. The forward (sense) and reverse (anti-sense) cross primers (PFs and PRa, respectively) are designed along with displacement primers, which are present upstream to cross-primers and are present in relatively lower concentration.

The primers form hybrids with the DNA template at sites where the DNA temporarily separates into bubbles at 63°C. Moreover, the high primer concentration compared to DNA template allows the primer to bind more readily than the DNA strands can re-join (Xu *et al.*, 2012).

A DNA polymerase, usually Bst, causes the cross and displacement primers to be extended. The displacement primer extends and pushes the cross-primer extension strand aside, resulting in a distinct 5' end. This extension/displacement process also creates a different priming site known as PFa, on the opposite strand and specifies a new end. After each extension/displacement cycle, a new priming site is introduced, resulting in multiple sites for primers to attach to, which speeds up the amplification.

To detect the target sequence, detector probes or primers are used. In some cases, a lateral flow strip detection platform is used as the preferred detection method.

8.8.2 Applications in Disease Detection

The exceptional specificity and sensitivity of this technique allows detection of very low concentration of pathogen in plant tissues, thereby generating amplicons from four genomic DNA copies (Xu et al., 2012). This allows timely detection of plant pathogens and prevention of further spread of the infection.

8.9 EXPONENTIAL AMPLIFICATION REACTION (EXPAR)

The Exponential Amplification Reaction (EXPAR) is a recently developed technique for detecting both nucleic acid and non nucleic acid targets. This advanced technique operates at a constant temperature of 50-60°C and produces amplification yields that are comparable to other methods. It uses a specific polymerase's strand displacement activity to convert dsDNA to ssDNA (Reid et al., 2018).

8.9.1 Mechanism

Exponential amplification reaction is an extremely efficient amplification technique. Reid et al. (2018) found that it can achieve amplification rates of 10^6 to 10^8 in under 30 minutes, making it suitable for signal amplification. EXPAR can identify a variety of substances, including DNA, proteins, and microRNAs. It achieves exponential amplification of a target sequence by combining a DNA template, deoxynucleotide triphosphates (dNTPs), and two enzymatic reactions. The amplification mechanism uses DNA polymerase and a nick endonuclease (NEase) in the same reaction mixture.

The EXPAR assay employs a specialized DNA template made up of two repeat regions. A short NEase recognition sequence separates these regions, which are intended to be complementary to a trigger sequence. This template is known as a T'-T' template, with T' representing the trigger sequence's complement and a brief NEase recognition sequence between them. The trigger sequence is a short nucleotide strand that can either be derived from larger targets or be the target itself, as in the case of miRNA. The length of the T' region is specifically designed to ensure a melting temperature (Tm) near the reaction temperature. This ensures both the trigger sequence and the template to have an effective hybridization. Once the trigger sequence binds to the template, a DNA polymerase having strand displacement activity expands the trigger along the template. After extension, NEase cleaves the extended trigger DNA strand, resulting in a nick. It is important to note that the NEase recognition sequence is found in dsDNA. A newly synthesized strand is displaced as the 3'-end of the nicked site is extended by the DNA polymerase. Because the template has two T' regions, the displaced DNA strand follows the same sequence as the trigger DNA. This displaced strand can then integrate with a different template molecule, starting a new amplification cycle. As a result, each cycle of the reaction generates two copies of the trigger DNA from a target molecule, yielding exponential amplification (2n).

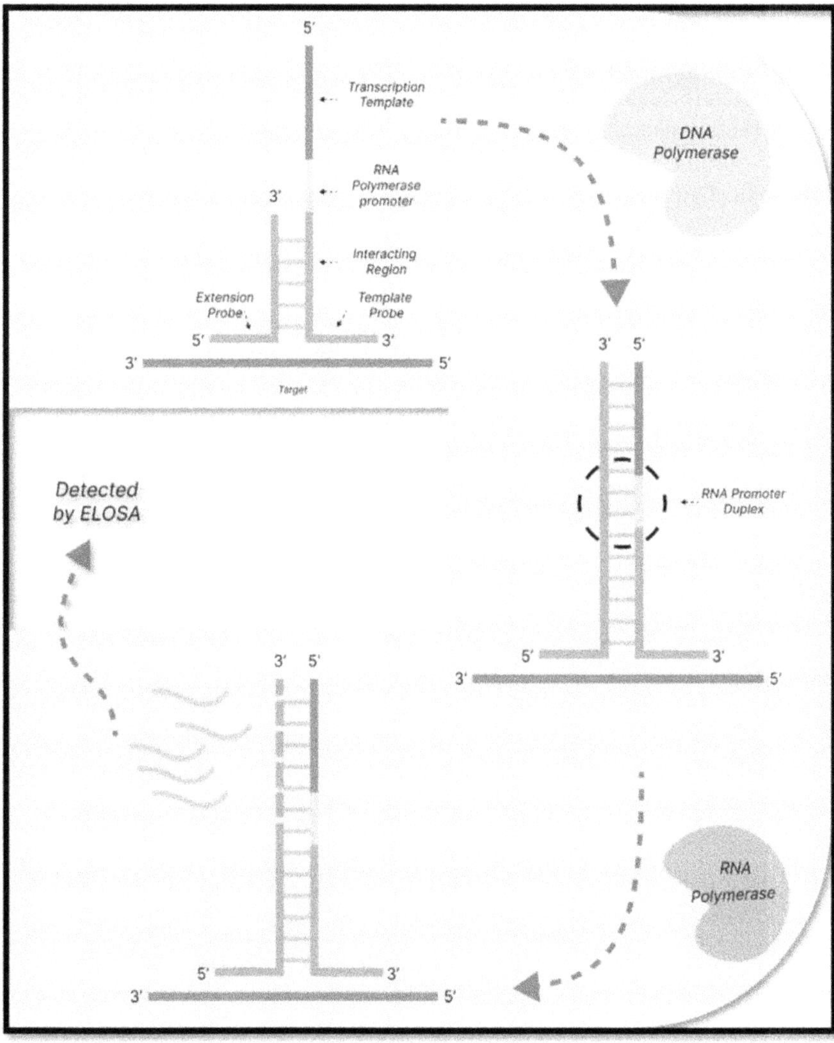

FIGURE 8.5 Schematic representation of signal-mediated amplification of RNA technology (SMART) assay. A three-way junction forms when the specific target sequence is present in the nucleic acid sample. DNA polymerase extends from the 3' end of the extension probe, creating a double-stranded functional RNA polymerase promoter. This promoter is used to generate multiple copies of RNA as a signal. The RNA signal is subsequently detected and quantified using an ELOSA method.

One notable feature of EXPAR is its ability to function at a relatively low temperature of 37°C. This lower temperature is advantageous for specific applications, such as protein detection, where antibodies are used to bind to the target proteins. Operating at lower temperatures makes the interactions between the target and the template more stable, which improves the detection process (Figure 8.5).

8.9.2 Advantages

EXPAR has a stable temperature of 37°C, making it suitable for applications where high-temperature denaturation steps are not desirable. Its isothermal nature reduces the need for complex and costly thermal cycling equipment. EXPAR is capable of achieving extremely high amplification rates, ranging from 10^6 to 10^8 within a short period, typically less than 30 minutes. This rapid amplification is particularly useful for time-sensitive assays and field-based testing.

8.9.3 Disadvantages

EXPAR necessitates the creation of a specialized DNA template with repeat regions that complement the trigger sequence. It can be time-consuming and may require a prior understanding of the target sequence. Like any nucleic acid amplification method, there is potential for contamination, leading to false-positive results. The effectiveness of EXPAR can also be affected by conditions of the reaction and the choice of enzymes, with suboptimal conditions impacting the success of assay.

8.9.4 Applications in Plant Disease Detection

EXPAR has an ability to function at a lower temperature (37°C) and can be used to detect specific nucleic acid markers associated with plant pathogens or diseases. The rapid amplification capability of EXPAR is valuable for quickly identifying the presence of pathogens, allowing timely disease management and control in agricultural field. Furthermore, the versatility of this technique allows for the detection of a variety of nucleic acid targets relevant to plant health.

8.10 SIGNAL-MEDIATED AMPLIFICATION OF RNA TECHNOLOGY (SMART)

SMART is an isothermal amplification assay that is easy to run, fast, and does not require heat cycling or duplication of lengthy target sequences. It is based on signal amplification produced only when a certain target is present. The target sequence can be distinguished between base variations using this technique (Hall et al., 2002).

8.10.1 Mechanism

SMART makes use of two single-stranded oligonucleotide probes: the extension probe and the template probe (Hall et al., 2002). These probes are structured around two key regions: the target-binding region and the interacting region. The target-binding region complements a specific target sequence, allowing probes to hybridize with it. The interacting region allows both probes to hybridize with one another. When exposed to the specific target sequence, the extension and template probes hybridize to the target, bringing both probes together to form a three-way junction (3WJ) structure (Yan et al., 2014). The template probe's single-stranded promoter sequence is then converted into a double-stranded functional promoter using DNA polymerase. This functional promoter allows RNA polymerase to produce multiple copies of an RNA signal in response to the presence of the target (Fig. 8.5). If necessary, the RNA signal can be amplified further by repeating the extension and transcription steps. This amplification step improves the sensitivity of the detection technique. An enzyme-linked oligosorbent assay (ELOSA) is used to detect and quantify the presence and amount of the RNA signal.

The assay allows for the precise measurement of the signal produced in response to the presence of the target. SMART technology is easily adaptable for detecting various target sequences. SMART is known for its sensitivity and ability to detect target analytes at low concentrations, such as 50 pM (pico Molar), in a relatively short period of time (Hall et al., 2002). SMART technology has demonstrated successful applications in detecting diverse targets, including genomic DNA and ribosomal RNA of *E. coli* (Hall et al.,2002), making it a valuable tool in molecular biology and diagnostics.

8.10.2 Advantages

SMART is designed to produce signal amplification only when a specific target is present, which enhances its specificity and minimizes false positives. It can distinguish between base variations within the target sequence, providing valuable information about genetic variations or mutations. SMART adaptability makes it versatile for detecting various target sequences with minimal probe adjustments, while its higher sensitivity allows the detection of target analytes at lowerconcentration (50 pM) in a shorter time.

8.10.3 Disadvantages

The design of the extension and template probes requires careful consideration, which may be a limitation for those without prior expertise in molecular biology techniques. In some cases, additional rounds of extension and transcription may be necessary to achieve the desired sensitivity, which can extend the overall assay time. The need for an ELOSA step in the detection process can add complexity to the assay and may require additional equipment and expertise.

8.10.4 Application in Plant Disease Detection

SMART technology has the potential to significantly improve plant disease detection. It uses specific genetic markers to identify disease-causing agents such as bacteria, viruses, and fungi in plants. Its rapid and sensitive detection capabilities allow for early intervention, helping to prevent the spread of diseases and timely disease management. SMART's ability to distinguish between base variations within the target sequence is valuable for assessing genetic variations or mutations in plant pathogens. This information is valuable for effective disease management in agriculture.

8.11 HELICASE-DEPENDENT AMPLIFICATION (HDA)

Helicase-Dependent Amplification (HDA) is a method of nucleic acid amplification that operates at a constant temperature and closely mimics the natural process of DNA replication. In HDA, a helicase enzyme is employed to separate the dsDNA, enabling the binding and extension of primers by a strand-displacing DNA polymerase. Vincent and coworkers (2004) devised this approach, which has found application in several FDA-approved diagnostic assays. Similar to PCR, the HDA system utilizes only two primers and yields concise DNA products.

8.11.1 Mechanism

HDA begins with the separation of duplex DNA strands using a DNA helicase, followed by the stabilization of ssDNA-binding proteins. Then, two specific primers are added to each end of the target DNA sequence. DNA polymerases extend these primers to produce dsDNA. In the subsequent cycle of reactions, the newly synthesized dsDNA products serve as substrates for DNA helicases. As a result, a chain reaction occurs concurrently, resulting in the desired target sequence being amplified exponentially (Vincent et al., 2004) (Figure 8.6).

Since the DNA helicase enzymatically unwinds dsDNA, the initial denaturation or thermocycling steps are not required. As a result, HDA offers a straightforward

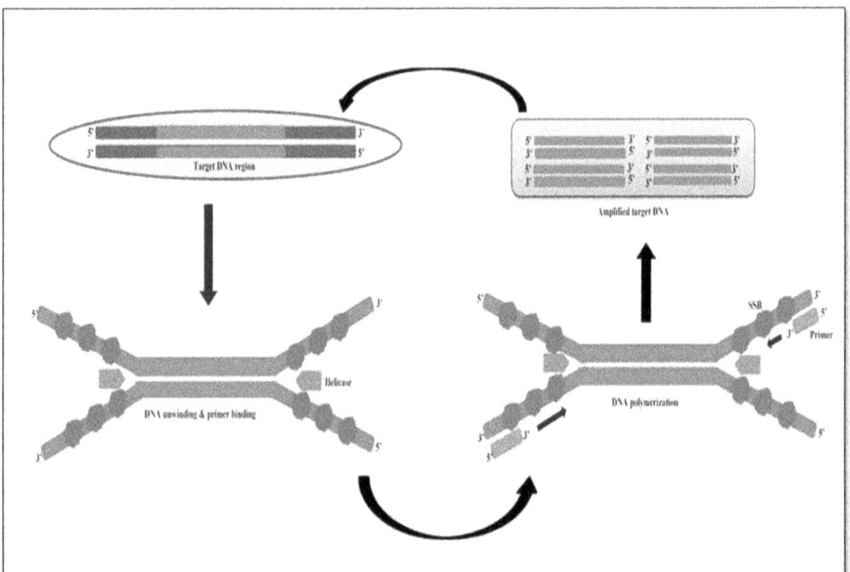

FIGURE 8.6 Helicase-dependent amplification (HAD) assay:The separation of DNA takes place with the assistance of a helicase in the presence of single-stranded binding proteins. This facilitates primer annealing and enables extension through the action of a strand-displacing DNA polymerase.

DNA amplification process with a consistent temperature maintained throughout the reaction from start to finish.

8.11.2 Applications in Plant Disease Detection

Helicase-dependent isothermal DNA amplification uses a constant temperature, eliminating the need for a thermocycler. Its simplicity and ease of use make it an ideal tool for creating portable DNA diagnostic devices for detecting plant pathogens in field samples. HDA assays have been used to detect citrus leprosis virus C, tobacco mosaic virus, and tomato spotted wilt virus (TSWV) in tomato plant samples (Wu et al., 2016).

8.12 DUAL PRIMING ISOTHERMAL AMPLIFICATION (DAMP)

Dual Priming Isothermal Amplification (DAMP) is a method of amplifying target DNA sequences with specialized primers. DAMP uses six primers: two outer primers (FO and RO), two inner primers (FI and RI), and two pairing-competition primers (FC and RC). These six primers are intended to recognize specific sites within the target DNA sequence. To accelerate and recycle the process, DAMP uses a unique "dual-priming" extension mechanism and pairing-competition primers (Ding et al. 2019).

8.12.1 Mechanism

DAMP employs six primers: two outer (FO and RO), two inner (FI and RI), and two pairing-competition (FC and RC) primers. DAMP's six primers recognize six distinct sites in target DNA sequences: (i) F1, R2, and R3 sites on the forward target sequence and (ii) F3c, F2c, and R1c sites on the reverse sequence. Sites with a lowercase "c" denote "complementary" sites (Ding et al., 2019).

DAMP's inner primers (FI and RI) are intended to recognize two target sites that are close together (less than 40 nucleotides), allowing for efficient "dual-priming" extension. Two pairing-competition primers (FC and RC) helps in the "dual-priming" extension and accelerate amplification. Surprisingly, the sequences of these primers match those of the inner primers' 5' parts (F1/R1).

Basic structure generation and cycling amplification. DNA synthesis begins in the basic structure production step when the FI primer binds to the F2c and the FO primer binds to the F3c sites. As DNA polymerase extends the FO primer, the strand synthesized by the FI primer is displaced and released. The RI and RO primers recognize the released strand, which contains R2 and R3 sites. The released strand then establishes a basic structure 1. The cycling amplification step consists of two types of extension events: self-priming extension and pairing-priming extension. Self-priming extension competes against three pairing-priming extension events. The first pairing-priming extension takes place when the F2 site of the FI primer anneals to the F2c site. The second type of "pairing-priming" extension results from a competition between the FC primer and the F1 site. Meanwhile, the third type of "pairing-priming" extension occurs when the first and second events occur concurrently. This competition limits the "self-priming" extension while encouraging the "pairing-priming" extension, which allows the formation of a variety of basic structures such as duplex and complementary structures. These structures are formed repeatedly, effectively recycling the amplification process (Ding et al., 2019).

8.12.2 Advantages

The DAMP dual-priming extension mechanism enables efficient and rapid amplification of target DNA sequences. With six different primers recognizing specific sites on the target DNA, DAMP enhances specificity in amplification. Its unique ability to recycle the process of amplification produces multiple basic structures, resulting in significant target DNA amplification.

8.12.3 Disadvantages

Designing and optimizing six distinct primers, which includes pairing-competition primers, can be a complex process that requires careful planning. DAMP involves multiple steps, including "self-priming" extension and "pairing-priming" extension events, which may increase the risk of experimental errors.

8.12.4 Application in Plant Disease Diagnosis

DAMP can be applied to plant disease diagnosis by selectively amplifying specific DNA sequences associated with plant pathogens. Its efficiency, specificity, and recycling amplification process make it a valuable tool for quick and accurate detection of plant diseases. This is critical for managing plant diseases effectively, as it enables timely actions to stop the spread of pathogens and reduce harm to crops. DAMP capability to spot low levels of pathogens is also helpful for keeping an eye on and managing plant diseases in agriculture. DAMP ability to detect low pathogen concentrations is also beneficial in monitoring and controlling plant diseases in agriculture.

8.13 SINGLE PRIMER ISOTHERMAL AMPLIFICATION (SPIA)

Single Primer Isothermal Amplification (SPIA) is a novel approach for selectively amplifying specific DNA sequences. This technology employs a specialized primer containing both DNA and RNA segments and operates at a constant temperature, making it an isothermal amplification method (Yang et al., 2018).

8.13.1 Mechanism

SPIA uses a single primer (which contains both DNA and RNA sequences), a short nucleic acid fragment (oligonucleotide) designed to complement a specific target DNA sequence for amplification (Kurn et al., 2005). The reaction in this technique begins with the addition of the DNA/RNA primer to the reaction mixture, which also contains the target DNA. A blocker is introduced as part of the amplification process. The blocker is intended to prevent or "block" the primer extension process. It is made up of a short strand of single-stranded oligonucleotides that are located downstream from the primer binding region in the target sequence of DNA. The blocker controls the amplification process. Once the primer and blocker have been added, DNA polymerase is incorporated into the reaction mixture in order to extend the DNA strand by adding complementary nucleotides to the target DNA sequence. Because the primer has an RNA component at its 5' end, after the first round of DNA amplification, an enzyme called RNase H is used to digest the primer's RNA portion (Kurn et al., 2005). This step exposes a portion of the primer-binding region on the desired DNA sequence. After the RNA is removed, a new DNA/RNA primer can bind to the exposed region of the desired DNA sequence. The DNA polymerase extends this primer, leading to additional amplification of the target sequence. The cycle of primer binding, extension, and digestion is repeated as the amplification process progresses. DNA polymerase continues to replace the previous product, ensuring the generation of ssDNA, which is an important output of the SPIA process.

The 3' end of the blocker is intended to have a high binding affinity for the target DNA sequence, making it resistant to displacement. This blocker feature terminates the amplification process at a specific point, preventing further DNA strand extension (Yang et al., 2018). Overall, the SPIA method enables the selective amplifying of a target DNA sequence in a constant temperature reaction, resulting in a substantial amount of ssDNA. It is a useful tool in molecular biology and diagnostics for analyzing gene expression and detecting particular DNAsequences.

8.13.2 Advantages

SPIA is designed to selectively amplify specific DNA sequences, making it an effective tool for targeted DNA amplification. It may identify and amplify trace amounts of target DNA, allowing the identification of specific DNA sequences. SPIA employs a single primer design, which simplifies the experimental setup and reduces the chance of primer-primer interactions, making it user-friendly.

8.13.3 Disadvantages

Designing aspecialized primer containing both DNA and RNA segments may require additional expertise and careful consideration. The design and optimization of the blocker, a critical component of SPIA, can be a complex task. The SPIA process involves multiple steps, which includes the introduction of the primer, blocker, and enzymes, which may increase the risk of experimental errors.

8.13.4 Applications in Plant Disease Detection

SPIA can be applied to plant disease diagnosis by detecting specific DNA sequences associated with plant pathogens. This selective amplification method allows for early and precise disease identification in plants, a critical aspect of effective agriculture disease management. It enables targeted and timely intervention to prevent pathogen spread and reduce crop damage. SPIA's sensitivity allows it to detect low pathogen concentrations, which is useful for monitoring and controlling plant diseases.

8.14 CONCLUSION

Isothermal amplification assays are important for in-field applications. However, each isothermal method has its own advantages and disadvantages. The different assays (i.e., NASBA, HDA, EXPAR, RPA, SDA, and SPIA) require one or more enzymes, accurate optimization and special reagents making them more tedious in carrying out the test. However, the other assays (i.e., LAMP, RCA, and RPA) require one enzyme at a fixed temperature. The most popular isothermal amplification, LAMP has been used to detect wide range of plant pathogens as it is cheap and easy to perform. But LAMP requires more than one set of primers which makes it complex and false-positive reactions also occur due to primer cross-dimer formations. RPA is a recently developed fastest amplification method and is tolerant to temperature fluctuation and efficient at 35-42°C. LAMP and RPA require user-friendly equipment as a block heater and can be performed in resource constrained laboratories and on-site conditions. However, LFA are recommended for rapid and sensitive detection of plant pathogens. RPA aligned with LFA, can be designed with CRISPR/Cas technology, making them suitable for onsite detection in the field and it can give the quickest results. Additives such as polyethylene glycol, dextran, proline, betaine, pullulan, and helicases have been used to improve the specificity of isothermal amplification methods. The combination of isothermal amplification techniques with other methods like colorimetry, nanoparticles, lateral flow biosensors, etc., results in the development of novel plant pathogen detection techniques. However, some techniques must be optimized in order to produce better results and increase their use for diagnosis at point-of-care sites. Notably, each amplification technique has a niche that expands when combined with others thereby reducing the need for PCR thermal cycler machines.

CONFLICT OF INTEREST

The authors declare that there is no conflict of interest.

REFERENCES

Ahrabi, S.M., Farajnia, S., Rahimi-Mianji, G., Saheb, S.M., & Nejati-Javaremi, A. (2010). Whole genome amplification: use of advanced isothermal method. *African Journal of Biotechnology*, *9*(54), 9248–9254.

Ali, M.M., Li, F., Zhang, Z., Zhang, K., Kang, D.K., & Ankrum, J.A. (2014). Rolling circle amplification: a versatile tool for chemical biology, materials science and medicine. *Chemical Society Reviews*, *43*, 3324–3341.

Babu, B., Ochoa-Corona, F.M., & Paret, M.L. (2018). Recombinase polymerase amplification applied to plant virus detection and potential implications. *Analytical Biochemistry*, *546*, 72–77.

Baner, J., Nilsson, M., Mendel-Hartvig, M., & Landegren, U. (1998). Signal amplification of padlock probes by rolling circle replication.*NucleicAcids Research*, *26*, 5073–5078.

Bruce, K.L., Leterme, S.C., Ellis, A.V., & Lenehan, C.E. (2015). Approaches for the detection of harmful algal blooms using oligonucleotide interactions. *Analytical and Bioanalytical Chemistry*, *407*, 95–116. doi: 10.1007/s00216-014-8193-x

Compton, J.(1991). Nucleic acid sequence-based amplification. *Nature*, *350*(6313), 91–92.

Daher, R.K., Stewart, G., Boissinot, M., & Bergeron, M.G. (2016). Recombinase polymerase amplification for diagnostic applications. *Clinical Chemistry*,*62*(7), 947–958.

Dai, T., Hu, T., Yang, X., Shen, D., Jiao, B., Tian, W., & Xu, Y. (2019a). A recombinase polymerase amplification-lateral flow dipstick assay for rapid detection of the quarantine citrus pathogen in China, *Phytophthora hibernalis*. *Peer J*, *7*, e8083.

Dai, T., Yang, X., Hu, T., Jiao, B., Xu, Y., Zheng, X., & Shen, D. (2019b). Comparative evaluation of a novel recombinase polymerase amplification-lateral flow dipstick (RPA-LFD) assay, LAMP, conventional PCR, and leaf-disc baiting methods for detection of Phytophthora sojae. *Frontiers in Microbiology*, *10*, 1884.

Davari, M., Van Diepeningen, A.D., Babai-Ahari, A., Arzanlou, M., Najafzadeh, M. J., & Van Der Lee, T.A.J. (2012). Rapid identification of Fusarium graminearum species complex using Rolling Circle Amplification (RCA). *Journal of Microbiological Methods*, *89*, 63–70. doi: 10.1016/j.mimet.2012.01.017.

Dean, F., Nelson, J., Giesler, T., & Lasken, R. (2001). Rapid amplification of plasmid and phage DNA using Phi29 DNA polymerase and multiply primed rolling circle amplification. *Genome Research*, *11*(7), 1095–1099.

Deiman, B., van Aarle, P., & Sillekens, P.(2002). Characteristics and applications of nucleic acid sequence-based amplification (NASBA). *Molecular Biotechnology*, *20*(2), 163–179.

Denschlag, C., Rieder, J., Vogel, R.F., & Niessen, L. (2014). Real-time loop-mediated isothermal amplification (LAMP) assay for group specific detection of important trichothecene producing *Fusarium* species in wheat. *International Journal of Food Microbiology*, *177*, 117–127. doi: 10.1016/j.ijfoodmicro.2014.02.010

Devi, H.C., Devi, K.S., Sanabam, R., Pathaw, N., Devi, O.P., Chanu, Ng.T., Maibam, A. Chanu, W.T., Sanasam, J., Roy, S.S., Devi, Ch.P., Singh, A.R., Devi, Ph. S., Gupta, N., & Sharma, S.K. (2023). Simplified extraction protocol for plant tissues and reverse transcription RPA assay for quick and reliable diagnosis and its application in resistance screening of chilliveinal mottle virus. *Crop Protection*, *170*, 106280.

Ding, X., Xu, Z., Yin, K., Sfeir, M., & Liu, C. (2019). Dual-priming isothermal amplification (DAMP) for highly sensitive and specific molecular detection with ultralow nonspecific signals. *Analytical Chemistry*, *91*(20), 12852–12858. doi: 10.3389/fpls.2022.857673

Ereku, L.T., Mackay, R.E., Craw, P., Naveenathayalan, A., Stead, T., & Branavan, M. (2018). RPA using a multiplexed cartridge for low cost point of care diagnostics in the field. *Analytical Biochemistry*,*547*, 84–88. doi: 10.1016/j.ab.2018.02.010

Fernandez-Ortuno, D., Tores, J.A., de Vicente, A., & Perez-Garcia, A. (2007). Multiple displacement amplification, a powerful tool for molecular genetic analysis of powdery mildew fungi. *Current Genetics*, *51*, 209–219.

Fischbach, J., Xander, N.C., Frohme, M., & Glökler, J.F. (2015). Shining a light on LAMP assays: A comparison of LAMP visualization methods including the novel use of berberine. *Biotechniques*, *58*(4), 189–194.

Foster, S.J., & Monahan, B.J. (2005). Whole genome amplification from filamentous fungi using Phi29-mediated multiple displacement amplification. *Fungal Genetics and Biology*, *42*(5), 367–375.

Gansen, A., Herrick, A.M., Dimov, I.K., Lee, L.P., & Chiu, D.T. (2012). Digital LAMP in a sample self-digitization (SD) chip. *Lab on a Chip*, *12*, 2247–2254. doi: 10.1039/c2lc21247a

Grigoras, I., Timchenko, T., Katul, L., Grande-Pérez, A., Vetten, H.J., & Gronenborn, B. (2009). Reconstitution of authentic nanovirus from multiple cloned DNAs. *Journal of Virology*, *83*,10778–10787.

Goto, M., Honda, E., Ogura, A., Nomoto, A., & Hanaki, K.I. (2009). Colorimetric detection of loop-mediated isothermal amplification reaction by using hydroxy naphthol blue. *BioTechniques*,*46*, 167–172. doi: 10.2144/000113072

Gupta, S., Saharan, M.S., Gurjar, M.S., Singh, J., Bashyal, B.M., & Aggarwal, R. (2020). Molecular detection of *Fusarium graminearum*causing head blight of wheat by loop mediated isothermal amplification (LAMP) assay. *Indian Phytopathology*, *73*, 667–672. doi: 10.1007/s42360-020-00261-8.

Haible, D., Kober, S., & Jeske, H. (2006). Rolling circle amplification revolutionizes diagnosis and genomics of geminiviruses. *Journal of Virological Methods*, *135*, 9–16.

Hall, M.J., Wharam, S.D., Weston, A., Cardy, D.L.N., & Wilson, W.H. (2002). Use of signal-mediated amplification of RNA technology (SMART) to detect marine cyanophage DNA. *Biotechniques*, *32*(3), 604–611.

Hao, X., Wang, L., Zhang, X., Zhong, Q., Hajano, J.U.D., & Xu, L. (2021). A real-time loop-mediated isothermal amplification for detection of the wheat dwarf virus in wheat and the insect vector *Psammotettixalienus*. *Plant Disease*, *105*, 4113–4120. doi: 10.1094/pdis-10-20-2279-re

Hellyer, T.J., & Nadeau, J.G. (2004). Strand displacement amplification: a versatile tool for molecular diagnostics. *Expert Review of Molecular Diagnostics*, *4*(2), 251–261.

Hu, S., Yan, C., Yu, H., Zhang, Y., & Zhang, C.Q. (2023). Establishment of the recombinase polymerase amplification-lateral flow dipstick (RPA-LFD) detection technique for Fusarium oxysporum. *Plant Disease*, *107*(9), 2665–2672.

Inoue-Nagata, A.K., Albuquerque, L.C., Rocha, W.B., & Nagata, T. (2004). A simple method for cloning the complete begomovirus genome using the bacteriophage φ29 DNA polymerase. *Journal of Virological Methods*, *116*, 209–211.

Iwamoto, T., Sonobe, T., & Hayashi, K. (2003). Loop-mediated isothermal amplification for direct detection of *Mycobacterium tuberculosis* complex, *M. avium*, and *M. intracellulare*in sputum samples. *Journal of Clinical Microbiology*, *41*, 2616–2622. doi: 10.1128/JCM.41.6.2616-2622.2003

James, A., & Macdonald, J. (2015). Recombinase polymerase amplification: emergence as a critical molecular technology for rapid, low-resource diagnostics. *Expert Review of Molecular Diagnostics*, *15*(11), 1475–1489. doi:10.1586/14737159.2015.1090877.

James, A.P., Geijskes, R.J., Dale, J.L., & Harding, R.M. (2011). Development of a novel rolling-circle amplification technique to detect banana streak virus that also discriminates between integrated and episomal virus sequences. *Plant Disease*, *95*, 57–62.

Ji, H.X., Yan, F., Lei, J.P., & Ju, H.X. (2012). Ultrasensitive electrochemical detection of nucleic acids by template-enhanced hybridization followed with rolling circle amplification. *Analytical Chemistry*, *84*(16), 7166–7171. doi: 10.1021/ac3015356.

Johne, R., Müller, H., Rector, A., van Ranst, M., & Stevens, H. (2009). Rolling-circle amplification of viral DNA genomes using phi29 polymerase. *Trends in Microbiology*, *17*(4), 205–211.

Karthik, K., Rathore, R., Thomas, P., Arun, T.R., Viswas, K.N., Agarwal, R.K., & Dhama, K. (2014). Loop-mediated isothermal amplification (LAMP) test for specific and rapid detection of Brucella abortus in cattle. *Veterinary Quarterly*, *34*(4), 174–179.

Kapoor, R., Srivastava, N., Kumar, S.,Saritha, R.K., Sharma, S.K., Jain, R.K., & Baranwal, V.K. (2017). Development of a recombinase polymerase amplification assay for the diagnosis of banana bunchy top virus in different banana cultivars. *Archives of Virology*, *162*, 2791–2796.

Kapoor, R., Srivastava, N., Kumar, R., Sharma, S.K., Rai, R., Kumar, S., & Baranwal, V.K. (2020). Detection of episomal banana streak Mysore virus by reverse transcription-recombinase polymerase amplification assay. *Journal of Plant Pathology*, *102*, 499–503.

Kishan, G., Kumar, R., Sharma, S.K., Srivastava, N., Gupta, N., Kumar, A., & Baranwal, V.K. (2023a). Development and application of crude sap-based recombinase polymerase amplification assay for the detection and occurrence of grapevine geminivirus A in Indian grapevine cultivars. *Frontiers in Plant Science*, *14*, 1151471.

Kishan, G., Kumar, R., Sharma, S.K., Srivastava, N., Gupta, N., Kumar, A., & Baranwal, V.K. (2023b). Trouble-free detection of grapevine leafroll-associated virus-3 employing reverse transcription-recombinase polymerase amplification assay. *Journal of Plant Diseases and Protection*, *131*(1), 1–13.

Kokane, A., Kokane, S., Warghane, A., Gubyad, M., Sharma, A., Reddy, M., & Ghosh, D.(2021). A rapid and sensitive reverse transcription–loop-mediated isothermalamplification (RT-LAMP) assay for the detection of Indian citrus ringspot virus. *Plant Disease*, *105*(5), 1346–1355.

Kumar, R., Pant, R.P., Kapoor, S., Khar, A., & Baranwal, V.K. (2021). Development of a reverse transcription-recombinase polymerase amplification (RT-RPA) assay for the detection of onion yellow dwarf virus (OYDV) in onion cultivars. *Indian Phytopathology*, *74*, 201–207.

Kurn, N., Chen, P., Heath, J.D., Kopf-Sill, A., Stephens, K.M., & Wang, S. (2005). Novel isothermal, linear nucleic acid amplification systems for highly multiplexed applications.*Clinical Chemistry*, *51*(10), 1973–1981.

Lau, H.Y., & Botella, J.R. (2017). Advanced DNA-based point-of-care diagnostic methods for plant diseases detection. *Frontiers in Plant Science*, *8*, 2016.

Le, D.T., & Vu, N.T. (2017). Progress of loop-mediated isothermal amplification technique in molecular diagnosis of plant diseases. *Applied Biological Chemistry*, *60*, 169–180. doi: 10.1007/s13765-017-0267-y

Li, J., & Macdonald, J. (2015). Advances in isothermal amplification: novel strategies inspired by biological processes. *Biosensors and Bioelectronics*, *64*, 196–211. doi: 10.1016/j.bios.2014.08.069

Li, T.T., Wang, J.L., Zhang, N.Z., Li, W.H., Yan, H.B., Li, L., Jia, W.Z., & Fu, B.Q. (2019). Rapid and visual detection of *Trichinella* spp. using a lateral flow strip-based recombinase polymerase amplification (LF-RPA) assay. *Frontiers in Cellular and Infection Microbiology*, *9*, 1.

Liang, C., Chu, Y., Cheng, S., Wu, H., Kajiyama, T., & Kambara, H. (2012). Multiplex loop-mediated isothermal amplification detection by sequence-based barcodes coupled with nicking endonuclease-mediated pyrosequencing. *Analytical Chemistry*, *84*, 3758–3763.

Liu, Y., Hao, J., Guo, Q., Yan, J., & Yao, Q. (2023). Establishment of a recombinase polymerase amplification detection method for *Puccinia striiformis* f. *sp.* tritici. *Scientific Reports*, *13*, 16133. https://doi.org/10.1038/s41598-023-42663-4

Lu, X., Zheng, Y., Zhang, F., Yu, J., Dai, T., Wang, R., Tian, Y., Xu, H., Shen, D., & Dou, D. (2020). A rapid, equipment-free method for detecting *Phytophthora infestans* in the field using a lateral flow strip-based recombinase polymerase amplification assay. *Plant Disease*, *104*, 2774–2778. https://doi.org/10.1094/PDIS-01-20-0203-SC

Malek, L., Sooknanan, R., & Compton, J. (1994). Nucleic acid sequence-based amplification (NASBA™). *Protocols for nucleic acid analysis by nonradioactive probes*, pp.253–260.

Mori, Y., Kitao, M., Tomita, N., & Notomi, T. (2004). Real-time turbidimetry of LAMP reaction for quantifying template DNA. *Journal of Biochemical and Biophysical Methods*, *59*, 145–157. doi: 10.1016/j.jbbm.2003.12.005

Nagamine, K., Hase, T., & Notomi, T. (2002). Accelerated reaction by loop-mediated isothermal amplification using loop primers. *Molecular and Cellular Probes*, *16*, 223–229. doi: 10.1006/mcpr.2002.0415

Najafzadeh, M.J., Sun, J., Vicente, V.A., & DeHoog, G.S. (2011). Rapid identification of fungal pathogens by rolling circle amplification using Fonsecaea as a model. *Mycoses*, *54*, E577–E582. doi: 10.1111/j.1439-0507.2010.01995.x.

Notomi, T., Okayama, H., Masubuchi, H., Yonekawa, T., Watanabe, K., & Amino, N. (2000). Loop-mediated isothermal amplification of DNA. *Nucleic Acids Research*,*28*, e63. doi: 10.1093/nar/28.12.e63

Panno, S., Matić, S., Tiberini, A., Caruso, A.G., Bella, P., & Torta, L. (2020). Loop mediated isothermal amplification: principles and applications in plant virology. *PlanTheory*, *9*, 461. doi: 10.3390/plants9040461

Parida, M., Horioke, K., Ishida, H., Dash, P.K., Saxena, P., Jana, A.M., & Morita, K. (2005). Rapid detection and differentiation of dengue virus serotypes by a real-time reverse transcription-loop-mediated isothermal amplification assay. *Journal of Clinical Microbiology*, *43*(6), 2895–2903.

Parida, M., Sannarangaiah, S., Dash, P.K., Rao, P.V.L., & Morita, K. (2008). Loop mediated isothermal amplification (LAMP): a new generation of innovative gene amplification technique; perspectives in clinical diagnosis of infectious diseases. *Reviews in Medical Virology*, *18*, 407–421. doi: 10.1002/rmv.593

Pieczul, K., Perek, A., & Kubiak, K. (2018). Detection of *Tilletia caries*, *Tilletia laevis* and *Tilletiacontroversa*wheat grain contamination using loop-mediated isothermal DNA amplification (LAMP). *Journal of Microbiological Methods*, *154*, 141–146. doi: 10.1016/j.mimet.2018.10.018

Piepenburg, O., Williams, C.H., Stemple, D.L., & Armes, N.A. (2006). DNA detection using recombination proteins. *PLOS Biology*, *4*(7), e204.

Raigond, B., Pathania, S.,Verma, G., Bhardwaj, P.,Kochhar, T., & Chakrabarti, S.K.(2023). Development and application of reverse transcription-loop mediated isothermal amplification assay for sensitive detection of groundnut bud necrosis virus infecting potato. *Virology*, *587*, 109872.

Raigond, B., Verma, A., Pathania, S., Sridhar J, Kochhar, T., & Chakrabarti, S.K. (2020). Development of a reverse transcription loop-mediated isothermal amplification for detection of *potato virus a* in potato and in insect vector aphids. *Crop Protection*, *137*, 105296.

Raigond, B., Verma, A., Roach, S., Kochhar, T., Shilpa, Arjunan, J., Kumar, R., Sharma, S., & Chakrabarti, S.K. (2019). One-step reverse transcription loop-mediated isothermal amplification: a simple, sensitive and rapid assay for detection of potato virus X in potato leaves and tubers. *Indian Phytopathology*, *72*, 321–328.

Reid, M.S., Le, X.C., & Zhang, H. (2018). Exponential isothermal amplification of nucleic acids and amplified assays for proteins, cells, and enzyme activities.Angewandte. *Chemie International Edition*, *57*, 11856–11866.

Russell, C., Welch, K., Jarvius, J., Cai, Y.X., Brucas, R., & Nikolajeff, F. (2014). Gold nanowire-based electrical DNA detection using rolling circle amplification. *ACS Nano*, *8*(2), 1147–1153. doi: 10.1021/nn4058825.

Schubert, J., Habekuss, A., Kazmaier, K., & Jeske, H. (2007). Surveying cereal infecting geminiviruses in Germany–Diagnostics and direct sequencing using rolling circle amplification. *Virus Research*, *127*, 61–70. doi: 10.1016/j.virusres.2007.03.018.

Sharma, S.K., Pathaw, N., Wangkhem, B., Jackson, K.S., Devi, K.S., Roy, S.S., Singh, A.R., Singh, R., Banerjee, A., Kumar, S., & Ningombam, A. (2022). Simple template-based reverse transcription-recombinase polymerase amplification assay for routine diagnosis of citrus tristeza virus. *Letters in Applied Microbiology*, *76*(1), p.ovac060.

Srivastava, N., Kapoor, R., Kumar, R., Kumar, S., Saritha, R.K., Kumar, S., & Baranwal, V.K. (2019). Rapid diagnosis of Cucumber mosaic virus in banana plants using a fluorescence-based real-time isothermal reverse transcription-recombinase polymerase amplification assay. *Journal of Virological Methods*, *270*, 52–58.

Sukal, A.C., KidanemariamD.B., Dale, J.L., Harding, R.M., & James, A.P. (2019).Assessment and optimization of rolling circle amplification protocols for the detection and characterization of badnaviruses. *Virology*, *529*, 73–80.

Tomita, N., Mori, Y., Kanda, H., & Notomi, T. (2008). Loop-mediated isothermal amplification (LAMP) of gene sequences and simple visual detection of products. *Nature Protocols*, *3*, 877–882. doi: 10.1038/nprot.2008.57

Tseng, Y., Wu, C., Lee, C., Chang, C., Chen, Y., & Jan, F. (2021). Universal primers for rapiddetection of six pospiviroids in solanaceae plants using one-step reverse-transcriptionPCR and reverse-transcription loop-mediated isothermal amplification. *Plant Disease*, *105*(10), 2867–2872.

Vincent, M., Xu, Y., & Kong, H.(2004). Helicase-dependent isothermal DNA amplification. *EMBO Reports*, *5*(8), 795–800.

Walker, G.T., Fraiser, M.S., Schram, J.L., Little, M.C., Nadeau, J.G., & Malinowski, D.P.(1992). Strand displacement amplification—an isothermal, in vitro DNA amplification technique. *Nucleic Acids Research*, *20*(7), 1691–1696.

Wambulwa, M.C., Wachira, F.N., Karanja, L.S., & Muturi, S.M. (2012). Rolling circle amplification is more sensitive than PCR and serology-based methods in detection of Banana streak virus in Musa germplasm. *American Journal of Plant Science*, *3*(11), 1581–1587.

Wong, Y.P., Othman, S., Lau, Y.L., Radu, S., & Chee, H.Y. (2018). Loop-mediated isothermal amplification (LAMP): a versatile technique for detection of micro-organisms. *Journal of Applied Microbiology*, *124*, 626–643. doi: 10.1111/ jam.13647.

Wu, X., Chen, C., Xiao, X., & Deng, M.J. (2016). Development of reverse transcription thermostable helicase-dependent DNA amplification for the detection of tomato spotted wilt virus. *Journal of AOAC International*, *99*(6), 1596–1599.

Xu, G., Hu, L., Zhong, H., Wang, H., Yusa, S.I., Weiss, T.C., Romaniuk, P.J., Pickerill, S., & You, Q. (2012). Cross priming amplification: Mechanism and optimization for isothermal DNA amplification. *Scientific Reports*, *2*(1), 246.

Yan, L., Zhou, J., Zheng, Y., Gamson, A.S., Roembke, B.T., Nakayama, S., & Sintim, H.O. (2014). Isothermal amplified detection of DNA and RNA. *Molecular BioSystems*, *10*, 970–1003.

Yang, Q., Zhang, Y., Li, C., Zhao, Y., Ma, X., Zhang, W., & Zhang, S. (2018). A rapid and visual single primer isothermal amplification-based method for the detection of staphylococcus aureus in raw pork products. *Food Analytical Method*, *11*, 3113–3120.

Yasuhara-Bell, J., Valent, B., & Stack, J.P. (2018). Specific detection of the wheat blast pathogen (*Magnaporthe oryzae Triticum*) by loop-mediated isothermal amplification. *Plant*Disease, *102*, 2550–2559.

Zhang, X., Zhang, H., Pu, J., Qi, Y., Yu, Q., & Xie, Y.(2013). Development of a real-time fluorescence loop-mediated isothermal amplification assay for rapid and quantitative detection of *Fusarium oxysporum*f. sp. *cubense*tropical race 4 in soil. *PLoS One*, *8*, e82841. doi: 10.1371/journal.pone.0082841

Zhao, W., Chi, Y., Ye, M., Wang, T., Xu, A., & Qi, R. (2021). Development and application of recombinase polymerase amplification assay for detection of *Bipolarissorokiniana*. *Crop Protection*, *145*(1). https://doi.org/10.1016/j.cropro.2021.105619.

9 Next-Generation Sequencing Approaches

Detection of Plant Pathogens with a Special Focus on Exploring Plant Virome

Naorem Loya Mangang, Pooja Bhardwaj, Pooja Thapa, Nishant Srivastava, Damini Diksha, Yazing S. Mailem, Nitika Gupta, Sajad U. Nabi, Virendra K. Baranwal and Susheel Kumar Sharma

9.1 INTRODUCTION

Plant pathogens, encompassing a diverse array of viruses, bacteria, fungi, nematodes, and other microorganisms, have formidable adversaries in agriculture and the natural world. These insidious agents stealthily infiltrate crops, jeopardizing food security, biodiversity, and economic stability (Strange & Scott, 2005). In an era characterized by a growing global population and increasing demands on agricultural productivity, the significance of robust and timely plant pathogen detection cannot be overstated. The plant pathogens remain a formidable challenge to global agriculture. They have the potential to cause substantial economic losses by reducing crop yields. In a world striving to feed a growing population, plant pathogen detection is indispensable for safeguarding food security (Savary et al., 2019). Recent outbreaks of emerging plant pathogens, such as *Xylella fastidiosa* and citrus greening, emphasize the urgency of early detection and management (Sicard et al., 2018). Swift identification of new threats is crucial to prevent their spread and minimize the impact on crops and ecosystems. Additionally, it helps to mitigate the spread of pathogens to neighbouringfields, preserving the health of agricultural ecosystems. Moreover, the consequences of undetected plant pathogens extend beyond agriculture. These stealthy adversaries can infiltrate natural ecosystems, threatening biodiversity by harming native plant species (Gilbert & Webb, 2007), so pathogen detection is essential for conserving biodiversity and maintaining the ecological balance. Its detection empowers farmers and agricultural stakeholders with data-driven decision-making capabilities. Advances in sensor technologies and data analytics provide opportunities for real-time monitoring and precision disease management (Mahlein et al., 2013), but on the other hand climate change is altering the distribution and

DOI: 10.1201/9781003474036-9

behaviour of plant pathogens. Timely detection and mitigation strategies are vital for managing the impact of climate-driven changes on plant health (Anderson et al., 2004). The paramount importance of plant pathogen detection lies in its pivotal role in safeguarding agricultural sustainability (Agrios, 2005). Sustainable agriculture practices, including reduced pesticide use, are gaining prominence. Recent research underscores the role of plant pathogen detection in optimizing sustainable farming by enabling targeted interventions and minimizing environmental impacts (Popp et al., 2013). Now international trade in agricultural products is subject to stringent quarantine regulations and the role of plant pathogen detection in ensuring compliance with these regulations and maintaining trade relationships (FAO, 2006). The ongoing battle against plant pathogens has been reshaped and invigorated by the advent of Next-Generation Sequencing (NGS) technologies. These revolutionary platforms have empowered researchers to decode the genetic blueprints of organisms rapidly and cost-effectively (Mardis, 2008). This transformative capability extends far beyond traditional detection methods, unlocking new dimensions of plant-pathogen interactions and fundamentally changing our approach to understanding and combatting these adversaries.

In the present chapter, we will delve into the profound importance of early and precise plant pathogen detection and how NGS technologies have emerged as powerful tools in this endeavor. We will also shine a spotlight on the often-overlooked dimension of the plant virome a complex network of viruses that coexist within plant ecosystems, influencing host health and ecological balance. Effective detection is thus essential for conserving the delicate balance of ecosystems, where plant pathogens can significantly alter the composition and dynamics of native flora. The emergence of NGS technologies has revolutionized our ability to detect and understand plant pathogens. NGS allows for the rapid and cost-effective sequencing of entire genomes, enabling researchers to identify pathogens with unprecedented accuracy (Mardis, 2008). It has vastly expanded the toolkit available to plant pathologists, enabling them to probe deeper into the genetic makeup of pathogens, monitor their evolution, and assess their virulence. This transformative technology transcends the confines of traditional detection methods, such as PCR and serological assays, which often rely on prior knowledge of specific pathogens. NGS empowers researchers to perform metagenomic analyses, where entire communities of pathogens can be identified and characterized simultaneously, without the need for prior assumptions about the pathogen's identity (Loman & Pallen, 2015). This capability is particularly advantageous when dealing with emerging or unknown pathogens. While the bulk of plant pathogen research has traditionally focused on a subset of pathogens with direct negative effects on crops, recent years have witnessed a growing appreciation for the plant virome a complex assemblage of viruses coexisting within plant hosts (Roossinck, 2015). The virome plays a multifaceted role in plant health, affecting plant physiology, interaction with other organisms, and even resilience to stressors. Exploring the plant virome is not only about understanding the potential threats posed by pathogenic viruses but also about uncovering the beneficial symbiotic relationships between plants and viruses (Roossinck, 2015). Some viruses can confer resistance to abiotic stressors or herbivores, enhance nutrient uptake, or modify host plant phenotypes. Consequently, a comprehensive understanding of the plant virome can open doors to innovative strategies for improving crop resilience and health.

The primary objective of this chapter is to illuminate the intricate tapestry of plant pathogen detection, NGS technologies, and the plant virome, with in-depth knowledge about the fundamental principles, methodologies, and practical applications underpinning these subjects. Additionally, the challenges and limitations inherent in these approaches and showcase real-world case studies that exemplify their significance will also be discussed in the present chapter. Our intent is to equip researchers, plant scientists, policymakers, and students with a profound understanding of the critical importance of early plant pathogen detection, the transformative potential of NGS technologies, and the untapped richness of the plant virome. By comprehending the far-reaching consequences of these insidious adversaries and the pivotal role of detection, we aim to inspire a renewed commitment to safeguard agriculture, enhance food security, and advance ecological stewardship in a dynamic and interconnected world. Join us on this illuminating journey through the nexus of plant pathology, cutting-edge technology, and the mysterious realm of the plant virome a journey that promises to reshape our approach to plant health and ecological sustainability, and ultimately, enhance our capacity to address the challenges of the future.

9.2 OVERVIEW OF PLANT PATHOGENS WITH SPECIAL FOCUS ON PLANT VIROME

Plant pathogens represent a diverse collection of microorganisms, and comprehending the distinct categories of these pathogens, which encompass viruses, bacteria, fungi, and nematodes, is fundamental for the efficient control of diseases and for the promotion of sustainable agriculture. However, in this specific context, our attention is directed towards viruses.

Plant viruses are genetic entities consisting of either DNA or RNA enclosed in a protein coat. They are obligate parasites that rely on host cellular machinery for their survival and multiplication. Viruses can be transmitted through various means, including vectors (insects, nematodes, fungi), through vegetative propagation (like budding, grafting etc.), mechanically or in some cases through pollen to the seeds. Infection with plant viruses can lead to a wide range of symptoms, such as leaf mottling, chlorosis, necrosis, chlorotic and necrotic spots, stunted growth, fruit and leaf deformations, reduced yields, etc. (Marwal & Gaur, 2020). Some well-known plant viruses include tobacco mosaic virus (tobamovirus), potato virus Y (potyvirus), tomato yellow leaf curl virus (begomovirus), plum pox virus, cucumber mosaic virus, etc. (Roossinck, 2015).

The plant virome, a complex and dynamic network of viruses coexisting within plant ecosystems, has recently garnered significant attention due to advances in high-throughput sequencing technologies (Roossinck, 2015). This intricate web of viral diversity, encompassing both known and novel species, is unveiling new dimensions of the interactions between plants and their viral inhabitants. The composition of the plant virome within a single plant can be astonishingly diverse, with numerous viral families, including *Potyviridae*, *Geminiviridae*, and *Tombusviridae*, etc. (Dolja &

Koonin, 2018). The diversity of viral species within the plant virome emphasizes the complexity of these interactions and the need for further exploration. Beyond its mere existence, the plant virome plays multifaceted roles in the realm of plant health and ecological balance. It encompasses both pathogenic viruses responsible for diseases and viruses that establish mutualistic relationships with their host plants, influencing various aspects of their biology (Roossinck, 2015). Recent research has illuminated several crucial functions and implications of the plant virome, shedding light on its significance. Some plant viruses, rather than causing harm, establish mutually beneficial relationships with their host plants. These viruses can enhance nutrient uptake, confer resistance to abiotic stressors, or modify host phenotypes, offering innovative strategies to improve crop resilience and productivity (Roossinck, 2015). The plant virome extends its influence beyond individual plants to shape entire ecosystems. Viruses can significantly impact plant communities by affecting native plant species and altering species diversity. These findings underscore the interconnectedness of the plant virome with broader ecological processes, emphasizing the need for comprehensive ecological studies.

The plant virome, like any ecological system, is subject to change. Climate change and human activities can influence the distribution and behavior of plant viruses (Roossinck, 2015; Anderson et al., 2004). This dynamic nature of the plant virome accentuates the importance of ongoing monitoring and research to understand and mitigate emerging threats. In the realm of agriculture, the plant virome presents both challenges and opportunities. Recent studies emphasize the need to consider the virome when developing disease management strategies (Roossinck, 2015). Understanding the intricate interactions between pathogenic and beneficial viruses within crop ecosystems could lead to more targeted approaches for disease control and sustainable agriculture (Kreuze et al., 2009). Further, continued exploration of the plant virome, coupled with innovative research and management strategies, promises to reshape our approach to plant health, agricultural sustainability, and ecosystem conservation. The intricate web of viral diversity within the plant virome invites further investigation, offering the potential to unlock new insights into plant-virus interactions and their roles in shaping the ecological and agricultural landscapes.

9.3 NGS TECHNOLOGIES FOR PLANT PATHOGEN DETECTION

Next-Generation Sequencing (NGS), also known as high-throughput sequencing, stands as a transformative technological leap in the realm of genomics and molecular biology. This introduction provides an overview of NGS, its fundamental principles, and its profound impact on various fields, including genetics, environmental science, agriculture and medicine with reference to recent research findings. NGS, characterized by its ability to sequence millions to billions of DNA or RNA fragments simultaneously, has revolutionized the way we approach DNA and RNA sequencing, enabling rapid, cost-effective, and comprehensive analysis of genetic material (Goodwin et al., 2016). It involves library preparation, where DNA or RNA samples are fragmented and adapters attached to facilitate identification and sequencing, followed by high-throughput sequencing and advanced bioinformatics analysis.

The applications and impact of NGS span a wide spectrum. In genomics and personalized medicine, NGS has enabled the rapid sequencing of whole genomes, facilitating personalized treatment strategies. It has also played a pivotal role in cancer genomics, allowing comprehensive profiling of genomic alterations in tumors and targeted therapies (Kandoth et al., 2013). In agriculture, NGS has advanced crop genomics, leading to improved yields and precision breeding (Thottathil et al., 2016). Moreover, NGS has transformed microbiome analysis, enabling the study of microbial communities in various environments (Shreiner et al., 2015). It has also facilitated environmental metagenomics, contributing to biodiversity assessment and ecosystem monitoring (Taberlet et al., 2012). NGS represents a paradigm shift in genomics, offering unparalleled insights into the genetic fabric of organisms and ecosystems. Recent advancements continue to expand its applications, from unraveling the intricacies of the human genome to revolutionizing agriculture, medicine, and environmental science. As we delve deeper into the genomic universe enabled by NGS, its impact on science, medicine, and society at large becomes increasingly profound.

9.4 ADVANTAGES OF NGS IN PLANT PATHOGEN DETECTION

NGS has ushered in a transformative era in plant pathogen detection, offering a multitude of advantages over conventional methods, with recent research findings underscoring its significance. NGS unparalleled sensitivity enables the detection of low-abundance pathogens in complex samples, a critical capability for early identification and precise characterization of emerging or rare pathogens. It empowers comprehensive pathogen profiling, allowing simultaneous detection and in-depth characterization of multiple pathogens within a single sample, shedding light on pathogen diversity and co-infections, which can significantly influence disease dynamics (Massart et al., 2017; Kreuze et al., 2009). Furthermore, NGS boasts a rapid turnaround time, essential for timely responses to disease outbreaks and effective containment measures (Massart et al., 2017). One of the distinctive advantages of NGS lies in its independence from prior knowledge of specific pathogens. This makes it particularly advantageous for the detection and identification of novel or previously unknown pathogens, enhancing our preparedness for emerging threats (Adams et al., 2009). In addition to detection, NGS provides comprehensive genomic insights into detected pathogens, enabling the study of their evolution, virulence factors, and potential for adaptation. Recent research has highlighted NGS role in unraveling the genomics of plant pathogens, advancing our understanding of pathogen-host interactions and contributing to more effective disease management strategies. The high-throughput capacity of NGS allows for the efficient screening of large plant populations or multiple samples concurrently, a valuable asset for epidemiological studies and monitoring disease prevalence (Adams et al., 2009; Roossinck et al., 2010). Importantly, NGS significantly reduces the risk of false-positive results, relying on multiple sequence reads and advanced bioinformatic tools for accurate pathogen identification (Massart et al., 2017). Furthermore, NGS extends its utility to environmental surveillance (Table 9.1), enabling the monitoring of pathogens in diverse environmental matrices such as soil, water, and air, bolstering early warning systems

TABLE 9.1
Advantages of NGS in Plant Pathogen Detection

Advantage	Description
High Sensitivity	NGS can detect even low-abundance pathogens in complex samples, enabling early detection.
Detection of Novel Pathogens	It can identify previously unknown or emerging pathogens without prior knowledge.
Rapid Turnaround Time	It offers a quick response to disease outbreaks, allowing timely containment measures.
High-Throughput Screening	It can screen multiple samples concurrently, making it suitable for large-scale studies.
Comprehensive Pathogen Profiling	It allows simultaneous detection and characterization of multiple pathogens in a single sample.
Genomic Insights	It provides in-depth genomic information on detected pathogens, aiding disease management.
Reduced Risk of False Positives	It relies on multiple sequence reads and advanced bioinformatics for accurate identification.
Environmental Surveillance	It can be used for monitoring pathogens in diverse environmental matrices (plant samples, soil, water, air).
Metagenomic Analysis	It enables the study of entire pathogen communities without prior knowledge about identity.
Targeted Genome Sequencing (TG-Seq)	TG-Seq enhances sensitivity and cost-effectiveness in plant virology diagnostics.

and risk assessments (Roossinck et al., 2010). Its metagenomic analysis capability is a boon for studying complex ecosystems, including plant viromes and microbiomes, and identifying potential pathogen reservoirs (Roossinck et al., 2010). In a holistic approach, NGS data integration with other omics data, such as transcriptomics and proteomics, enhances our understanding of host-pathogen interactions, disease mechanisms, and potential targets for intervention (Sen et al., 2023; Mansoor et al., 2024).

One important type of NGS technique is the Targeted Genome Sequencing (TG-Seq), an innovative approach in plant virology that holds significant promise for the sensitive and cost-effective detection of multiple plant viruses simultaneously. In the context of agricultural and horticultural crops, where plant viruses can cause substantial damage, TG-Seq offers a targeted solution to the challenges faced by traditional methods. Unlike metagenomics, TG-Seq focuses on specific viral genome regions, minimizing the presence of host-derived sequences in sequencing data, thereby reducing both sequencing cost and bioinformatic complexity. This approach has demonstrated higher sensitivity, outperforming conventional Multiplex PCR (mPCR) techniques and gel electrophoresis in terms of virus detection, especially for low-titre pathogens. The use of transposome beads-based library chemistry further enhances TG-Seq's robustness and coverage uniformity, making it a valuable tool for plant virus diagnostics. As TG-Seq technology is continuously evolving, it has the potential to become a routine high-throughput sequencing-based diagnostic method, thus supporting activities such as germplasm certification, biosecurity investigations, and baseline surveillance in agriculture, ultimately contributing towards improved crop health and food production (Maina et al., 2021).

9.4.1 Sample Preparation and Library Construction for NGS

Sample preparation plays a pivotal role in NGS, encompassing several crucial stages, including sample pre-processing, library preparation, sequencing, and subsequent bioinformatics analysis (Fig. 9.1) (Goodwin et al., 2016; Metzker, 2010). Irrespective of the sequencing technology employed, each step yields a sequencing library ready for instrument loading (Goodwin et al., 2016; Metzker, 2010). These libraries consist of DNA fragments with defined length, and oligomer adapters at the 5′ and 3′ ends for barcoding and facilitating the sequencing process. Post-sequencing, data analysis relies mainly on bioinformatic tools and resources.

9.4.2 Sample Preparation and Sequencing Protocol for Illumina NGS

NGS on Illumina platforms is a powerful and widely used method for genomics research, offering high-throughput and accuracy. The following is a step-by-step protocol for preparing and sequencing samples on Illumina instruments:

1. **Sample Collection and Extraction:** Begin by collecting the biological samples of interest (i.e., DNA, RNA, or other nucleic acids) from the experimental system. Samples include leaves, roots, fruits, and seeds. The choice of sample depends on the research objectives, the specific plant species of interest, and the nature of the study, such as investigating the presence of viruses, understanding viral diversity, or studying interactions between the plant and its virome. The quality and quantity of the starting material need to be determined as theyplay a crucial role in successful sequencing.
2. **Library Preparation:** Library preparation involves converting the extracted nucleic acid into a sequencing library. This step includes DNA fragmentation (if necessary), end repair, A-tailing, and adapter ligation. During this process, unique barcode sequences may be added to individual samples for multiplexing.
3. **Library Amplification:** To increase the quantity of the sequencing library, Polymerase Chain Reaction (PCR) amplification is performed. This step ensures there is enough material for sequencing and that it is representative of the original sample.
4. **Library Quantification:** Accurate quantification of the prepared library is essential to determine the amount of DNA present. Various methods, such as quantitative PCR (qPCR) or fluorometric assays, can be used for quantification.
5. **Library Normalization:** If working with multiple libraries, it is important to normalize them to an equal concentration to ensure even representation during sequencing.

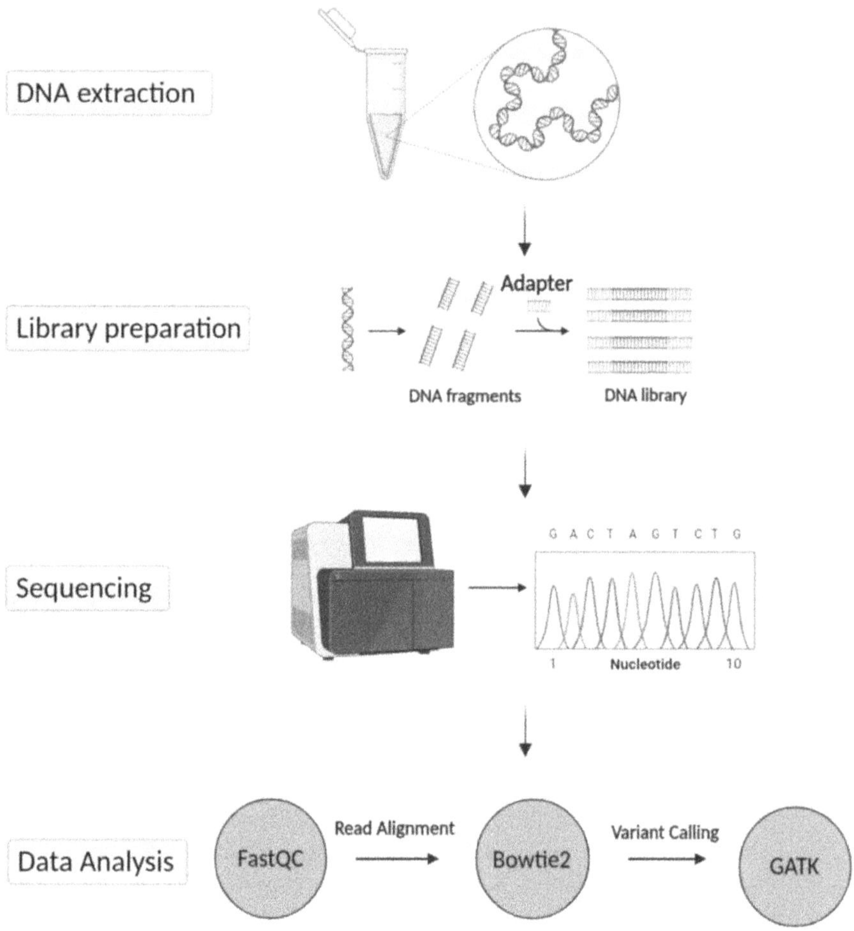

FIGURE 9.1 Simplified Workflow of Next-Generation Sequencing (NGS). Overview of the core steps in the NGS process, beginning with DNA extraction from the sample followed by library preparation, high-throughput sequencing, and bioinformatics analysis.

6. **Sequencing Run Setup:** Preparation of the Illumina sequencing instrument for the run by loading the samples in the flow cell, which contains millions of nanowells for individual sequencing reactions. The libraries are immobilized on the flow cell surface.
7. **Sequencing:** The sequencing process is initiated by introducing fluorescently labeled nucleotides into the flow cell. As the complementary bases are incorporated into the growing DNA strands, fluorescence signals are recorded and converted into sequencing data. Illumina instruments generate short reads, typically ranging from 50 to 300 base pairs, with high accuracy.
8. **Data Analysis:** Once the sequencing run is complete, the raw data, in the form of a FASTQ file, is generated. Data analysis involves several steps,

including quality control, read alignment (if applicable), variant calling, and downstream data interpretation. A variety of bioinformatic tools and software packages are available for these tasks.

9. **Data Interpretation:** The final step is to interpret the sequencing data in the context of our research goals. This may involve identifying genetic variants, quantifying gene expression, characterizing microbial communities, or other specific analyses.

Illumina sequencing technology is evolving, with improvement in read length, throughput, and chemistry. Researchers should follow the manufacturer's guidelines and protocols specific to the Illumina platform they are using, as there may be variations in library preparation and sequencing procedures among different instruments and generations of sequencers.

9.4.3 High-throughput Sequencing (HTS)

High-throughput sequencing (HTS), also known as NGS, deep sequencing, and third-generation sequencing has revolutionized the field of biology (i.e., epigenomics, genomics, metagenomics, transcriptomics), thereby playing a pivotal role in different areas of research by generating a vast amount of genetic data at a much faster rate as compared to the traditional techniques (Lee, 2023). HTS technologies have the potential to sequence multiple DNA molecules in parallel, thus generating billions of reads, resulting in gigabytes of data in a shorter duration of time and at much lesser cost (Kulski et al. 2016). In recent years (i.e., in the early 2000s), numerous HTS sequencing platforms have been developed including Ion torrent technology, Roche/454 pyrosequencing, Illumina/Solexa, ABI's SOLiD, Heliscope and sequencing by Oligonucleotide ligation and detection (Margulies et al., 2005; Harris et al., 2008; Valouev et al., 2008; Lee, 2023). Though there has been great improvement in many aspects in NGS techniques with regard to cost and speed in comparison to first-generation sequencing, they still depend upon PCR amplification, which could result in amplification bias and generation of shorter reads (~20-200 bp), thus resulting in gaps and misassemblies. In order to overcome the above mentioned issues, further optimization was done that led to the development of third and fourth generation platforms including single-molecule real-time sequencing by Pacific Biosciences (PacBio) and Nanopore sequencing (Krishna et al., 2019). The PacBio sequencing platform generates longer reads of about ~1.5 kb (recently new sequencers produce 10-15 kb reads), while Nanopore sequencing results in the generation of reads of around ~4 Mb length, thereby eliminating the requirement for clonal amplification of the fragmented DNA template (Lee et al., 2023). These platforms are different in terms of their accuracy, preparatory procedures, read type, length, price, sequencing technology, the size of generated data, and nucleotide incorporation technology (Ibaba & Gubba, 2020). The application of HTS in combination with rapid advancement in developing bioinformatics tools (so as to simplify the data analysis) has led to a great revolution in plant virology throughout the world, thus serving as an important tool in plant virus discovery and diagnosis (Barba et al., 2014; Hadidi et al., 2016).

HTS does not require any prior knowledge of the suspected disease-causing pathogen and it has a distinct property of unbiased and hypothesis-free testing of plant samples. Thus, this method has the potential to simultaneously detect multiple viruses regardless of their genome structure and nature (Jones et al., 2017). Hence, various HTS protocols have been developed with specific bioinformatics workflow using total RNAs, double-stranded RNAs, ribosomal-RNA-depleted RNAs, messenger RNAs, small interfering RNAs, or virion-associated nucleic acids for the detection of plant viruses, thereby serving as an important tool in the discovery of viruses (Blawid et al., 2017; Pecman et al., 2017; Chalupowicz et al., 2019; Massart et al., 2019; Villamor et al., 2019).

9.5 BIOINFORMATIC TOOLS AND PIPELINES FOR PATHOGEN IDENTIFICATION

Bioinformatic tools and pipelines play a crucial role in identifying pathogens in plant virome data generated through NGS technologies. These tools help analyze the vast amount of sequencing data to pinpoint viral sequences, detect novel pathogens, and understand their interactions within the plant virome. An overview of bioinformatic tools and pipelines commonly used for pathogen identification in plant virome NGS data is presented in Table 9.2.

9.6 CHALLENGES AND STRATEGIES IN NGS FOR PLANT VIROME DISCOVERY AND LIBRARY PREPARATION

NGS is a powerful tool for uncovering plant viromes, but its effectiveness can be influenced by the level of host background, which limits sensitivity in detecting pathogen-derived sequences. To enhance sensitivity for identifying novel agents in plant virome studies, researchers have adopted several strategies. For viruses, capsid purification methods have been developed to enrich samples for infectious particles. Additionally, techniques to deplete host DNA, such as methylation-specific DNase treatment, removal of host ribosomal RNA, and Duplex-Specific Nuclease (DSN) normalization, have been implemented. Target enrichment using biotinylated probes has also been employed to enrich NGS libraries for pathogen sequences. Choosing the right NGS platform for plant virome discovery depends on two key factors: read length and read depth. Longer reads (typically at least 100 to 300 nucleotides) are essential for identifying novel pathogens and distinguishing them from host or background flora. Sufficient read depth, or the number of sequence reads generated per run, is crucial for achieving higher sensitivity. The Roche 454 GS-FLX+pyrosequencing platform has been widely used due to its long reads and high accuracy. More recently, Illumina NGS platforms (GAIIx, HiSeq, and MiSeq) have gained popularity for their improved read depth and gradually increasing read lengths. Paired-end sequencing, which generates forward and reverse reads from each DNA fragment, aids in designing PCR primers for confirming potential sequence findings and de novo genome assembly. Other NGS technologies, such as Ion Torrent and Pacific Biosciences, have yet to see extensive use in plant virome studies but hold promise

TABLE 9.2
Bioinformatic Tools and Pipelines for Pathogen Identification in Plant Virome Analysis

Bioinformatics Step	Tools	Purpose
Sequence Quality Control	FastQC, Trimmomatic, Cutadapt	Assess sequencing read quality, remove low-quality bases, and trim adapters.
Read Alignment	Bowtie2, BWA, HISAT2	Map cleaned reads to reference genomes of known plant viruses to identify their presence.
De Novo Assembly	SPAdes, Velvet, Trinity	Assemble sequencing reads into longer contigs for novel virus discovery or complete viral genome assembly.
Virus Detection	ViromeScan, VirFind, ViraMiner	Detects viral sequences in the virome data, including known and novel viruses.
Taxonomic Classification	Kraken, Kaiju, Centrifuge	Assign taxonomic classifications to viral sequences, identifying genera, families, or species.
Variant Calling	GATK, Samtools, VarScan	Identify single nucleotide polymorphisms (SNPs) and variations within viral genomes to assess genetic diversity.
Phylogenetic Analysis	RAxML, PhyML, BEAST	Construct phylogenetic trees to understand the evolutionary relationships among detected viruses.
Functional Annotation	Prokka, RAST, Blast2GO	Annotate viral genes and predict their functions.
Visualization	Tablet, IGV, Artemis	Visualize and explore aligned reads, assemblies, and phylogenetic trees.
Integration and Interpretation	Custom pipelines or scripting	Integrate results, filter false positives, and generate comprehensive reports.

for specific applications. However, contamination remains a concern in unbiased NGS technologies, emphasizing the need for rigorous quality control measures (Chiu, 2013). Ensuring the reliability and standardization of every stage within the NGS process is paramount in routine laboratory practice (Endrullat et al., 2016; Gargis et al., 2012). Challenges arise at various workflow stages, demanding attention to maintain high-quality sequencing outcomes. For instance, the extraction of sufficient DNA from the sample input, free of inhibitory substances, can be intricate, especially when dealing with diverse sample materials. During sequencing runs, preventing carry-over contamination is critical to avoiding significant errors (Trombetta et al., 2014). In the realm of bioinformatics, managing the vast data generated by high-throughput NGS necessitates the substantial use of IT resources and comprehensive analysis solutions (Korneliussen et al., 2014; Scholz et al., 2012).

Library preparation specifically presents three primary challenges: (i) protocol complexity, (ii) contamination risks, and (iii) cost considerations.

(i) Protocol complexity: The protocols themselves can be complex, often requiring numerous steps for attaching adapters and barcodes to nucleotides. Notably, bead-based purification steps, involving magnets and magnetic particles, pose a potential source of user-generated errors (Meyer & Kircher, 2010).
(ii) Contamination risks: Sample contamination is another concern, since libraries are frequently prepared in parallel, with pre-amplifications needed for low initial nucleic acid concentrations thereby representing a prominent source of contamination (Salter et al., 2014).
(iii) Cost considerations: Cost associated with library preparation include expenses for laboratory equipments, reagents, and labor, as trained personnel are needed to perform the procedures.

However, the cost of library preparation on overall sequencing expenses can vary, and it remains a significant factor, particularly in scenarios involving extensive multiplexing, such as bacterial whole genome sequencing. Notably, the introduction of tagmentation reactions, which combine nucleotide fragmentation with adapter ligation, marked a substantial reduction in protocol costs (Marine et al., 2011).

9.6.1 NGS: Quality Control, Computational Resources, Validation, and Ethics

Quality control is paramount, as it involves verifying the accuracy and dependability of sequencing data. Challenges encompass managing sequencing errors, addressing adapter contamination, and handling low-quality reads. Tools like FastQC have been instrumental in identifying and filtering out problematic data (Andrews et al., 2010). Additionally, NGS generates vast data, necessitating significant computational power and storage resources. Analyzing these large datasets efficiently requires high-performance computing clusters, with researchers increasingly turning to cloud-based solutions like AWS and Google Cloud to scale resources as needed, alleviating the burden of maintaining local infrastructure (Stein, 2010). Ensuring the accuracy of NGS results with validation methods such as PCR-based assays, Sanger sequencing, and cross-referencing with independent sequencing platforms is essential and of paramount importance. Rigorous validation is particularly critical in clinical applications (Auwera et al., 2013). Moreover, NGS presents ethical and regulatory challenges, especially in the context of human genomics. Concerns regarding privacy, data sharing, and informed consent are pivotal issues. The regulatory landscape varies per country, with adherence to data protection laws like GDPR, which is of prime importance, and ethical review boards playing a crucial role in ensuring NGS research (Knoppers et al., 2014). Addressing the above mentioned multifaceted challenges is integral for harnessing the full potential of NGS technology in genomics and its broader applications.

9.7 CASE STUDIES AND SUCCESS STORIES

In an innovative study, NGS was employed to investigate the virome of sweet potato crops in East Africa, where the co-infection of sweet potato feathery mottle virus (SPFMV) and sweet potato chlorotic stunt virus (SPCSV) leads to the severe synergistic sweet potato virus disease (SPVD), a major threat to sweet potato cultivation (Karyeija et al., 1998). The experiment initially aimed to understand the effects of SPVD on RNA silencing and small interfering RNA (siRNA) accumulation in sweet potato. However, the data obtained was repurposed for diagnostic purposes. Small RNAs (sRNAs) were isolated and sequenced from infected sweet potato plants using Illumina deep sequencing. The majority of sequences were 21 to 24 nucleotides in size. Assembly of sRNA sequences into contigs revealed the complete genomic sequence of SPFMV and identified contigs related to badnaviruses and mastreviruses. These symptomless viruses were successfully characterized, and their genomes were deposited in GenBank (FJ155666, FJ560943, FJ560944, FJ560945, and FJ560946). The method also demonstrated its sensitivity by detecting viruses even in the case of latent and low-titre infections. Moreover, the technique showed the potential to simultaneously analyze a large number of samples, making it valuable for germplasm indexing and improving the safety of internationally exchanged plant material (Kreuze et al., 2009). In another study of grapevines affected by Grapevine Leafroll Disease (GLRD) NGS played an important role in uncovering the intricate dynamics of sRNAs in virus and viroid-infected grapevines. This research expanded our understanding of how various pathogens interact within the plant, particularly when cultivated in the field. The study revealed that GLRD-affected grapevines exhibited significant modulation in sRNA profiles, notably among miRNA families, which are known to play a critical role in plant development. For instance, miR162 and miR168, which regulate the miRNA pathway, showed altered expression levels in GLRD-affected leaves. Furthermore, miRNA families like miR156, miR166, and miR167, which target transcription factors affecting plant development, were also affected. The findings indicated that GLRD might disrupt miRNA biology, potentially contributing to disease symptoms. Additionally, the study highlighted the modulation of sRNAs specific to viroids in grapevines co-infected with multiple pathogens, shedding light on the complex interplay between different pathogens in the same host (Alabi et al., 2012). This case study demonstrates how NGS can provide valuable insights into the molecular mechanisms underlying complex plant diseases, such as GLRD. Some of the case studies are listed in Table 9.3.

9.8 FUTURE PERSPECTIVES

NGS has ushered a new era in plant virome research, offering unparalleled insights into the intricate world of plant viruses and viroids. As we look ahead, several promising avenues emerge. NGS is poised to significantly enhance disease management strategies by enabling precise interventions, including the development of resistant cultivars and site-specific control measures. The future holds the promise of rapid, on-site virome analysis tools, facilitating early detection and continuous monitoring of viral and viroid infections. Moreover, the complex synergistic interactions

TABLE 9.3
Some of the Published HTS Studies for the Detection of Viruses Infecting Plants

Sl. no.	Plant species	HTS Technology used	Viruses and viroids identified	References
1	Grapevine	Illumina	GVE, GRSPaV, GVA, GLRaV-3	Coetzee et al. 2010
2	Maize	Roche 454 GS-FLX+	MCMV, SCMV	Adams et al. 2013
3	Sugarcane	Illumina HiSeq; Roche 454 GS FLX Titanium	SSEV, SCWSV	Candresse et al. 2014
4	Maize	Illumina MiSeq Illumina HiSeq	MCMV, SCMV	Mahuku et al. 2015
5	Cassava	Illumina MiSeq	CBSV, UCBSV	Ndunguru et al. 2015
6	Potato	Illumina HiSeq	PVY	Ximba et al. 2017
7	Sweet potato	Illumina MiSeq	SPLCV, SPLCSPV, SPMV	Mulabisana et al. 2018
8	Maize	Illumina MiSeq	MaBV, MCMV, SCMV, MSV, ZMCV1, MPtV-1, MYDV-RMV2, MaYMV, MYDV-like, BYDV, MATV,	Asiimwe et al. 2020
9	Grapevine	Illumina NextSeq500	GFkV, GRVFV, GVA, GVB, GVE, GVF, GVL, TSV, GLRaV3, GLRaV4, GLRaV 5, 6, GLRaV2, GR5LaV, AaMV1, EnMV1, EnMV3, AaCV1, AGVd, GYSVd1, GYSVd2, GLVd, HSVd, GGVA	Sidharthan et al. 2020

10	Apple	Illumina HiSeq 2500	ApNMV, ApMV, ASGV, ASPV, ACLSV, AHVd	Nabi et al. 2022
11	Grapevine	Illumina	GVB, GVD, GVE, GVF, GVH, GVL, GLRaV-1, GLRaV-3, GLRaV-4, GLRaV-2, GRSPaV, GRVFV, GFkV, GFLV, GGVA, AGVd, GYSVd1, GYSVd2 GLVd, HSVd, GVA, GVK, GLRaV-7	Kishan et al. 2023

*Virus names and abbreviations: Grapevine leafroll associated virus-1 (GLRaV-1, GLRaV-2, GLRaV-3, GLRaV-4, GLRaV5, GLRaV6, GLRaV-7); grapevine virus A (GVA), grapevine virus B (GVB), grapevine virus D (GVD), grapevine rupestris stem pitting associated virus (GRSPaV); grapevine fanleaf virus (GFLV), Alternaria alternatachrysovirus 1 (AaCV1), Alternaria arborescensmitovirus 1 (AaMV1), Australian grapevine viroid (AGVd), Erysiphe necatormitovirus 1 (EnMV1), Erysiphe necatormitovirus 3 (EnMV3), grapevine fleck virus (GFkV), grapevine latent viroid (GLVd), grapevine virus L (GVL), grapevine virus E (GVE), grapevine virus F (GVF), grapevine yellow speckle viroid (GYSVd1), Grapevine yellow speckle viroid-2 (GYSVd2), hop stunt viroid (HSVd), tobacco streak virus (TSV) Grapevine rupestris stem pitting-associated virus (GRSPaV), Maize chlorotic mottle virus (MCMV), Sugarcane mosaic virus (SCMV), Cassava brown streak virus (CBSV), Ugandan cassava brown streak virus (UCBSV), Sugarcane streak Egypt Virus (SSEV), Sugarcane white streak Virus (SCWSV), Potato Virus Y (PVY), Sweet potato leaf curl virus (SPLCV), Sweet potato leaf curl Sao Paulo virus (SPLCSPV), Sweet potato mosaic virus (SPMV), Maize chlorotic mottle virus (MCMV), Maize streak virus (MSV), Maize yellow dwarf virus (RMV2) (MYDV-RMV2), Maize yellow mosaic virus (MaYMV), Barley yellow dwarf virus (BYDV), Maize-associated totivirus (MATV), Maize pteridovirus 1 (MPtV-1), Zea mays chrysovirus 1 (ZMCV1), Maize-associated betaflexivirus (MaBV), Apple necrotic mosaic virus (ApNMV), apple mosaic virus (ApMV), apple stem grooving virus (ASGV) and apple stem pitting virus (ASPV), apple chlorotic leaf spot virus (ACLSV), apple hammerhead viroid (AHVd) observed in co-infected plants highlight the need for further research in this regard, unravelling the mechanisms and implications of synergism in disease development. Functional characterization of sRNAs specific to various plant pathogens is essential for a deeper understanding of plant immunity and innovative antiviral strategies. Expanding the catalog of viromes across diverse plant hosts and ecosystems will provide a more comprehensive view of viral diversity. Finally, integrating NGS with other omics technologies, such as metagenomics, proteomics, and metabolomics promises a holistic understanding of host-pathogen interactions, paving the way for comprehensive disease management strategies. The future of NGS in plant virome research is bright, with potential benefits for disease management, early detection, and a deeper understanding of the intricate relationships between plants and their viral inhabitants, ultimately contributing to food security and ecosystem preservation in a changing world.

9.9 CONCLUSION

The intricate world of plant pathogens, including viruses, bacteria, fungi, nematodes, and other microorganisms, presents formidable challenges to agriculture, biodiversity, and economic stability. The significance of robust and timely plant pathogen detection cannot be overstated, especially in the face of a growing global population and increasing demands on agricultural productivity. Recent outbreaks of emerging plant pathogens underscore the urgency of early detection and management, not only to safeguard food security but also to preserve the health of agricultural ecosystems and conserve biodiversity. The advent of NGS technologies has revolutionized our ability to detect and understand plant pathogens. NGS enables rapid and cost-effective sequencing of entire genomes, allowing for the identification of pathogens with unprecedented accuracy. It goes beyond traditional detection methods by empowering researchers to perform metagenomic analyses, identify multiple pathogens simultaneously, and explore the complex plant virome a network of viruses co-existing within plant ecosystems. NGS has numerous advantages in plant pathogen detection, including its sensitivity, ability to detect novel pathogens, rapid turnaround time, and capacity for high-throughput screening. Moreover, NGS extends its utility to environmental surveillance, enabling the monitoring of pathogens in diverse matrices. TG-Seq is an innovative approach that enhances sensitivity and cost-effectiveness in plant virology, offering valuable diagnostic capabilities. However, challenges exist, such as protocol complexity, contamination risks during library preparation, and the need for computational resources. Quality control, validation, and ethical considerations are also crucial aspects of NGS in genomics and beyond. Case studies and success stories highlight the practical applications of NGS in plant virology, demonstrating its effectiveness in detecting and characterizing plant pathogens, even in latent infections.

Looking to the future, NGS holds immense promise in enhancing disease management, early detection, and our understanding of plant-pathogen interactions. The integration of NGS with other omics technologies and the expansion of virome catalogs across diverse ecosystems will contribute to comprehensive disease management strategies, ultimately benefiting food security and ecosystem preservation in a changing world. The journey through the nexus of plant pathology, cutting-edge technology, and the mysterious realm of the plant virome promises to reshape our approach to plant health and ecological sustainability, inspiring a renewed commitment to safeguard agriculture and advancement in ecological stewardship.

CONFLICT OF INTEREST

The authors declare that there is no conflict of interest.

REFERENCES

Adams, I. P., Glover, R. H., Monger, W. A., Mumford, R., Jackeviciene, E., Navalinskiene, M., & Boonham, N. (2009). Next-generation sequencing and metagenomic analysis: a universal diagnostic tool in plant virology. *Molecular Plant Pathology*, *10* (4), 537–545.

Adams, I.P., Miano, D.W., Kinyua, Z. M., Wangai, A., Kimani, E., Phiri, N., Reeder, R., Harju, V., Glover, R., Hany, U., et al. (2013). Use of next-generation sequencing for the identification and characterization of Maize chlorotic mottle virus and Sugarcane mosaic virus causing maize lethal necrosis in Kenya. *Plant Pathology*, 62, 741–749.

Agrios, G. N. (2005). *Plant pathology* (5th ed.). Academic Press.

Alabi, O. J., Zheng, Y., Jagadeeswaran, G., Sunkar, R., & Naidu, R. A. (2012). High-throughput sequence analysis of small RNAs in grapevine (Vitis vinifera L.) affected by grapevine leafroll disease. *Molecular Plant Pathology*, *13* (9), 1060–1076. https://doi.org/10.1111/j.1364-3703.2012.00815.x

Anderson, P. K., Cunningham, A. A., Patel, N. G., Morales, F. J., Epstein, P. R., & Daszak, P. (2004). Emerging infectious diseases of plants: pathogen pollution, climate change and agrotechnology drivers. *Trends in Ecology & Evolution*, *19*(10), 535–544.https://doi.org/10.1016/j.tree.2004.07.021\

Andrews, S. (2010). *FastQC: A Quality Control Tool for High Throughput Sequence Data.* Available at: https://www.bioinformatics.babraham.ac.uk/projects/fastqc/

Asiimwe, T., Stewart, L.R., Willie, K., Massawe, D.P., Kamatenesi, J., & Redinbaugh, M.G. (2020). Maize lethal necrosis viruses and other maize viruses in Rwanda. *Plant Pathology*, 69, 585–597.

Auwera, G.A. van, Carneiro, M.O., Hartl, C., et al. (2013). From FastQ Data to High-Confidence Variant Calls: The Genome Analysis Toolkit Best Practices Pipeline. *Current Protocols in Bioinformatics*, 43(1), 11.10.1–11.10.33. doi: 10.1002/0471250953.bi1110s43

Barba, M., Czosnek, H., & Hadidi, A. (2014). Historical perspective, development and applications of next-generation sequencing in plant virology. *Viruses*, *6* (1), 106–136.

Blawid, R., Silva, J.M.F., & Nagata, T. (2017). Discovering and sequencing new plant viral genomes by next-generation sequencing: Description of a practical pipeline. *Annals of Applied Biology*, 170, 301–314.

Candresse, T., Filloux, D., Muhire, B., Julian, C., Galzi, S., Fort, G., Bernardo, P., Daugrois, J. H., Fernandez, E., Martin, D. P., et al. (2014). Appearances can be deceptive: revealing a hidden viral infection with deep sequencing in a plant quarantine context. *PLoS ONE*, 9, e102945.

Chalupowicz, L., Dombrovsky, A., Gaba, V., Luria, N., Reuven, M., Beerman, A., Lachman, O., Dror, O., Nissan, G., & Manulis-Sasson, S. (2019). Diagnosis of plant diseases using the Nanopore sequencing platform. *Plant Pathology*, 68, 229–238.

Chiu, C. Y. (2013). Viral pathogen discovery. *Current Opinion in Microbiology*, *16* (4), 468–478.

Coetzee, B., Freeborough, M.J., Maree, H.J., Celton, J. M., Rees, D.J.G., & Burger, J.T. (2010). Deep sequencing analysis of viruses infecting grapevines: Virome of a vineyard. *Virology*, 400, 157–163.

Dolja, V. V., & Koonin, E. V. (2018). Metagenomics reshapes the concepts of RNA virus evolution by revealing extensive horizontal virus transfer. *Virus Research*, 244, 36–52.

Endrullat, C., Glökler, J., Franke, P., & Frohme, M. (2016). Standardization and quality management in next-generation sequencing. *Applied & Translational Genomics*, 10, 2–9.

Food and Agriculture Organization of the United Nations (FAO). (2006). International standards for phytosanitary measures. FAO.

Gargis, A. S., Kalman, L., Berry, M. W., Bick, D. P., Dimmock, D. P., Hambuch, T., Lu, F., Lyon, E., Voelkerding, K. V., Zehnbauer, B. A., Agarwala, R., Bennett, S. F., Chen, B., Chin, E. L., Compton, J. G., Das, S., Farkas, D. H., Ferber, M. J., Funke, B. H., Furtado, M. R., … Lubin, I. M. (2012). Assuring the quality of next-generation sequencing in clinical laboratory practice. *Nature biotechnology*, *30* (11), 1033-1036. https://doi.org/10.1038/nbt.2403

Gilbert, G. S., & Webb, C. O. (2007). Phylogenetic signal in plant pathogen-host range. *Proceedings of the National Academy of Sciences of the United States of America*, 104 (12), 4979–4983.https://doi.org/10.1073/pnas.0607968104

Goodwin, S., McPherson, J. D., & McCombie, W. R. (2016). Coming of age: ten years of next-generation sequencing technologies. *Nature Reviews. Genetics*, 17 (6), 333–351.https://doi.org/10.1038/nrg.2016.49

Hadidi, A., Flores, R., Candresse, T., & Barba, M. (2016). Next-generation sequencing and genome editing in plant virology. *Frontiers in Microbiology*, 7, 1325.

Harris, T. D., Buzby, P. R., Babcock, H., Beer, E., Bowers, J., Braslavsky, I., Causey, M., Colonell, J., DiMeo, J., Efcavitch, J.W., Giladi, E., Gill, J., Healy, J., Jarosz, M., Lapen, D., Moulton, K., Quake, S.R., Steinmann, K., Thayer, E., Tyurina, A., Ward, R., Weiss, H., & Xie, Z. (2008). Single-molecule DNA sequencing of a viral genome. *Science*, 320, 106–109.

Ibaba, J. D. & Gubba, A. (2020). High-throughput sequencing application in the diagnosis and discovery of plant-infecting viruses in Africa, a decade later. *Plants*, 9, 1376.

Jones, S., Baizan-Edge, A., Macfarlane, S., & Torrance, L. (2017). Viral diagnostics in plants using next generation sequencing: computational analysis in practice. *Frontiers in Plant Science*, 8, 1770.

Kandoth, C., McLellan, M. D., Vandin, F., Ye, K., Niu, B., Lu, C., & Ding, L. (2013). Mutational landscape and significance across 12 major cancer types. *Nature*, 502 (7471), 333–339.

Karyeija, R. F., Gibson, R. W., & Valkonen, J. P. T. (1998). The significance of sweet potato feathery mottle virus in subsistence sweet potato production in Africa. *Plant Disease*, 82 (1), 4–15.https://doi.org/10.1094/PDIS.1998.82.1.4

Kishan, G., Sharma, S. K., Holkar, S. K., Gupta, N., Khan, Z. A., Singh, S. K., & Baranwal, V. K., (2023). Diverse spectra of virus infection identified through high throughput sequencing in nursery plants of two Indian grapevine cultivars. *Physiological and Molecular Plant Pathology*, 128, 102135.

Knoppers, B. M. (2014). Framework for responsible sharing of genomic and health-related data. *The HUGO Journal*, 8 (1), 3.

Korneliussen, T. S., Albrechtsen, A., & Nielsen, R. (2014). ANGSD: analysis of next generation sequencing data. *BMC Bioinformatics*, 15 (1), 1–13.

Kreuze, J. F., Perez, A., Untiveros, M., Quispe, D., Fuentes, S., Barker, I., & Simon, R. (2009). Complete viral genome sequence and discovery of novel viruses by deep sequencing of small RNAs: a generic method for diagnosis, discovery and sequencing of viruses. *Virology*, *388* (1), 1–7. https://doi.org/10.1016/j.virol.2009.03.024

Krishna, B.M., Khan, M.A., & Khan, S.T. (2019). Next-Generation Sequencing (NGS) Platforms: An Exciting Era of Genome Sequence Analysis. In *Microbial Genomics in Sustainable Agroecosystems*; Tripathi, V., Kumar, P., Tripathi, P., Kishore, A., Kamle, M., Eds.; Springer Nature: Singapore, 2, pp. 89–109.

Kulski, J. K. (2016). Next-Generation Sequencing—An Overview of the History, Tools, and "Omic" Applications. In *Next Generation Sequencing—Advances, Applications and Challenges*; Kulski, J.K., Ed.; InTechOpen: London, UK, pp. 3–60.

Lee, J. Y. (2023). The principles and applications of high-throughput sequencing technologies. *Development & Reproduction*, 27 (1), 9–24. https://doi.org/10.12717/DR.2023.27.1.9

Loman, N. J., & Pallen, M. J. (2015). Twenty years of bacterial genome sequencing. *Nature Reviews. Microbiology*, 13 (12), 787–794.

Mahlein, A. K., Rumpf, T., Welke, P., Dehne, H. W., Plümer, L., Steiner, U., & Oerke, E. C. (2013). Development of spectral indices for detecting and identifying plant diseases. *Remote Sensing of Environment*, 128, 21–30.

Mahuku, G., Wangai, A., Sadessa, K., Teklewold, A., Wegary, D., Ayalneh, D., Adams, I., Smith, J., Bottomley, E., Bryce, S., et al. (2015). First report of maize chlorotic mottle virus and maize lethal necrosis on maize in Ethiopia. *Plant Disease*, 99, 1870.

Maina, S., Zheng, L., & Rodoni, B. C. (2021). Targeted genome sequencing (TG-Seq) approaches to detect plant viruses. *Viruses*, 13 (4), 583.

Mansoor, S., Tripathi, P., Ghimire, A., Hamid, S., Abd El-moniem, D., Chung, Y. S., & Kim, Y. (2024). Comparative transcriptomic analysis of the nodulation-competent zone and inference of transcription regulatory network in silicon applied Glycine max [L.]-Merr. *Roots. Plant Cell Reports*, 43(7), 169.

Mardis, E. R. (2008). Next-generation DNA sequencing methods. *Annual Review of Genomics and Human Genetics*, 9, 387–402.

Margulies, M., Egholm, M., Altman, W. E., Attiya, S., Bader, J. S., Bemben, L. A., et al. (2005). Genome sequencing in microfabricated high-density picolitre reactors. *Nature*, 437, 376–380.

Marine, R., Polson, S. W., Ravel, J., Hatfull, G., Russell, D., Sullivan, M., & Wommack, K. E. (2011). Evaluation of a transposase protocol for rapid generation of shotgun high-throughput sequencing libraries from nanogram quantities of DNA. *Applied and Environmental Microbiology*, 77 (22), 8071–8079.

Marwal, A., & Gaur, R. K. (2020). Host plant strategies to combat against viruses effector proteins. *Current Genomics*, 21 (6), 401–410.

Massart, S., Candresse, T., Gil, J., Lacomme, C., Predajna, L., Ravnikar, M., & Wetzel, T. (2017). A framework for the evaluation of biosecurity, commercial, regulatory, and scientific impacts of plant viruses and viroids identified by NGS technologies. *Frontiers in Microbiology*, 8, 45.

Massart, S., Chiumenti, M., De Jonghe, K., Glover, R., Haegeman, A., Koloniuk, I., Komínek, P., Kreuze, J., Kutnjak, D., Lotos, L., et al. (2019). Virus detection by high-throughput sequencing of small RNAs: large-scale performance testing of sequence analysis strategies. *Phytopathology*, 109, 488–497.

Metzker, M. L. (2010). Sequencing technologies—the next generation. *Nature Reviews Genetics*, 11(1), 31–46.

Meyer, M., & Kircher, M. (2010). Illumina sequencing library preparation for highly multiplexed target capture and sequencing. *Cold Spring Harbor Protocol, 2010* (6), t5448.

Mulabisana, M., Cloete, M., Mabasa, K., Laurie, S., Oelofse, D., Esterhuizen, L., & Rey, M. (2018). Surveys in the Gauteng, Limpopo and Mpumalanga provinces of South Africa reveal novel isolates of sweet potato viruses. *South African Journal of Botany*, 114, 280–294.

Nabi, S. U., Baranwal, V. K., Rao, G. P., Mansoor, S., Vladulescu, C., Raja, W. H., ... & Alansi, S. (2022). High-throughput RNA sequencing of mosaic infected and non-infected apple (Malus× domestica Borkh.) cultivars: from detection to the reconstruction of whole genome of viruses and viroid. *Plants*, 11 (5), 675.

Ndunguru, J., Sseruwagi, P., Tairo, F., Stomeo, F., Maina, S., Djinkeng, A., Kehoe, M., & Boykin, L.M. (2015). Analyses of twelve new whole genome sequences of cassava brown streak viruses and Ugandan cassava brown streak viruses from East Africa: diversity, supercomputing and evidence for further speciation. *PLoS ONE*, 10, e0139321.

Pecman, A., Kutnjak, D., Gutiérrez-Aguirre, I., Adams, I., Fox, A., Boonham, N., & Ravnikar, M. (2017). Next generation sequencing for detection and discovery of plant viruses and viroids: comparison of two approaches. *Frontiers in Microbiology*, 8, 1998.

Popp, J., Pető, K., & Nagy, J. (2013). Pesticide productivity and food security. A review. *Agronomy for Sustainable Development*, 33, 243–255.

Roossinck, M. J. (2015). Plants, viruses and the environment: ecology and mutualism. *Virology*, 479, 271–277.

Roossinck, M. J., Saha, P., Wiley, G. B., Quan, J., White, J. D., Lai, H., ... & Roe, B. A. (2010). Ecogenomics: using massively parallel pyrosequencing to understand virus ecology. *Molecular Ecology*, 19, 81–88.

Salter, S. J., Cox, M. J., Turek, E. M., Calus, S. T., Cookson, W. O., Moffatt, M. F., ... & Walker, A. W. (2014). Reagent and laboratory contamination can critically impact sequence-based microbiome analyses. *BMC Biology*, 12, 1–12.

Savary, S., Willocquet, L., Pethybridge, S. J., Esker, P., McRoberts, N., & Nelson, A. (2019). The global burden of pathogens and pests on major food crops. *Nature Ecology & Evolution*, 3 (3), 430–439.

Scholz, M. B., Lo, C. C., & Chain, P. S. (2012). Next generation sequencing and bioinformatic bottlenecks: the current state of metagenomic data analysis. *Current Opinion in Biotechnology*, 23 (1), 9–15.

Sen, P., & Orešič, M. (2023). Integrating omics data in genome-scale metabolic modeling: a methodological perspective for precision medicine. *Metabolites*, 13(7), 855. https://doi.org/10.3390/metabo13070855

Shreiner, A. B., Kao, J. Y., & Young, V. B. (2015). The gut microbiome in health and in disease. *Current Opinion in Gastroenterology*, 31 (1), 69.

Sicard, A., Zeilinger, A. R., Vanhove, M., Schartel, T. E., Beal, D. J., Daugherty, M. P., & Almeida, R. P. (2018). Xylella fastidiosa: insights into an emerging plant pathogen. *Annual Review of Phytopathology*, 56, 181–202.

Sidharthan, V.K., Sevanthi, A.M., Jaiswal, S., & Baranwal, V.K. (2020). Robust virome profiling and whole genome reconstruction of viruses and viroids enabled by use of available mRNA and sRNA-seq datasets in grapevine (Vitis vinifera L.). *Frontiers in Microbiology*,11, 1232. doi: 10.3389/fmicb.2020.01232.

Stein, L. D. (2010). The case for cloud computing in genome informatics. *Genome Biology*, 11, 1–7.

Strange, R. N., & Scott, P. R. (2005). Plant disease: a threat to global food security. *Annual Review of Phytopathology*, 43, 83–116.

Taberlet, P., Coissac, E., Pompanon, F., Brochmann, C., & Willerslev, E. (2012). Towards next-generation biodiversity assessment using DNA metabarcoding. *Molecular Ecology*, 21 (8), 2045–2050.

Thottathil, G. P., Jayasekaran, K., & Othman, A. S. (2016). Sequencing crop genomes: a gateway to improve tropical agriculture. *Tropical Life Sciences Research*, 27 (1), 93–114.

Trombetta, J. J., Gennert, D., Lu, D., Satija, R., Shalek, A. K., & Regev, A. (2014). Preparation of single-cell RNA-seq libraries for next generation sequencing. *Current Protocols in Molecular Biology*, 107 (1), 4–22.

Valouev, A., Ichikawa, J., Tonthat, T., Stuart, J., Ranade, S., Peckham, H., Zeng, K., Malek, J. A., Costa, G., McKernan, K., Sidow, A., Fire, A., & Johnson, S. M. (2008). A high-resolution, nucleosome position map of *C. elegans* reveals a lack of universal sequence-dictated positioning. *Genome Research*, 18, 1051–1063.

Villamor, D.E., Ho, T., Al Rwahnih, M., Martin, R.R., & Tzanetakis, I. (2019). High throughput sequencing for plant virus detection and discovery. *Phytopathology*, 109, 716–725.

Ximba, S., Ibaba, J.D., & Gubba, A. (2017). Potato virus Y strains infecting potatoes in the Msingå district in the province of KwaZulu-Natal, South Africa. *Crop Protection*, 96, 188–194.

10 Lab-on-Chip Platform for Field-based Diagnosis of Plant Pathogens

Devendra Kumar Choudhary and Sunil Kumar Sunani

10.1 INTRODUCTION

In the agricultural domain, the persistent threat posed by plant pathogens represents an enduring challenge, with the potential to compromise both food security and economic stability. Traditional diagnostic methods, characterized by time delays and logistical constraints, impede the swift identification and containment of plant diseases (Buja et al., 2021). The emergence of Lab-on-Chip (LoC) platforms marks a paradigm shift, promising to revolutionize plant pathogen diagnostics by transporting the capabilities of advanced laboratories directly to the agricultural fields. This introduction delves into the significance of LoC platforms in the realm of field-based diagnosis, illuminating their technological foundations, applications, and the transformative impact they hold for precision agriculture. In navigating this landscape of innovation, the LoC platform stands out as a beacon of hope, providing a powerful tool to mitigate the pervasive threats to global crop health and agricultural sustainability. The swift and precise detection of plant diseases is paramount for mitigating their impact on crops and ensuring global food security. The LoC platform, with its capacity for on-site diagnostics, addresses the imperative of identifying and managing plant pathogens in agricultural settings (He et al., 2024). By enabling early detection, the LoC platform facilitates timely decision-making, empowering farmers and agricultural professionals to implement targeted interventions to contain disease spread. The significance of the LoC platform is heightened by its potential to revolutionize conventional diagnostic methods, offering a portable and efficient alternative to time-consuming laboratory analyses (Marudamuthu et al., 2023). Given the rapid escalation of plant diseases, especially in the field where environmental conditions play a crucial role, the LoC platform's speed and precision provide immediate information to farmers. This capability proves instrumental in mitigating economic losses associated with crop damage and reducing the necessity for broad-spectrum treatments, thereby contributing to more sustainable agricultural practices. Furthermore, the importance of plant diseases in the context of the LoC platform extends its role in advancing our understanding of disease dynamics. The platform facilitates the monitoring of specific biomarkers linked to plant pathogens, shedding light on the real-time prevalence and distribution of diseases (Kumari et al., 2023). This data not only aids in immediate

 DOI: 10.1201/9781003474036-10

decision making but also provides valuable insights for researchers studying the epidemiology and evolution of plant pathogens, ultimately informing the development of more resilient crop varieties and targeted disease management strategies.

10.2 IMPORTANCE OF PLANT DISEASE

Plant diseases wield a profound influence on both natural ecosystems and agriculture, serving as crucial components in the delicate balance of these environments. Fungi, bacteria, viruses, and nematodes are among the diverse pathogens responsible for these diseases, impacting plant populations by influencing growth, reproduction, and survival (Singh et al., 2023). Losses due to plant disease inflict significant economic losses worldwide, impacting agricultural productivity and food security. These losses result from reduced crop yields, diminished quality of harvested produce, and increased costs associated with disease management (Singh et al., 2023). Plant pathogens, including fungi, bacteria, viruses, and nematodes, can devastate entire crops, leading to substantial financial setbacks for farmers and affecting global food supplies (Strange and Scott, 2005). Various plant diseases significantly impact crop quality and reduce food availability, especially when the affected crop serves as a staple food in resource-limited regions with large populations. Historical evidence demonstrates that a single pathogen affecting a staple crop can lead to widespread famine or migration (Gai and Wang, 2024). A recent illustration is the Ug99 strain of *Puccinia graminis tritici*, posing a significant threat to global wheat production, particularly in South Asia, with 90% of initially tested wheat varieties highly susceptible (Singh et al., 2015). Despite extensive efforts to develop resistance and manage the disease, an estimated annual global productivity loss of 10–16% persists due to plant pathogens (Zhan et al., 2015). Plant pathogens encompass a diverse array of microorganisms, including bacteria (e.g., Xanthomonas, Pseudomonas), fungi (e.g., Fusarium, Phytophthora), viruses (e.g., Tobacco mosaic virus, Potato virus Y), and nematodes (e.g., Root-knot nematodes, cyst nematodes), collectively posing a significant threat to global agriculture Zhan et al., 2015). These pathogens induce various diseases in plants, such as leaf spots, root rot, and mosaic patterns, leading to substantial economic losses by reducing crop yields and compromising quality. Detection methods range from visual inspection to sophisticated molecular techniques like Polymerase Chain Reaction (PCR) and remote sensing using satellite imagery.

Managing plant diseases involves specific agricultural practices such as crop rotation, clean seed usage, employing resistant varieties, and applying fungicides to prevent and control diseases (Gupta and Kumar, 2020). However, excessive or prolonged fungicide use has led to resistance development within fungal populations against specific modes of action (Lucas et al., 2015). Hence, there is a critical need to develop tools for accurate, sensitive, and rapid detection of pathogenic causes. This advancement will enable more effective preemptive application of appropriate chemicals, reducing unnecessary use in cropping systems, alleviating selective pressures on pathogen populations, and mitigating potential secondary environmental impacts.

The importance of plant disease detection is paramount in agriculture and ecosystem health. Rapid and accurate detection serves as an early warning system, enabling timely and targeted interventions critical for mitigating the economic impact of diseases on crop yields and quality. Identifying diseases at their incipient stages allows farmers to adopt precise measures, such as fungicide application or removal of infected plants, preventing further pathogen spread within crops or to neighboring fields. Plant disease detection is not solely about immediate economic concerns; it is also vital for preserving biodiversity and ecological balance. Early detection allows for the implementation of strategies to contain disease spread, preserving the ecological integrity of both agricultural and natural environments. Furthermore, plant disease detection plays a pivotal role in global food security. As the world's population continues to grow, the demand for food intensifies, making it imperative to safeguard crop production from pathogenic threats. Implementing effective detection methods ensures a resilient and sustainable agricultural system, providing a foundation for stable food supplies and reducing the need for excessive and environmentally detrimental pesticide use (Gai and Wang, 2024). In the context of LoC platforms and other advanced diagnostic technologies, the importance of plant disease detection is heightened. These technologies offer rapid, on-site analyses, empowering farmers with real-time information and facilitating prompt decision making. This not only reduces economic losses associated with diseases but also contributes to sustainable and environmentally friendly agricultural practices (Zhang et al., 2020).

10.3 EVOLUTION OF DETECTION TECHNIQUES

The evolution of plant pathogen detection methods has undergone a significant transformation, transitioning from traditional approaches to sophisticated molecular technologies. Initially, visual inspection and symptom-based identification were the primary methods for detecting plant diseases, where farmers relied on observable signs such as wilting or discoloration (Negi and Anand, 2024). However, this approach had limitations due to nonspecific symptoms that could be confused with other stress factors. As scientific knowledge advanced, the mid-20th century saw the adoption of culturing techniques to isolate and identify pathogens, although this process was time consuming and required expertise in microbial cultures. The late 20th century witnessed a revolutionary shift with the advent of molecular biology, particularly PCR, enabling the amplification and identification of specific DNA or RNA sequences associated with pathogens (Váradi et al., 2017). Real-time PCR and quantitative PCR further improved sensitivity and precision. In recent years, advanced technologies like Next-Generation Sequencing (NGS) have played a crucial role, allowing high-throughput sequencing of entire genomes for rapid identification of multiple pathogens (Marudamuthu et al., 2023). The integration of bioinformatics and data analytics has enhanced our ability to interpret complex genomic data, improving accuracy in pathogen identification. Ongoing developments in portable and field deployable diagnostic tools, such as handheld devices and biosensors, highlight the current trend towards decentralized and rapid on-site detection.

10.4 LAB-ON-CHIP APPROACH

Lab-on-Chip (LoC) technology is a groundbreaking innovation that miniaturizes and integrates laboratory functions onto a single microchip. Its development traces back to the late 20th century, with researchers in the early 1990s exploring the potential of creating microscale systems capable of performing key laboratory processes such as sample preparation and analysis. By 1996, the concept of a "micro total analysis system" (μTAS) was introduced, marking a significant advancement in making laboratory procedures more compact and efficient (Shoji, 1999). The μTAS concept aimed to shrink entire laboratory workflows onto a small chip, thereby reducing costs, improving efficiency, and enabling point-of-care diagnostics. Microfabrication techniques borrowed from the semiconductor industry played a pivotal role in the development of these microfluidic devices. The ability to manipulate fluids on a microscale opened the door to precise control of reactions and analyses within miniature channels. Over the following years, significant strides were made in microfluidics, materials science, and biochemistry, leading to the realization of functional LoC devices. Researchers developed methods for integrating components like pumps, valves, mixers, and detectors onto small platforms. These devices found applications in various fields, including medical diagnostics, environmental monitoring, and chemical analysis (Hajam and Khan, 2024).

The evolution of LoC technology was accelerated by advancements in nanotechnology, allowing for even greater miniaturization and increased sensitivity. As the 21st century unfolded, the field witnessed a surge in interdisciplinary research, bringing together experts from physics, chemistry, engineering, and biology to collaborate on refining and expanding the capabilities of LoC devices. Today, LoC technology continues to advance rapidly, with ongoing efforts to improve sensitivity, scalability, and usability. The impact of these devices is evident in the development of portable diagnostic tools, rapid point-of-care testing, and the exploration of novel applications in personalized medicine. The history of LoC reflects not only technological innovation but also the transformative potential of converging scientific disciplines to address complex challenges in diagnostics and analytical chemistry. The LoC applications in various fields are listed in Table 10.1.

10.4.1 Advantages

1. Portability and On-site Testing: One of the primary advantages of LoC platforms is their portability. These miniaturized systems can be easily transported to field locations, allowing for on-site testing without the need to transport samples to a central laboratory (Puiu and Bala, 2020).
2. Rapid and Real-time Results: LoC devices enable rapid analysis of samples, providing real-time results. This quick turnaround is essential for making timely decisions in the field, such as implementing control measures or adjusting agricultural practices (Marudamuthu et al., 2023).
3. Reduced Sample Volume and Reagent Consumption: The micro-scale nature of LoC systems allows for the use of significantly smaller sample volumes and reduced amounts of reagents.

TABLE 10.1
Lab-on-Chip Technologies and Their Applications in Plant Pathogen Detection and Diagnosis

Technique	Uses	Reference
Lab-on-a-Chip	For detection of *Phytophthora* species	Julich et al., 2011
Real-Time Microchip PCR System	For portable plant disease diagnosis of six different fungal and bacterial plant pathogens with 100% success rate to a detection limit of 5 ng/8 μl sample	Koo et al., 2013
Lab-on-a-Chip	For rapid detection assay of *Xylella fastidiosa* subsp. *pauca* strain CoDiRO	Chiriacò et al., 2018
Lab-on-a-Chip	For detection of *Ampelovirus* and *Nepovirus* for large scale health screening of Grapevine	Buja et al., 2022
Lab-on-a-Chip	For the detection of the quarantine potato cyst nematode *Globodera pallida*	Camacho et al., 2023
LAMP on a microfluidic chip	For rapid diagnosis of four important tomato pathogens	Shymanovich et al., 2024
Isothermal nucleic acid amplification based microfluidic lab-on-a-chip	For the detection of pathogenic bacteria and viruses in agri-foods	Lu et al., 2024

4. Multiplexing Capabilities: LoC platforms can be designed to incorporate multiplexed assays, allowing for the simultaneous detection of multiple pathogens in a single analysis (Jagannath et al., 2022).
5. High Sensitivity and Specificity: The microfluidic components of LoC platforms enable precise control over reaction conditions, resulting in high sensitivity and specificity (Koo et al., 2017).
6. Integration with Molecular Techniques: LoC technology seamlessly integrates with molecular techniques such as PCR or LAMP. This allows for the detection of specific DNA or RNA sequences associated with plant pathogens, providing a molecular-level understanding of the infection (Lau and Botella, 2017).
7. User Friendly Operation: Many modern LoC platforms are designed to be user friendly, requiring minimal technical expertise to operate (Lonchamps et al., 2022).

10.4.2 Types of LoC

10.4.2.1 LoC PCR (Lab-on-Chip Polymerase Chain Reaction)

PCR involves a cyclic process to amplify DNA. Initially, the DNA sample undergoes denaturation, separating it into single strands. Primers then anneal to the target sequences, marking the starting points for DNA synthesis. DNA polymerase extends the primers, synthesizing complementary strands. This cycling process, involving denaturation, primer annealing, and extension, is repeated, doubling the DNA content with each cycle (Zhu et al., 2020). A final extension ensures complete strand synthesis. PCR, reliant on specific components like primers, DNA polymerase, and nucleotides, is a versatile molecular biology tool used in genetic testing, research, forensics, and diagnostics. Conventional PCR has notable drawbacks, including time inefficiency due to lengthy temperature cycles, the requirement for substantial chemical amounts, and elevated costs (Singh et al., 2014). The extended reaction times, often taking hours, can be a limitation in applications where rapid results are crucial. The necessity for larger quantities of reagents, including DNA primers and polymerase, contributes to increased costs and resource consumption, particularly concerning limited or precious samples. The lack of portability in conventional PCR machines further hinders their use in fieldwork or point of care settings. These drawbacks have prompted the development and adoption of alternative PCR technologies addressing these limitations.

LoC-PCR refers to the integration of PCR technology onto LoC platforms. These microfluidic devices typically consist of a network of tiny channels and chambers etched onto a small chip. The chip may contain all the necessary components for a PCR reaction, including microvalves, pumps, and reaction chambers. The microscale design allows for precise manipulation of small volumes of reagents and samples (Lee et al., 2008). LoC-PCR offers numerous advantages over conventional PCR methods. Its miniaturized format allows for portability, making it suitable for point-of-care applications and fieldwork. With reduced sample and reagent volumes, LoC-PCR is cost effective, especially with precious samples, and it contributes to faster reaction times due to improved heat transfer efficiency. The integration and automation of multiple PCR steps in a single device enhance efficiency, reduce contamination risks, and enable multiplexing. Additionally, LoC-PCR provides precise temperature control, increased sensitivity, and environmental benefits through minimized waste and energy consumption (Foglieni et al., 2010). These features collectively make LoC-PCR a valuable and versatile tool in various applications, ranging from diagnostics to research. LoC PCR finds diverse applications in various fields due to its ability to miniaturize and integrate the PCR process onto a microscale platform. LoC PCR has found diverse applications in various fields, including medical diagnostics, environmental monitoring, and agriculture. In the context of plant pathology, LoC PCR is employed for the rapid and on-site detection of plant pathogens. This technology enables the amplification and identification of specific DNA sequences associated with plant pathogens, allowing for quick and accurate diagnosis. The use of microscale devices reduces the time required for amplification and analysis, making it suitable for field based applications (Baldi and Porta, 2020).

The structure of a LoC-PCR device is designed to miniaturize and integrate the various components of the PCR process onto a small chip. The microscale structure of LoC PCR devices is aimed at enhancing the efficiency, speed, and portability of the PCR reaction. Here are the key components and features of the structure:

1. Microfluidic Channels: LoC PCR devices contain a network of microfluidic channels that guide the flow of reagents, samples, and reaction mixtures.
2. Reaction Chambers: Microscale reaction chambers are incorporated into the chip to contain the PCR reaction.
3. Heating and Cooling Elements: LoC PCR devices include miniature heating and cooling elements to control the temperature cycling required for the denaturation, annealing, and extension steps of PCR.
4. Microvalves and Pumps: To control the movement of fluids within the microchannels, LoC PCR devices may incorporate microvalves and pumps.
5. Detection Systems: Some LoC PCR devices integrate detection systems for real-time monitoring of the amplification process.
6. Integrated Sample Preparation: In advanced LoC PCR designs, sample preparation steps may be integrated into the chip, allowing for a more comprehensive and streamlined diagnostic process.
7. Materials: The materials used in the fabrication of LoC PCR devices are often selected for their compatibility with biological samples and their ability to withstand the thermal cycling conditions of PCR. Materials such as polymers, glass, or silicon are commonly used.

10.4.2.2 LoC RT-PCR

In 1993, around the time of the Nobel Prize for PCR, Higuchi et al. recognized the potential for monitoring PCR through the addition of a fluorescent label binding to the accumulating PCR product (Higuchi et al., 1993). This insight led to the development of quantitative real-time PCR (qPCR). In current qPCR technology, fluorescence signals are generated using either fluorescent DNAbinding dyes or oligonucleotide probes. DNAbinding dyes, like SYBR® Green I, are added to the PCR buffer and exhibit increased fluorescence when bound to the amplifying DNA product. An alternative approach involves using labeled primers or labeled, often quenched, oligo probes between the two primers. Common probe options include Dual Labeled Probes (TaqMan®), Molecular Beacons, Scorpions® Probes, and LightCycler® Probes. These advancements offer precise monitoring and quantification of PCR reactions in real-time. Conventional real-time PCR typically demands larger sample volumes and involves higher reagent consumption due to larger reaction volumes. The longer reaction times in conventional PCR, attributed to the need for larger thermal masses, contrast with the faster and more efficient thermal cycling of LoC systems.

10.4.2.3 LoC-LAMP (Lab-on-Chip Loop-Mediated Isothermal Amplification)

Loop-Mediated Isothermal Amplification (LAMP) is a cost-effective DNA amplification technique developed in 2000 at the University of Tokyo, offering a low-cost alternative for disease detection. Its variant, Reverse transcription loop mediated isothermal amplification (RTLAMP), for RNA detection, operates at a constant temperature (60–65 °C), eliminating the need for a thermal cycler. The LAMP employs two to three sets of primers, including "loop primers," to amplify six distinct regions on the target gene, enhancing specificity. Detection methods include turbidity measurement, fluorescence, or visible colour change (Safavieh et al., 2016). LoC-LAMP technology offers advantages over conventional LAMP by allowing reduced sample and reagent volumes, enhancing cost effectiveness and portability. Integration of multiple LAMP steps automates the process, reducing contamination risks and enabling faster reaction times (Li et al., 2017). The platforms may also offer improved multiplexing capabilities, real-time monitoring, and enhanced precision, crucial for reliable results in diagnostics and research. Moreover, efficient sample handling and reduced reagent consumption contribute to environmental sustainability. Overall, LoC-LAMP is particularly advantageous in resource limited settings, POC diagnostics, and applications requiring portability and streamlined workflows (Das and Chuang, 2022).

The structure of LoC-LAMP devices integrates microfluidic components and reaction chambers to perform LAMP on a miniaturized scale. These devices incorporate several key structural elements to facilitate efficient amplification and detection processes. Microfluidic channels form pathways for the controlled movement of reagents and samples throughout the chip, enabling precise fluid manipulation. Within the chip, microscale reaction chambers contain small volumes of reaction mixtures, ensuring efficient amplification (Tao and Cai,, 2013). While LAMP is an isothermal technique, some LoC LAMP devices may include integrated temperature control elements for precise regulation. Additionally, mixing and dispensing components such as microvalves and micropumps aid in thorough reagent mixing. Real-time detection systems, including fluorescence detection components, allow for the monitoring of amplification in real time. Integrated sample loading ports facilitate the introduction of samples and reagents onto the chip. Materials chosen for LoC LAMP devices are compatible with the LAMP reaction and microfluidic processes, often including biocompatible polymers. Many LoC LAMP devices are designed for single use to minimize contamination risks (Safavieh et al., 2016). Advanced designs may integrate sample preparation modules directly onto the chip. Overall, the compact and portable form factor of LoC LAMP devices enables their use in diverse settings, including POC diagnostics, field testing, and resource limited environments.

The LoC LAMP technology holds significant promise in various plant related applications within agriculture and plant pathology. Rapid Onsite Detection of RNA based Plant Pathogens. The LoC-RT-LAMP is instrumental in swiftly detecting RNA based plant pathogens, such as viruses and viroids, directly in the field environment, facilitating timely management strategies (Aman et al., 2020). Early Disease Detection: Timely identification of plant pathogens is crucial for effective disease management. LoC LAMP's rapid and sensitive nature allows for the detection of infections at early stages, even before visible symptoms manifest, aiding in proactive disease control (Zhang et al., 2014). Multiplexed assays for pathogen detection: The microscale design of LoC LAMP devices enables the integration of multiplexed assays, permitting simultaneous detection of multiple pathogens in a single analysis (Sasaki et al., 2014). This capability enhances efficiency and throughput in plant disease diagnostics, allowing for comprehensive screening of plant samples for various pathogens in a single testing run. Screening for disease resistance: Researchers and plant breeders leverage LoC LAMP technology to screen plants for disease resistance genes or markers, facilitating the development of disease resistant crop varieties and aiding in breeding programs (Tomlinson et al., 2010).

10.4.2.4 LoC Microarray

Microarrays enable the early diagnosis of plant diseases by analyzing expression profiles, aiding in prompt control measures (Fowler and Thomashow, 2002). However, challenges like limited dynamic range, cross-hybridization, and high production costs constrain their efficacy (Kuo et al., 2002; Li and Wong, 2001; Shalon et al., 1996). LoC Microarrays are part of the broader history of microarray technology and LoC devices. The concept of microarrays, also known as DNA or gene chips, emerged in the 1990s (Schena et al., 1995). The technique involves immobilizing thousands of DNA probes on a solid surface to enable the simultaneous analysis of gene expression levels. In the early 2000s, researchers began exploring ways to integrate microarrays with LoC technology (Whitesides, 2006). The goal was to miniaturize and automate microarray-based assays, enabling high-throughput and multiplexed analyses on a compact platform. LoC Microarrays were designed to enable multiplexed analysis, allowing the simultaneous detection of multiple targets in a single experiment (Yager et al., 2006).

In contrast, LoC microarrays present a paradigm shift in genomic research, offering enhanced efficiency, miniaturization, and multifunctionality. LoC microarrays operate with reduced sample volumes, crucial for limited biological samples, and integrate multiple functions onto a single chip, streamlining analyses (Lagally et al., 2001). Their compact design and multifunctionality make them ideal for point of care applications, facilitating on-site testing without complex setups (Srinivasan et al., 2004). Advancements in surface chemistry and probe designs reduce cross-hybridization, enhancing assay specificity (Lin et al., 2004). Moreover, LoC microarrays offer high-throughput capability, enabling parallel analysis of multiple samples and targets, and multiplexed detection of analytes within a single assay (Hatch et al., 2001). These features prove beneficial for studying complex biological systems and largescale screenings (Holliger et al., 2002). The portability of LoC microarrays extends their utility to on-site testing and field applications, such as plant disease detection in agriculture, enabling real-time monitoring and data collection (Soper et al., 2006). Thus, LoC microarrays represent a significant advancement in gene expression analysis, offering versatile solutions for diverse research and diagnostic needs in plant science and agriculture.

The LoC Microarray structures are meticulously designed to merge microarray technology onto compact chips, harnessing the advantages of micro-fluidics alongside high throughput multiplexed analysis. At the core of LoC Microarrays lie several pivotal components. Firstly, microfluidic channels, akin to other LoC devices, guide the flow of samples and reagents across the chip, ensuring precise manipulation (Dhanpala et al., 2020). The microarray surface, a flat or planar area on the chip, serves as the canvas for deposition, typically functionalized with probes or capture molecules like DNA, RNA, proteins, or antibodies, selectively binding to target analytes. The fabrication process employs spotting technology, meticulously printing microscopic droplets of probe molecules onto the microarray surface in an array format, ensuring accuracy and precision in deposition. This structure allows for the spatial separation of different probes, facilitating multiplexed assays and parallel analysis of numerous targets in a single experiment.

Sample loading ports, strategically positioned, enable controlled delivery of biological samples onto the chip surface, while integrated microvalves and microfluidic control components regulate the flow of samples and reagents. Reaction chambers may also be integrated into the chip structure, providing a controlled environment for reactions to occur. Furthermore, fluorescence detectors or imaging systems serve as detection systems, aiding in visualizing and quantifying signals generated by interactions on the microarray surface. In some applications, temperature control elements are included to regulate temperature during specific reactions. Common materials like polydimethylsiloxane (PDMS) or glass are utilized for microfabrication, ensuring biocompatibility and suitability.

LoC Microarrays have diverse applications in plant research and agriculture, offering powerful tools for studying gene expression, genetic variation, and interactions between plants and their environment (Curtis et al., 2010). Here are specific ways in which LoC microarrays are used in plant-related contexts:

- Gene Expression Profiling: LoC Microarrays enable comprehensive gene expression profiling in plants. Researchers can simultaneously analyze the expression levels of thousands of genes (Lim et al., 2022).
- Pathogen Detection and Resistance Mechanisms: Researchers use LoC Microarrays to study plant-pathogen interactions by analyzing gene expression changes during infection (Windram et al., 2012).
- On-Site Diagnostics in Agriculture: The portable nature of LoC Microarrays makes them suitable for on-site diagnostics in agriculture. Farmers and plant health professionals can use these devices to quickly assess the health of crops, detect pathogens, and make informed decisions about disease management (Lim et al., 2022).

10.5 CONCLUSIONS AND FUTURE PROSPECTS

The LoC platform represents a transformative technology in the field of plant pathology, providing rapid, on-site diagnosis of plant pathogens that contributes significantly to early disease detection and effective management. As the technology continues to advance, the integration of LoC platforms with precision agriculture holds the key to a more sustainable and resilient agricultural future. The historical journey of LoC technology, from its inception as a tool to streamline laboratory methods to its current role in enabling field-based diagnostics, demonstrates its revolutionary potential. The ability of LoC platforms to address the critical need for rapid, on-site detection of plant pathogens is highlighted, emphasizing their role in mitigating economic and environmental consequences associated with delayed diagnoses. The future of LoC technology holds exciting prospects across various fields, driven by ongoing advancements in microfluidics, nanotechnology, biosensors, and other related areas. Here are some potential directions for the future of LoC:

- Continued efforts toward miniaturization and integration of multiple functions on a single chip are expected, leading to more compact and portable LoC devices with enhanced capabilities, suitable for point-of-care diagnostics and on-site field testing (Soper et al., 2006).
- Multiplexed Assays and High-Throughput Screening: The future of LoC involves expanding the capacity for multiplexed assays, allowing the simultaneous analysis of multiple analytes. This is particularly crucial in fields like genomics, proteomics, and diagnostics, where high-throughput screening is essential for comprehensive analyses (Zarei, 2017).
- Advanced Sensing Technologies: Integration of advanced sensing technologies, such as improved biosensors and detectors, will enhance the sensitivity and specificity of LoC devices (Wang et al., 2021).
- Smart and Autonomous Devices: Future LoC devices are likely to incorporate smart and autonomous features, enabling real-time monitoring, data analysis, and decision-making (Sackmann et al., 2014).
- Point-of-Care Diagnostics: The development of user-friendly, rapid, and cost-effective LoC platforms will enable healthcare professionals and even individuals to perform sophisticated diagnostic tests outside traditional laboratory settings (Nayak et al., 2017).
- Integration with Other Technologies: Synergies with other emerging technologies, such as artificial intelligence (AI) and Internet of Things (IoT), will likely shape the future of LoC. Integration with AI can improve data analysis and interpretation, while connectivity to IoT networks can enable remote monitoring and data sharing (Zarei, 2017).

The future of LoC holds immense potential to revolutionize how we approach diagnostics, research, and various applications across healthcare, agriculture, environmental science, and beyond. Continued interdisciplinary collaboration and advancements in materials science, engineering, and biological sciences will drive the evolution of LoC technology in the coming years.

REFERENCES

Aman, R., Mahas, A., Marsic, T., Hassan, N., & Mahfouz, M. M. (2020). Efficient, rapid, and sensitive detection of plant RNA viruses with one-pot RT-RPA–CRISPR/Cas12a assay. *Frontiers in Microbiology*, *11*, 610872.

Baldi, P., & La Porta, N. (2020). Molecular approaches for low-cost point-of-care pathogen detection in agriculture and forestry. *Frontiers in Plant Science*, *11*, 570862.

Buja, I., Sabella, E., Monteduro, A. G., Chiriacò, M. S., De Bellis, L., Luvisi, A., & Maruccio, G. (2021). Advances in plant disease detection and monitoring: From traditional assays to in-field diagnostics. *Sensors*, *21*(6), 2129.

Buja, I., Sabella, E., Monteduro, A. G., Rizzato, S., Bellis, L. De, Elicio, V., Formica, L., Luvisi, A., & Maruccio, G. (2022). Detection of Ampelovirus and Nepovirus by Lab-on-a-Chip: A promising alternative to ELISA test for large scale health screening of grapevine. *Biosensors*, 12(3), 147.

Camacho, M. J., Albuquerque, D. C., de Andrade, E., Martins, V. C., Inácio, M. L., Mota, M., & Freitas, P. P. (2023). A lab-on-a-chip approach for the detection of the quarantine potato cyst nematode Globodera pallida. *Sensors*, 23(2), 647.

Chiriacò, M. S., Luvisi, A., Primiceri, E., Sabella, E., De Bellis, L., & Maruccio, G. (2018). Development of a lab-on-a-chip method for rapid assay of Xylella fastidiosa subsp. pauca strain CoDiRO. *Scientific Reports*, 8(1), 7376.

Curtis, T. Y., & Shewry, P. R. (2010). Improving the protein content and composition of cereal grain. *Journal of Cereal Science*, 51(3), 293-299.

Das, D., Lin, C. W., & Chuang, H. S. (2022). LAMP-based point-of-care biosensors for rapid pathogen detection. *Biosensors*, *12*(12), 1068.

Dhanapala, L., Krause, C. E., Jones, A. L., & Rusling, J. F. (2020). Printed electrodes in microfluidic arrays for cancer biomarker protein detection. *Biosensors*, *10*(9), 115.

Foglieni, B., Brisci, A., Biagio, F. S., Pietro, P. D., Petralia, S., Conoci, S., ... & Cremonesi, L. (2010). Integrated PCR amplification and detection processes on a Lab-on-Chip platform: a new advanced solution for molecular diagnostics. *Clinical Chemistry and Laboratory Medicine*, *48*(3), 329-336.

Fowler, S., & Thomashow, M. F. (2002). Arabidopsis transcriptome profiling indicates that multiple regulatory pathways are activated during cold acclimation in addition to the CBF cold response pathway. *The Plant Cell*, 14(8), 1675-1690.

Gai, Y., & Wang, H. (2024). Plant Disease: A Growing Threat to Global Food Security. *Agronomy*, *14*(8), 1615.

Gupta, A., & Kumar, R. (2020). Management of seed-borne diseases: an integrated approach. *Seed-borne Diseases of Agricultural Crops: Detection, Diagnosis & Management*, 717-745.

Hajam, M. I., & Khan, M. M. (2024). Microfluidics: a concise review of the history, principles, design, applications, and future outlook. *Biomaterials Science*, 12, 218–251.

Hatch, A., Ray, T., & Lunte, S. (2001). Development of a microchipbased device for electrochemical detection of in vivo neurotransmitters. *Electrophoresis*, 22(18), 3961–3968.

He, H. J., da Silva Ferreira, M. V., Wu, Q., Karami, H., & Kamruzzaman, M. (2024). Portable and miniature sensors in supply chain for food authentication: a review. *Critical Reviews in Food Science and Nutrition*, 1–21.

Higuchi, R., Dollinger, G., Walsh, P. S., & Griffith, R. (1993). Simultaneous amplification and detection of specific DNA sequences. *Bio/technology*, 11(9), 1026–1030.

Holliger, C., Hudson, P. J., & Tee, K. L. (2002). DNA, RNA and protein synthesis in the lab: perspectives on chemistry and its role in understanding biology. *Philosophical Transactions of the Royal Society of London. Series B: Biological Sciences*, 357(1422), 69–76.

Jagannath, A., Cong, H., Hassan, J., Gonzalez, G., Gilchrist, M. D., & Zhang, N. (2022). Pathogen detection on microfluidic platforms: Recent advances, challenges, and prospects. *Biosensors and Bioelectronics: X*, *10*, 100134.

Julich, S., Riedel, M., Kielpinski, M., Urban, M., Kretschmer, R., Wagner, S., Fritzsche, W., Henkel, T., Möller, R., & Werres, S. (2011). Development of a lab-on-a-chip device for diagnosis of plant pathogens. *Biosensors and Bioelectronics*, 26(10), 4070–4075.

Koo, C., Malapi-Wight, M., Kim, H. S., Cifci, O. S., Vaughn-Diaz, V. L., Ma, B., Kim, S., Abdel-Raziq, H., Ong, K., Jo, Y.-K., Gross, D. C., Shim, W.-B., & Han, A. (2013). Development of a real-time microchip PCR system for portable plant disease diagnosis. *PLoS ONE*, 8(12), e82704.

Koo, K. M., Wee, E. J., Wang, Y., & Trau, M. (2017). Enabling miniaturised personalised diagnostics: from lab-on-a-chip to lab-in-a-drop. *Lab on a Chip*, *17*(19), 3200-3220.

Kumari, S., Saha, U., Bose, M., Murugan, D., Pachauri, V., Sai, V. R., & Madaboosi, N. (2023). Microfluidic platforms for single cell analysis: applications in cellular manipulation and optical biosensing. *Chemosensors*, 11(2), 107.

Kuo, W. P., Jenssen, T. K., Butte, A. J., OhnoMachado, L., & Kohane, I. S. (2002). Analysis of matched mRNA measurements from two different microarray technologies. *Bioinformatics*, 18(3), 405412.

Lagally, E. T., Medintz, I. L., & Mathies, R. A. (2001). Single-molecule DNA amplification and analysis in an integrated microfluidic device. *Analytical Chemistry*, 73(3), 565570.

Lau, H. Y., & Botella, J. R. (2017). Advanced DNA-based point-of-care diagnostic methods for plant diseases detection. *Frontiers in Plant Science*, *8*, 2016.

Lee, S. H., Kim, S. W., Kang, J. Y., & Ahn, C. H. (2008). A polymer lab-on-a-chip for reverse transcription (RT)-PCR based point-of-care clinical diagnostics. *Lab on a Chip*, *8*(12), 2121-2127.

Li, C., & Wong, W. H. (2001). Modelbased analysis of oligonucleotide arrays: expression index computation and outlier detection. *Proceedings of the National Academy of Sciences*, 98(1), 3136.

Li, Y., Fan, P., Zhou, S., & Zhang, L. (2017). Loop-mediated isothermal amplification (LAMP): A novel rapid detection platform for pathogens. *Microbial Pathogenesis*, *107*, 54-61.

Lim, P. K., Zheng, X., Goh, J. C., & Mutwil, M. (2022). Exploiting plant transcriptomic databases: Resources, tools, and approaches. *Plant Communications*, *3*(4).

Lin, M., Johnson, S., Wang, Z., & Prasanna, H. A. (2004). Environmental DNA mapping and sensing. *Proceedings of the IEEE*, 92(5), 808845.

Lonchamps, P. L., He, Y., Wang, K., & Lu, X. (2022). Detection of pathogens in foods using microfluidic "lab-on-chip": A mini review. *Journal of Agriculture and Food Research*, *10*, 100430.

Lu, Y., Zhang, J., Lu, X., & Liu, Q. (2024). Isothermal nucleic acid amplification based microfluidic "lab-on-a-chip" for the detection of pathogenic bacteria and viruses in agri-foods. *Trends in Food Science & Technology*, 148, 104482.

Lucas, J. A., Hawkins, N. J., & Fraaije, B. A. (2015). The evolution of fungicide resistance. *Advances in Applied Microbiology*, *90*, 29-92.

Marudamuthu, B., Sharma, T., Purru, S. et al. (2023). Next-generation sequencing technology: a boon to agriculture. *Genetic Resources and Crop Evolution*, 70, 353–372. https://doi.org/10.1007/s10722-022-01512-5

Nayak, S., Blumenfeld, N. R., Laksanasopin, T., & Sia, S. K. (2017). Point-of-care diagnostics: recent developments in a connected age. *Analytical Chemistry*, 89(1), 102-123.

Negi, P., & Anand, S. (2024). Plant Disease Detection, Diagnosis, and Management: Recent Advances and Future Perspectives. *Artificial Intelligence and Smart Agriculture: Technology and Applications*, 413-436.

Puiu, M., & Bala, C. (2020). Microfluidics-integrated biosensing platforms as emergency tools for on-site field detection of foodborne pathogens. *TRAC Trends in Analytical Chemistry*, *125*, 115831.

Sackmann, E. K., Fulton, A. L., & Beebe, D. J. (2014). The present and future role of microfluidics in biomedical research. *Nature, 507*(7491), 181-189.

Safavieh, M., Kanakasababapathy, M. K., Tarlan, F., Ahmed, M. U., Zourob, M., Asghar, W., & Shafiee, H. (2016). Emerging loop-mediated isothermal amplification-based microchip and microdevice technologies for nucleic acid detection. *ACS Biomaterials Science & Engineering*, *2*(3), 278-294.

Sasaki, T., Tanaka, H., Takahashi, H., Kishino, H., & Nakamura, T. (2014). Development and application of loopmediated isothermal amplification (LAMP) assays for rapid detection of Erysiphe necator. *Journal of General Plant Pathology*, 80(6), 541547.

Schena, M., Shalon, D., Davis, R. W., & Brown, P. O. (1995). Quantitative monitoring of gene expression patterns with a complementary DNA microarray. *Science*, 270(5235), 467470.

Shalon, D., Smith, S. J., & Brown, P. O. (1996). A DNA microarray system for analyzing complex DNA samples using twocolor fluorescent probe hybridization. *Genome Research*, 6(7), 639645.

Shoji, S. (1999). Micro total analysis system (μTAS). *Electronics and Communications in Japan (Part II: Electronics)*, *82*(2), 21-29.

Shymanovich, T., Saville, A. C., Paul, R., Wei, Q., & Ristaino, J. B. (2024). Rapid detection of viral, bacterial, fungal, and oomycete pathogens on tomatoes with microneedles, LAMP on a microfluidic chip, and smartphone device. *Phytopathology*, 114(8), 1975–1983.

Singh, B. K., Delgado-Baquerizo, M., Egidi, E., Guirado, E., Leach, J. E., Liu, H., & Trivedi, P. (2023). Climate change impacts on plant pathogens, food security and paths forward. *Nature Reviews Microbiology*, *21*(10), 640-656.

Singh, J., Birbian, N., Sinha, S., & Goswami, A. (2014). A critical review on PCR, its types and applications. *International Journal of Advanced Research in Biological Sciences*, *1*(7), 65-80.

Singh, R. P., Hodson, D. P., Jin, Y., Lagudah, E. S., Ayliffe, M. A., Bhavani, S., ... & Hovmøller, M. S. (2015). Emergence and spread of new races of wheat stem rust fungus: continued threat to food security and prospects of genetic control. *Phytopathology*, *105*(7), 872-884.

Soper, S. A., Brown, K., Ellington, A., Frazier, B., Garcia-Manero, G., Gau, V., ... & Wilson, D. (2006). Point-of-care biosensor systems for cancer diagnostics/prognostics. *Biosensors and Bioelectronics*, *21*(10), 1932-1942.

Srinivasan, V., Pamula, V. K., & Fair, R. B. (2004). An integrated digital microfluidic labonchip for clinical diagnostics on human physiological fluids. *Lab on a Chip*, 4(4), 310-315.

Strange, R. N., & Scott, P. R. (2005). Plant disease: a threat to global food security. *Annual Review of Phytopathology*, *43*(1), 83-116.

Tao, G., & Cai, L. (2013). Molecular diagnosis of Colletotrichum kahawae by loop-mediated isothermal amplification (LAMP). *Journal of Plant Pathology*, 519-524.

Tomlinson, J. A., Dickinson, M. J., Boonham, N., & Sharp, P. M. (2010). Development and evaluation of a onehour DNA extraction and loopmediated isothermal amplification assay for rapid detection of phytoplasmas. *Plant Pathology*, 59(3), 465-471.

Váradi, L., Luo, J. L., Hibbs, D. E., Perry, J. D., Anderson, R. J., Orenga, S., & Groundwater, P. W. (2017). Methods for the detection and identification of pathogenic bacteria: past, present, and future. *Chemical Society Reviews*, *46*(16), 4818-4832.

Wang, C., Liu, M., Wang, Z., Li, S., Deng, Y., & He, N. (2021). Point-of-care diagnostics for infectious diseases: From methods to devices. *Nano Today*, *37*, 101092.

Whitesides, G. M. (2006). The origins and the future of microfluidics. *Nature*, 442(7101), 368-373.

Windram, O., Madhou, P., McHattie, S., Hill, C., Hickman, R., Cooke, E., ... & Denby, K. J. (2012). Arabidopsis defense against Botrytis cinerea: chronology and regulation deciphered by highresolution temporal transcriptomic analysis. *The Plant Cell*, 24(9), 3530-3557.

Yager, P., Domingo, G. J., Gerdes, J., et al. (2006). Pointofcare diagnostics for global health. *Annual Review of Biomedical Engineering*, 10(1), 107-144.

Zarei, M. (2017). Portable biosensing devices for point-of-care diagnostics: Recent developments and applications. *TrAC Trends in Analytical Chemistry, 91*, 26-41.

Zhan, J., Thrall, P. H., Papaïx, J., Xie, L., & Burdon, J. J. (2015). Playing on a pathogen's weakness: using evolution to guide sustainable plant disease control strategies. *Annual Review of Phytopathology*, *53*(1), 19-43.

Zhang, N., Yang, G., Pan, Y., Yang, X., Chen, L., & Zhao, C. (2020). A review of advanced technologies and development for hyperspectral-based plant disease detection in the past three decades. *Remote Sensing*, 12(19), 3188.

Zhang, X., Lowe, S. B., & Gooding, J. J. (2014). Brief review of monitoring methods for loopmediated isothermal amplification (LAMP). *Biosensors and Bioelectronics*, 61, 491–499.

Zhu, H., Fohlerová, Z., Pekárek, J., Basova, E., & Neužil, P. (2020). Recent advances in lab-on-a-chip technologies for viral diagnosis. *Biosensors and Bioelectronics*, *153*, 112041.

11 Potential Use of Biosensors in Rapid Detection and Quantification of Pathogens Associated with Plant Disease

Pankhuri Singhal and Marimuthu Elangovan

11.1 INTRODUCTION

Due to expanding of the global population, the agriculture ecosystem has expanded to around 40% of the planet surface in order to ensure the food supply/security (Foley et al., 2011). The threat to food security is consistently posed by a range of abiotic and biotic stresses. Abiotic stresses, such as drought, nutrient deficiency, and chemical toxicity, can be easily recognized and addressed. On the other hand, biotic stresses, including fungi, bacteria, phytoplasma, viruses, and virus-like pathogens, lead to numerous diseases in crops, thereby presenting a substantial risk to global food security and sustainable agriculture. To minimize the crop loss and yield reduction caused by a myriad of plant pathogens, specific management practices are needed. Plant diseases are controlled through the implementation of specific cultivation techniques and the application of chemicals (Mancini & Romanazzi, 2014). This includes practices like crop rotation, using clean seeds, utilizing resistant varieties, and treating seeds and foliage with fungicides. Chemical treatments are especially important for achieving effective disease control. However, the excessive and prolonged use of chemicals has resulted in the emergence of pests and pathogen populations that are resistant to certain modes of action (Donoso & Valenzuela, 2018). Thus, early identification of pathogens is vital in minimizing widespread crop damage, reducing economic losses, and curbing the excessive use of agrochemicals. Prompt recognition allows for the segregation of diseased plants, curbing the spread of infections and preserving overall crop yield and quality. Additionally, early intervention helps to prevent pathogen evolution and development of resistant strains and ensure the effectiveness of available treatment options.

The earliest traditional method of plant disease diagnosis relies on visually inspecting crops, which may indicate that the pathogen has already spread extensively

DOI: 10.1201/9781003474036-11

through the host populations. However, this method may have lower accuracy and sensitivity and requires experienced or trained technical persons. The development of molecular pathogen detection methods relies on protein and nucleic acid (NA)-based techniques such as enzyme linked immunosorbent assay (ELISA), polymerase chain reaction (PCR), and loop-mediated isothermal amplification (LAMP). In protein-based diagnostics, specific antibodies are designed to target antigens of pathogens. Traditional ELISA is considered the most reliable method for detecting various types of pathogens, but it has limitations in terms of sensitivity and accuracy and is not suitable for multiplexing. Additionally, antibody production for ELISA is more complex, resource-intensive, and expensive. On the other hand, NA-based pathogen detection methods rely on amplifying NA sequence followed by amplicon detection and visualization. However, these assays require well-equipped laboratories and skilled personnel.

In this context, biosensors have emerged as powerful tools revolutionizing pathogen detection and quantification. Integrating biological recognition elements with transducing components, biosensors offer swift and precise identification of specific pathogens. Their ability to measure pathogen load aids in assessing infection severity and implementing tailored management practices. Real-time monitoring enhances the efficacy of disease control measures while promoting the sustainable use of resources. Moreover, the portability, user-friendliness, and cost-effectiveness of biosensors democratize pathogen detection, making them accessible to a broader range of users, including farmers, agronomists, and researchers.

The objective of this chapter is to provide an in-depth understanding of the principles and mechanisms underlying biosensors and their specific applications in the context of plant pathogen detection. Furthermore, this chapter aims to explore the recent technological advancements and innovations in biosensor development, including the integration of nanotechnology, artificial intelligence, and machine learning, and their implications for enhancing the sensitivity and specificity of pathogen detection.

11.2 UNDERSTANDING BIOSENSORS AND THEIR OPERATION

11.2.1 Definition

A biosensor is an analytical device that detects changes in biological processes such as enzymes, tissues, microorganisms, cells, etc., and converts them into an electrical signal. It is a small, portable, analytical diagnostic piece of equipment made up of a biological or biomimetic sensing component that is either directly coupled to or integrated into a physicochemical transducer system. The selective binding of the target analyte to the complementary biorecognition component immobilized on an appropriate support platform constitutes the detection principle. An analyte of biological element quantitatively utilizing by a bioreceptor component (called a biorecognition component) and a signal transduction component.

The detection principle relies on the selective binding of a target analyte to its complementary bio-recognition component, which is immobilized on an appropriate support platform. This process involves a bioreceptor (the bio-recognition component) that interacts with the analyte and a signal transduction component that enables quantitative analysis.

The analytes of benefits can be biological, such as DNA, protein, and cells. When a bioreceptor component perceives a particular analyte and binds, a transducer component gives quantitative or semi-quantitative data about the objective by changing the physicochemical signal produced during the biorecognition into a quantifiable electrical signal (Ali et al., 2018; Chen & Shamsi, 2017). A change in one or more physico-chemical properties (such as pH change, electron transfer, mass change, heat transfer, uptake or release of gases, or specific ions) as a result of the particular interaction is recognized and may be measured by the transducer.

The term biosensor has been connected to numerous innovations. They can be both subjective or quantitative, encrypted or solid physical devices, giving concurrent time or time-fractioned information, *in-vivo* or *ex-vivo*, and of numerous different structures (Sadanandom & Napier, 2010). Fundamentally, a biosensor couples to a recognition element or a chemical response by a biological element with a transducer, which then converts this response into a readable form. The goal is typically to create an electronic signal whose size or frequency is proportionate to the concentration of a particular analyte or collection of analytes that the biosensing element binds.

The biorecognition mechanism of a biosensor can be used to categorize them. The enzyme/substrate, antibody/antigen, and nucleic acids/complementary sequences are the main biological components employed in biosensor technology. Additionally, the biosensing system can contain bacteria, complete animal or plant cells, and tissue slices. An alternate strategy that makes use of artificial biomimetic recognition systems has been made possible by recent advancements and innovations in the field of molecular imprinting. According to Haupt & Mosbach (2000), molecular imprinted polymers can theoretically be created for any analyte molecule and are capable of binding to target molecules with affinities and specificities comparable to those of biological recognition elements.

A biosensor comprises a biological or biomimetic sensing component integrated into a transducer system, enabling the selective binding of target analytes to the biorecognition component. The transducer measures resulting physicochemical changes, generating an electronic signal proportional to the analyte concentration. By harnessing the specificity of the biological recognition element and the sensitivity of the transducer, biosensors for plant pathogen detection offer a rapid and reliable method for monitoring and controlling plant diseases. Various biological components, including enzymes, antibodies, and nucleic acids, are employed in biosensor technology, allowing for diverse applications in plant disease diagnostics. Recent advances have also introduced biomimetic recognition systems, expanding the potential of biosensors in detecting and quantifying plant pathogens.

11.2.2 Components of Biosensors

(1) **Bioreceptor (Biological Recognition Element):** This component is responsible for recognizing the target analyte. It can be an enzyme, an antibody, a whole cell, or nucleic acids, depending on the nature of the substance to be

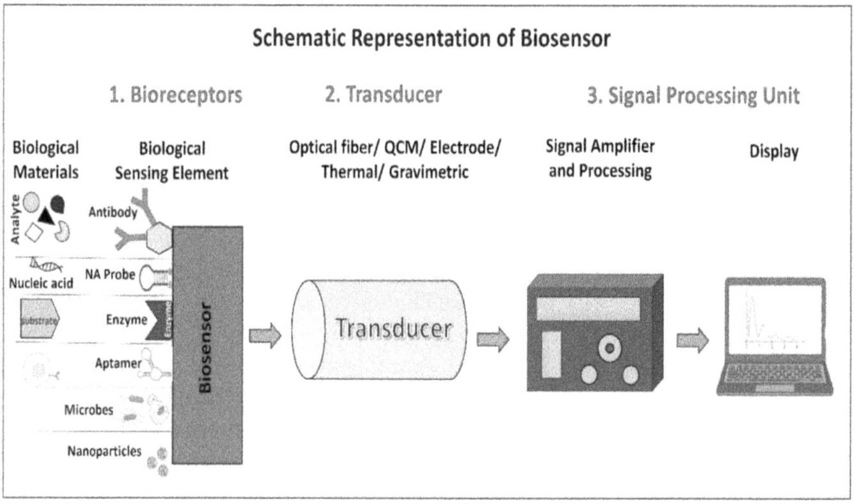

FIGURE 11.1 Schematic representation of biosensor components and their workflow.

detected. The selection of the biological recognition element is crucial, as it determines the specificity and sensitivity of the biosensor.

(2) **Transducer:** The transducer converts the biochemical signal generated by the interaction between the biological recognition element and the target analyte into a measurable signal, typically an electrical signal. Various types of transducers, such as electrochemical, optical, piezoelectric, or thermal transducers, can be employed based on the nature of the detection and the desired sensitivity.

(3) **Signal Processing System:** This component is responsible for processing the signal generated by the transducer and converting it into a readable and interpretable form. The signal processing system may include electronic circuits, amplifiers, and display units that enable the user to interpret and analyze the data obtained from the biosensor (Figure 11.1).

11.2.3 Working Principle

The working principle of biosensors for plant pathogen detection involves the integration of a biological recognition element and a transducer to facilitate the rapid and specific identification of plant pathogens.

11.2.4 Key Steps Involved in Detection via Biosensors

(i) Bioreceptor Interaction:
The biosensor's biological recognition element, such as antibodies or DNA probes, selectively binds to the specific target of pathogen's biomolecules, initiating a biochemical interaction that is highly specific to the pathogen of interest.

(ii) Transducer Conversion:
The transducer component of the biosensor converts the biochemical interaction between the biological recognition element and the pathogen into a measurable signal. Various transducer types, including electrochemical, optical, or piezoelectric transducers, are commonly employed to transform the biochemical event into an electrical, optical, or acoustic signal.

(iii) Signal Processing and Analysis:
The generated signal is processed and analyzed through electronic circuits and signal processing units within the biosensor. This processing enables the quantification and interpretation of the pathogen's presence or concentration, providing real-time and accurate data for disease diagnosis and management.

11.3 TYPES OF BIOSENSORS IN PLANT PATHOGEN DETECTION

Several types of biosensors are used in plant disease diagnostics to detect and monitor various pathogens. These biosensors utilize different transduction mechanisms and have distinct advantages in terms of sensitivity, specificity, and ease of use. Depending on the method of signal transduction, biosensors can also be divided into different groups: amperometric, electrochemical, optical, thermometric, piezoelectric, or magnetic. Amperometric devices are the most commonly reported on class of biosensors.

11.3.1 Amperometric Biosensors

Amperometric detection typically relies on an enzyme system that catalytically converts analytes into products that can be oxidized or reduced at a working electrode and maintained at a specific potential. The main advantage of this transducer is the low cost and the use of disposable electrodes. The high reproducibility of these single-use electrodes eliminates the requirement for repeated calibration. Limitations of amperometric transducers include interference from electroactive compounds.

11.3.2 Electrochemical Biosensors

These biosensors utilize electrochemical transduction methods to convert a biochemical interaction into an electrical signal. They are known for their high sensitivity and rapid response, making them suitable for the detection of specific plant pathogens in various environmental samples. Electrochemical biosensors are often used to measure changes in current, potential, or impedance caused by the interaction between the biological recognition element and the target pathogen.

11.3.3 Optical Biosensors

Optical biosensors rely on the detection of changes in light properties, such as absorbance, fluorescence, or refractive index, resulting from the binding events between the biological recognition element and the target pathogen. These biosensors offer high sensitivity and specificity, enabling real-time monitoring and quantitative analysis of plant pathogens in complex sample matrices. Surface plasmon resonance (SPR) and fluorescence-based biosensors are commonly used optical techniques in plant disease diagnostics. In recent years, a key stimulus for the development of optical biosensors has been the availability of high-quality fibers and optoelectronic components. The optical biosensor format may involve direct detection of the analyte of interest or indirect detection through optically labelled probes and the optical transducer may detect changes in the absorbance, luminescence, polarization or refractive index. The advantages of optical biosensors are their speed, the immunity of the signal to electrical or magnetic interference, and the potential for higher information content (spectrum of information available), but the main drawback can be the high cost of some instrumentation.

11.3.4 Thermal Biosensors

Thermal biosensors operate based on the measurement of heat changes resulting from the biochemical interaction between the target pathogen and the biological recognition element. These biosensors can detect subtle temperature variations caused by specific binding events, allowing for the sensitive and selective detection of plant pathogens. Thermal biosensors are particularly useful for monitoring pathogen interactions on plant surfaces and in other environmental samples.

11.3.5 Piezoelectric Biosensors

Piezoelectric biosensors rely on the measurement of changes in mass, typically resulting from the binding events between the target pathogen and the biological recognition element. The piezoelectric biosensor is based on measuring frequency changes directly related to mass change on the sensor surface. These biosensors convert the mass changes into electrical signals, enabling the detection and quantification of plant pathogens with high sensitivity and real-time monitoring capabilities. Piezoelectric biosensors are often used in the detection of pathogens in plant tissues, seeds, and other complex matrices. One of the main advantages of this technique is the real-time binding reaction detection, allowing kinetic evaluation of affinity interactions (a feature similar to surface plasmon resonance biosensors) and, in addition, the low cost of the instrumentation required. Limitations of this transduction method are the need for calibration of each crystal and the possible variability when coating the surface with the antigen or antibody.

11.4 APPLICATION OF BIOSENSORS IN PLANT DISEASE MANAGEMENT

11.4.1 Rapid Detection of Plant Diseases using Biosensors

Plant pathogens significantly cause reduced crop productivity and yield loss, posing a threat to food security for both animals and humans. According to the Food and Agricultural Organization (FAO), the world population is projected to reach approximately 10 billion by 2050, necessitating a 50% increase in food production to meet the demand (FAO, 2019). To address this challenge, much research in the field of plant pathology focuses on minimizing catastrophic crop losses directly or indirectly (Li et al., 2020). Traditionally, the management of plant diseases has relied on the application of recommended doses of fungicides through foliar and soil treatments. However, the excessive use of these chemicals had detrimental effects on ecosystems and has led to the development of resistance among pathogens. Consequently, there is a need for development of simple and rapid detection tools to aid in plant disease management. Currently, nano-biosensors have emerged as effective tools for pathogen detection and disease management. Researchers have developed various biosensing strategies, including electrochemical-, amperometric-, and microfluidic-based biosensors. A list of several plant pathogens that have been successfully detected using various nano-biosensors is given in Table 11.1.

TABLE 11.1
Various biosensor techniques developed for rapid detection of plant pathogens

S. No.	Type of Biosensor	Target pathogen	Sensitivity	Reference
		Fungi		
1.	Amperometric	*Botrytis cinerea*	3mg/L	Cinquanta et al. (2015)
2.	Optical	*Phakopsora pachyrhizi*	2.2 ng/mL	Miranda et al. (2013)
3.	-do-	*Phytophthora infestans*	0.4 ppm	Li et al. (2019)
4.	-do-	*Alternaria panax*	0.01 pg/µL	S. Wei et al. (2018)
5.	Aptameric	*Phytophthora palmivora*	0.3 pg/mL	Franco et al. (2019)
6.	Electrochemical	*Phytophthora cactorum*	35-62 nmol/L	Fang et al. (2014)
7.	-do-	*Ustiloginoida virens*	10 fmol/L	Rana et al. (2021)
		Bacteria		
1.	Electrochemical	*Pantoea stewarti*	7.8×10^3 CFU/mL	Zhao et al. (2014)
2.	-do-	*Xanthomonas axonopodis*	0.01 nM	Haji-Hashemi et al. (2018)
3.	-do-	*Pseudomonas syringea* pv. *lachrymans*	2.6 pg/mL	Cebula et al. (2019)
4.	-do-	*Ca. phytoplasma* Aurantifolia	1500 pg/mL	Ebrahimi et al. (2019)
5.	Microfluidic	*Xanthomonas arboricola*	0.8 pg/mL	Regiart et al. (2017)
6.	Optical	*Pantoea stewarti*	5.38-538 pg/mL	Feng et al. (2015)
7.	-do-	*Acidovorax avenea* sub sp. *Citruli*	3.5×10^3 pg/mL – 7.25×10^4 pg/mL	Charlermroj et al. (2013)

(continued)

TABLE 11.1 (Continued)
Various biosensor techniques developed for rapid detection of plant pathogens

S. No.	Type of Biosensor	Target pathogen	Sensitivity	Reference
		Virus		
1.	Electrochemical	*Plum pox virus*	10 pg/mL	Jarocka et al. (2011)
2.	-do-	*Cucumber mosaic virus*	10 ng/mL	Chartuprayoon et al. (2013)
3.	-do-	*Groundnut bud necrosis virus*	5-7 ng/mL	Chaudhary et al. (2021)
4.	QCM based	*Maize chlorotic mottle virus*	2.5×10^5 pg/mL	Huang et al. (2014)
5.	Amperometric	*Capsicum chlorosis virus*	-	Sharma et al. (2017)
6.	-do-	*Citrus tristiza virus*	0.3 fg/mL	Freitas et al. (2019)
7.	Impedimetric	*Citrus tristiza virus*	0.1-10μM	Khater et al. (2019)
8.	Optical	*Potato virus* X	2 × 103 pg/mL	Drygin et al. (2012)
9.	-do-	*Cymbidium mosaic virs*	48 pg/mL	Lin et al. (2014)
10.	-do-	*Banana bunchy top virus*	3.2×10^4 pg/mL	J. Wei et al. (2014)
11.	-do-	*Banana streak virus*	0.1 pg/mL	Tang et al. (2007)
12.	-do-	*Grapevine virus* A type protein	1 pg – 10ng / mL	Tereshchenko et al. (2016)

11.4.2 Significance of Rapid Detection in Preventing Disease Spread

Rapid detection of plant pathogens by biosensors plays a crucial role in preventing the spread of diseases in several ways as follows:

(i) **Non-destructiveness of sample/plants:** Biosensors offer a non-destructive method for detecting plant pathogens without causing any damage to the plants. While traditional methods often involve destructive sampling, which can harm the plant and affect its growth and productivity.

(ii) **Real-time monitoring:** Biosensors enable the real-time monitoring of plant pathogens, which enables researchers to track the presence and spread of pathogens continuously, and also provide valuable insights into the dynamics of pathogen infections. Real-time monitoring can help in implementing timely control measures and preventing the further spread of diseases.

(iii) **Early detection:** Biosensors can detect the infection of plant pathogens at an early stage, even before visible symptoms appear. This early detection allows for quick intervention and control measures to be implemented and can prevent the further spread of the pathogen and minimize crop loss.

(iv) **Rapid detection:** Biosensors offer rapid detection of plant pathogens, allowing farmers and agricultural authorities to respond promptly. This may help in implementing quarantine measures, removing infected plants, or initiating targeted treatments to prevent the disease from spreading to other plants or regions.

(v) **High sensitivity and specificity:** Biosensors provide accurate identification of specific plant pathogens. This helps in distinguishing between different pathogen strains or species, enabling tailored control strategies to be implemented for each specific pathogen. It also reduces false-positive results and unnecessary application of broad-spectrum pesticides and fungicides, reducing the ecological impact and cost.

(vi) **Point-of-care testing:** Biosensors can be designed as portable and user-friendly devices, enabling on-site or point-of-care testing. This eliminates the need for laboratory analysis, reduces the time required for diagnosis, and provides real-time results. Farmers and growers can quickly identify infected plants and take necessary actions, saving time, resources, and reducing the risk of further spread.

(vii) **Cost effective:** Biosensors can be cost effective compared to conventional detection techniques. They require minimal sample preparation, reduced reagent consumption, and often allow the detection of multiple pathogens in a single test, reducing overall costs.

11.5 ADVANTAGES OF BIOSENSORS OVER CONVENTIONAL METHODS OF PATHOGEN DETECTION

In the realm of plant disease diagnostics, biosensors offer several distinct advantages over conventional methods, revolutionizing the speed and accuracy of pathogen detection. Biosensors typically exhibit higher sensitivity and specificity compared to conventional methods. The precise and selective binding capabilities of biological recognition elements in biosensors allow for the accurate detection of low pathogen concentrations, reducing the risk of false positives or false negatives. In contrast, some conventional methods may lack the required sensitivity and specificity, leading to potential inaccuracies in pathogen identification. Biosensors enable real-time or rapid detection of plant pathogens, providing immediate results that allow for timely interventions. Conversely, conventional methods often involve time-consuming procedures, including sample preparation, culturing, and staining, which may delay the diagnosis and subsequent management decisions.

Biosensors are often portable and enable on-site testing, facilitating immediate decision-making in the field. This capability is particularly advantageous for detecting and monitoring plant pathogens in remote agricultural settings. In contrast, conventional methods typically require specialized laboratory equipment and trained personnel, limiting their applicability for on-site diagnostics. While the development and optimization of biosensors may incur initial costs, their long-term benefits often outweigh the investment, especially in terms of rapid, precise, and cost-effective diagnostics. Conventional methods may be more accessible and cost-effective initially, but they can be resource-intensive in terms of time, labour, and reagents, particularly when analyzing a large number of samples. Biosensors can be designed for multiplexing, enabling the simultaneous detection of multiple pathogens or biomolecules in a single sample. This capability streamlines the diagnostic process and enhances the efficiency of disease surveillance. Conventional methods may require separate tests for each pathogen, making the process more time-consuming and labour-intensive.

11.6 EMERGING TECHNOLOGIES AND INNOVATIONS IN BIOSENSOR DEVELOPMENT

Recent advancements in biosensor technology have led to the development of highly sensitive, specific, and rapid biosensors that have been instrumental in improving the efficiency and effectiveness of pathogen detection. Some notable advancements include the following.

11.6.1 Nanotechnology Integration

The association of nanotechnology with biosensor technology has ushered in a new era of heightened precision and sensitivity in the detection of plant pathogens. The integration of nanomaterials, such as nanoparticles, nanowires, and nanotubes, has substantially enhanced the sensitivity and selectivity of biosensors (Naresh & Lee, 2021). Nanomaterials possess large surface area, electronic conductivity, plasmonic characteristics and other unique physicochemical properties, serving as a platform for the construction of biorecognition molecules (Ramesh et al., 2022). These characteristics raise the detection limit enabling detection at extremely low concentrations. Many nanomaterials have been used for biosensor applications in the field of plant pathogen detection, including graphene, carbon nanotubes, quantum dots, and metal and metal oxide nanoparticles (Fang et al., 2014; Jain et al., 2021; Kumar & Arora, 2020; Li et al., 2020; Pudake et al., 2019).

There are several diverse methodologies made possible by nanotechnology and that can be utilized in the creation of (bio)sensors with improved characteristics. For example, carbon nanotubes or graphene derivatives can be used to increase the conductivity of the sensor (Li et al., 2019; Schroeder et al., 2019). Electro-spun nanofibers or metallic nanoparticles with tunable surface functionality allow binding sites for biospecific immobilization can be used to achieve a significant improvement in surface-to-volume ratio (Mercante et al., 2017; Chang et al., 2019; Huang et al., 2014). Molecularly imprinted polymer can be used to tune the selectivity of the sensor because it only interacts with the target analyte and prevents interference (Dickert et al., 2004).

A nanoparticle-based electrochemical biosensor designed to detect target pathogen DNA using recombinase polymerase amplification (RPA), combined with a gold electrophoresis method. Nanoparticle-based sensors exerted 10,000 times more sensitivity in identifying infected plants prior to visible disease symptoms (Lau et al., 2017). A rice fungal disease false smut affecting rice production is caused by *Ustilaginoidea virens.* To detect false smut, a study developed a graphene-based geno-biosensor on paper electrodes as a biological recognition element used as probe

DNA (cost-effective), and to enhance detection, oxidized graphene was used (up to 10 fM of target ssDNA) to determine the sensitivity of the sensor. In the presence of a pathogen, single-stranded DNA probe hybridized with pathogen target DNA was analyzed using linear sweep voltammetry and cyclic voltammetry (Rana et al., 2021). A multiwalled carbon nanotube-based zinc nanocomposite was used to detect the Chili leaf curl beta satellite (ChLCB). Zn-NPs are used for the immobilization of the probe ssDNA strand, and upon binding of the target DNA to ssDNA, leads to hybridization, which was analyzed using cyclic and differential pulse voltammetry. This nano DNA sensor was reported to be three times more specific compared to the conventional method (Tahir et al., 2017).

The utilization of gold nanoparticles in biosensors has improved the signal amplification and facilitated the detection of plant pathogens with higher sensitivity and specificity due to high surface-to-volume ratio, good stability, and high conductivity. A label-free impedimetric biosensor electrodeposited with AuNPs was developed for the detection of the DNA of the *Citrus tristeza virus*. Nano-sensors specifically detect target viruses in the presence of non-specific DNA with enhanced reproducibility (Khater et al., 2019). Similarly, the functionalization of carbon nanotubes with specific biomolecules has led to the development of highly selective biosensors capable of discriminating between closely related strains of plant-infecting viruses such as that of geminiviruses in cotton with a detection limit of 0.01 ng/μL (Tahir et al., 2018).

11.6.2 Lab-on-a-Chip Technology

The miniaturization of biosensors using lab-on-a-chip technology has enabled the development of portable and cost-effective diagnostic tools. These integrated microfluidic systems allow for the simultaneous analysis of multiple samples, reducing the time and resources required for pathogen detection. Lab-on-a-chip biosensors have facilitated the rapid on-site detection of plant pathogens, enabling real-time monitoring and timely decision-making in agricultural settings. Among the visible symptom-based plant disease diagnosis methods, lateral flow immunoassay (LFIA)-based test strips were reported to detect *Ralstonia solanacearum,* which is a quarantine organism that causes potato brown rot using enlarging gold nanoparticles (GNPs) (Borse & Srivastava, 2019).

11.6.3 Smartphone-Based Detection

The integration of biosensors with smartphone technology has facilitated the development of point-of-care diagnostic tools, allowing for rapid and convenient pathogen detection in the field. Smartphone-based biosensors utilize the device's camera, processing power, and connectivity to analyze and interpret the data obtained from the biosensor, enabling seamless data transmission and analysis. This advancement has democratized pathogen detection, making it accessible to a broader range of end-users, including farmers and agricultural workers. It has been employed for the detection of various human diseases (Pawar et al., 2023) and has the potential to be utilized for the detection of various plant pathogens.

11.6.4 Use of Artificial Intelligence and Machine Learning

The convergence of machine learning (ML) and biosensors presents a transformative synergy. Biosensors traditionally grapple with limitations in sensitivity and adaptability, hurdles that ML algorithms aim to overcome. ML, a subset of artificial intelligence, empowers computers to learn from data, potentially revolutionizing biosensors by extracting complex insights and automating decision-making processes. The pivotal advantage of ML lies in its adeptness with vast datasets, discerning subtle patterns, and correlations that traditional methods may miss. This capability significantly enhances biosensors' accuracy and sensitivity, particularly in managing intricate data. Additionally, ML optimizes biosensor performance by tailoring parameters, leading to more precise and adaptable biosensors suited for specific applications (Schackart & Yoon, 2021). Moreover, ML-equipped biosensors exhibit adaptability, swiftly adapting to changing conditions. These intelligent systems efficiently process rapidly generated data, mitigating the need for time-consuming manual analysis. Consequently, this integration holds promise for advancing biosensing technologies, promising improved data processing, analysis efficiency, and overall performance (Cui et al., 2020). AI-enabled biosensing for rapid pathogen detection has been employed for detecting human pathogen and pathogens in water and agricultural food (Zhang et al., 2021; Kang et al., 2021).

11.6.5 Internet of Things (IoT) Integration

The integration of biosensors with IoT technologies has enabled real-time data monitoring and remote disease management. IoT-enabled biosensors can continuously monitor environmental parameters and detect changes in pathogen concentrations, providing early warning systems for disease outbreaks. These interconnected biosensing networks facilitate data sharing and analysis, enabling proactive disease control measures and the optimization of agricultural practices to prevent crop losses. The integration of Cloud-IoT services in agriculture has streamlined sensor installation and connectivity, facilitating remote monitoring. Technological advancements, such as IoT systems and AI tools in "smart farming," analyze environmental data to optimize planting decisions based on factors like soil type and weather forecasts. 5G technology has further revolutionized agriculture by enabling faster communication networks. It supports drones equipped with advanced sensors and cameras, allowing real-time monitoring and diagnosis of plant issues. This technology drastically reduces costs by efficiently collecting high-quality data, especially through drones' ability to assess crop health and detect stress levels in plants. The IoT-integrated biosensors, driven by AI, drones, and 5G, detect plant stress and contribute to the development of superior nanosensor products. Companies like Razzberry, Nasys, Tracense, and nGageIT pioneer nanosensor innovations, monitoring water and soil chemistry, measuring air pollution, and tracing contaminants in food production. These advancements hold immense promise for optimizing agriculture practices and improving crop management through cutting-edge technologies and data-driven solutions.

11.7 CONCLUSION

By leveraging the unique capabilities of the different types of biosensors, researchers and plant pathologists can effectively detect and monitor various plant pathogens, enabling timely interventions and the implementation of targeted disease management strategies. The selection of the appropriate biosensor type depends on the specific requirements of the diagnostic application, the nature of the pathogen being targeted, and the environmental conditions of the sample matrix.

The integration of nanotechnology in biosensors has catalyzed a paradigm shift in pathogen detection, underscoring its pivotal role in ensuring sustainable agricultural practices and global food security. With its unprecedented sensitivity, selectivity, and portability, nanotechnology-integrated biosensors have emerged as indispensable tools for safeguarding crop health and mitigating the adverse impacts of plant diseases on agricultural productivity. Leveraging ongoing advancements and interdisciplinary collaborations, the widespread adoption of nanotechnology-integrated biosensors is poised to shape the future of precision agriculture, empowering stakeholders with effective and efficient strategies for plant disease management and crop protection.

11.8 FUTURE CHALLENGES AND PROSPECTIVES

Despite their remarkable potential, biosensors face challenges related to cost-effectiveness, standardization, and sample preparation, which need to be addressed through further research and technological innovations. By exploring potential strategies to overcome these challenges, this chapter emphasized the promising prospects and pivotal role of biosensors in ensuring sustainable and resilient agricultural practices in the face of emerging plant diseases and changing environmental conditions.

REFERENCES

Ali, S., Hassan, A., Hassan, G., Eun, C.-H., Bae, J., Lee, C. H., & Kim, I.-J. (2018). Disposable all-printed electronic biosensor for instantaneous detection and classification of pathogens. *Scientific Reports*, *8*(1), 5920. https://doi.org/10.1038/s41598-018-24208-2

Borse, V., & Srivastava, R. (2019). Process parameter optimization for lateral flow immunosensing. *Materials Science for Energy Technologies*, *2*(3), 434–441. https://doi.org/10.1016/j.mset.2019.04.003

Cebula, Z., Żołędowska, S., Dziąbowska, K., Skwarecka, M., Malinowska, N., Białobrzeska, W., Czaczyk, E., Siuzdak, K., Sawczak, M., Bogdanowicz, R., & Nidzworski, D. (2019). Detection of the plant pathogen *Pseudomonas Syringae* pv. *Lachrymans* on antibody-modified gold electrodes by electrochemical impedance spectroscopy. *Sensors*, *19*(24), 5411. https://doi.org/10.3390/s19245411

Chang, W., Liu, W., Liu, Y., Zhan, F., Chen, H., Lei, H., & Liu, Y. (2019). Colorimetric detection of nucleic acid sequences in plant pathogens based on CRISPR/Cas9 triggered signal amplification. *Microchimica Acta*, *186*(4), 243. https://doi.org/10.1007/s00604-019-3348-2

Charlermroj, R., Himananto, O., Seepiban, C., Kumpoosiri, M., Warin, N., Oplatowska, M., Gajanandana, O., Grant, I. R., Karoonuthaisiri, N., & Elliott, C. T. (2013). Multiplex detection of plant pathogens using a microsphere immunoassay technology. *PLoS ONE*, *8*(4), e62344. https://doi.org/10.1371/journal.pone.0062344

Chartuprayoon, N., Rheem, Y., Ng, J. C. K., Nam, J., Chen, W., & Myung, N. V. (2013). Polypyrrole nanoribbon based chemiresistive immunosensors for viral plant pathogen detection. *Analytical Methods*, *5*(14), 3497. https://doi.org/10.1039/c3ay40371h

Chaudhary, M., Verma, S., Kumar, A., Basavaraj, Y. B., Tiwari, P., Singh, S., Chauhan, S. K., Kumar, P., & Singh, S. P. (2021). Graphene oxide based electrochemical immunosensor for rapid detection of groundnut bud necrosis orthotospovirus in agricultural crops. *Talanta*, *235*, 122717. https://doi.org/10.1016/j.talanta.2021.122717

Chen, S., & Shamsi, M. H. (2017). Biosensors-on-chip: a topical review. *Journal of Micromechanics and Microengineering*, *27*(8), 083001. https://doi.org/10.1088/1361-6439/aa7117

Cinquanta, L., Albanese, D., De Curtis, F., Malvano, F., Crescitelli, A., & Di Matteo, M. (2015). Rapid assessment of gray mold (*Botrytis cinerea*) infection in grapes with a biosensor system. *American Journal of Enology and Viticulture*, *66*(4), 502–508. https://doi.org/10.5344/ajev.2015.15029

Cui, F., Yue, Y., Zhang, Y., Zhang, Z., & Zhou, H. S. (2020). Advancing biosensors with machine learning. *ACS Sensors*, *5*(11), 3346–3364.

Dickert, F. L., Hayden, O., Bindeus, R., Mann, K.-J., Blaas, D., & Waigmann, E. (2004). Bioimprinted QCM sensors for virus detection screening of plant sap. *Analytical and Bioanalytical Chemistry*, *378*(8), 1929–1934. https://doi.org/10.1007/s00216-004-2521-5

Donoso, A., & Valenzuela, S. (2018). In-field molecular diagnosis of plant pathogens: recent trends and future perspectives. *Plant Pathology*, *67*(7), 1451–1461.

Drygin, Y. F., Blintsov, A. N., Grigorenko, V. G., Andreeva, I. P., Osipov, A. P., Varitzev, Y. A., Uskov, A. I., Kravchenko, D. V., & Atabekov, J. G. (2012). Highly sensitive field test lateral flow immunodiagnostics of PVX infection. *Applied Microbiology and Biotechnology*, *93*(1), 179–189. https://doi.org/10.1007/s00253-011-3522-x

Ebrahimi, M., Norouzi, P., Safarnejad, M. R., Tabaei, O., & Haji-Hashemi, H. (2019). Fabrication of a label-free electrochemical immunosensor for direct detection of *Candidatus Phytoplasma Aurantifolia*. *Journal of Electroanalytical Chemistry*, *851*, 113451. https://doi.org/10.1016/j.jelechem.2019.113451

Fang, Y., Umasankar, Y., & Ramasamy, R. P. (2014). Electrochemical detection of p-ethylguaiacol, a fungi infected fruit volatile using metal oxide nanoparticles. *The Analyst*, *139*(15), 3804–3810. https://doi.org/10.1039/C4AN00384E

FAO. (2019). *The state of food and agriculture. 2019, Moving forward on food loss and waste reduction*. Retrieved February 11, 2024, from www.fao.org/3/ca6030en/CA6030EN.pdf

Feng, M., Kong, D., Wang, W., Liu, L., Song, S., & Xu, C. (2015). Development of an immunochromatographic strip for rapid detection of *Pantoea stewartii* subsp. *stewartii*. *Sensors*, *15*(2), 4291–4301. https://doi.org/10.3390/s150204291

Foley, J. A., Ramankutty, N., Brauman, K. A., Cassidy, E. S., Gerber, J. S., Johnston, M., Mueller, N. D., O'Connell, C., Ray, D. K., West, P. C., Balzer, C., Bennett, E. M., Carpenter, S. R., Hill, J., Monfreda, C., Polasky, S., Rockström, J., Sheehan, J., Siebert, S., … Zaks, D. P. M. (2011). Solutions for a cultivated planet. *Nature*, *478*(7369), 337–342. https://doi.org/10.1038/nature10452

Franco, A. J. DM., Merca, F. E., Rodriguez, M. S., Balidion, J. F., Migo, V. P., Amalin, D. M., Alocilja, E. C., & Fernando, L. M. (2019). DNA-based electrochemical nanobiosensor for the detection of Phytophthora palmivora (Butler) Butler, causing black pod rot in cacao (*Theobroma cacao* L.) pods. *Physiological and Molecular Plant Pathology*, *107*, 14–20. https://doi.org/10.1016/j.pmpp.2019.04.004

Freitas, T. A., Proença, C. A., Baldo, T. A., Materón, E. M., Wong, A., Magnani, R. F., & Faria, R. C. (2019). Ultrasensitive immunoassay for detection of Citrus tristeza virus in citrus sample using disposable microfluidic electrochemical device. *Talanta*, *205*, 120110. https://doi.org/10.1016/j.talanta.2019.07.005

Haji-Hashemi, H., Norouzi, P., Safarnejad, M. R., Larijani, B., Habibi, M. M., Raeisi, H., & Ganjali, M. R. (2018). Sensitive electrochemical immunosensor for citrus bacterial canker disease detection using fast Fourier transformation square-wave voltammetry method. *Journal of Electroanalytical Chemistry*, *820*, 111–117. https://doi.org/10.1016/j.jelechem.2018.04.062

Haupt, K., & Mosbach, K. (2000). Molecularly imprinted polymers and their use in biomimetic sensors. *Chemical Reviews*, *100*(7), 2495–2504. https://doi.org/10.1021/cr990099w

Huang, X., Xu, J., Ji, H.-F., Li, G., & Chen, H. (2014). Quartz crystal microbalance-based biosensor for rapid and sensitive detection of maize chlorotic mottle virus. *Analytical Methods*, *6*(13), 4530–4536. https://doi.org/10.1039/C4AY00292J

Jain, U., Pudake, R. N., Chauhan, N., & Pareek, S. (2021). Advancements in biosensors for fungal pathogen detection in plants. In: Pudake., R.N. Jain, U., Kole, C. (eds) Biosensors in Agriculture: Recent trends and future perspectives. Concepts and Strategies in Plant Sciences. Springer, Cham. (pp. 205–216). https://doi.org/10.1007/978-3-030-66165-6_11

Jarocka, U., Wąsowicz, M., Radecka, H., Malinowski, T., Michalczuk, L., & Radecki, J. (2011). Impedimetric immunosensor for detection of *Plum Pox Virus* in plant extracts. *Electroanalysis*, *23*(9), 2197–2204. https://doi.org/10.1002/elan.201100152

Kang, R., Park, B., Ouyang, Q., & Ren, N. (2021). Rapid identification of foodborne bacteria with hyperspectral microscopic imaging and artificial intelligence classification algorithms. *Food Control*, *130*, 108379. https://doi.org/10.1016/j.foodcont.2021.108379

Khater, M., de la Escosura-Muñiz, A., Quesada-González, D., & Merkoçi, A. (2019). Electrochemical detection of plant virus using gold nanoparticle-modified electrodes. *Analytica Chimica Acta*, *1046*, 123–131. https://doi.org/10.1016/j.aca.2018.09.031

Kumar, V., & Arora, K. (2020). Trends in nano-inspired biosensors for plants. *Materials Science for Energy Technologies*, *3*, 255–273. https://doi.org/10.1016/j.mset.2019.10.004

Lau, H. Y., Wu, H., Wee, E. J. H., Trau, M., Wang, Y., & Botella, J. R. (2017). Specific and sensitive isothermal electrochemical biosensor for plant pathogen DNA detection with colloidal gold nanoparticles as probes. *Scientific Reports*, *7*(1), 38896. https://doi.org/10.1038/srep38896

Li, Z., Paul, R., Ba Tis, T., Saville, A. C., Hansel, J. C., Yu, T., Ristaino, J. B., & Wei, Q. (2019). Non-invasive plant disease diagnostics enabled by smartphone-based fingerprinting of leaf volatiles. *Nature Plants*, *5*(8), 856–866. https://doi.org/10.1038/s41477-019-0476-y

Li, Z., Yu, T., Paul, R., Fan, J., Yang, Y., & Wei, Q. (2020). Agricultural nanodiagnostics for plant diseases: recent advances and challenges. *Nanoscale Advances*, *2*(8), 3083–3094. https://doi.org/10.1039/C9NA00724E

Lin, H.-Y., Huang, C.-H., Lu, S.-H., Kuo, I.-T., & Chau, L.-K. (2014). Direct detection of orchid viruses using nanorod-based fiber optic particle plasmon resonance immunosensor. *Biosensors and Bioelectronics*, *51*, 371–378. https://doi.org/10.1016/j.bios.2013.08.009

Mancini, V., & Romanazzi, G. (2014). Seed treatments to control seedborne fungal pathogens of vegetable crops. *Pest Management Science*, *70*(6), 860–868. https://doi.org/10.1002/ps.3693

Mercante, L. A., Scagion, V. P., Migliorini, F. L., Mattoso, L. H. C., & Correa, D. S. (2017). Electrospinning-based (bio)sensors for food and agricultural applications: A review. *TrAC Trends in Analytical Chemistry*, *91*, 91–103. https://doi.org/10.1016/j.trac.2017.04.004

Miranda, B. S., Linares, E. M., Thalhammer, S., & Kubota, L. T. (2013). Development of a disposable and highly sensitive paper-based immunosensor for early diagnosis of Asian soybean rust. *Biosensors and Bioelectronics*, *45*, 123–128. https://doi.org/10.1016/j.bios.2013.01.048

Naresh, V., & Lee, N. (2021). A review on biosensors and recent development of nanostructured materials-enabled biosensors. *Sensors*, *21*(4), 1109. https://doi.org/10.3390/s21041109

Pawar, A. A., Patwardhan, S. B., Barage, S., Raut, R., Lakkakula, J., Roy, A., Sharma, R., & Anand, J. (2023). Smartphone-based diagnostics for biosensing infectious human pathogens. *Progress in Biophysics and Molecular Biology*, *180–181*, 120–130. https://doi.org/10.1016/j.pbiomolbio.2023.05.002

Pudake, R. N., Chauhan, N., & Kole, C. (2019). *Nanoscience for Sustainable Agriculture* (R. N. Pudake, N. Chauhan, & C. Kole, Eds.). Springer International Publishing. https://doi.org/10.1007/978-3-319-97852-9

Ramesh, M., Janani, R., Deepa, C., & Rajeshkumar, L. (2022). Nanotechnology-enabled biosensors: A review of fundamentals, design principles, materials, and applications. *Biosensors*, *13*(1), 40. https://doi.org/10.3390/bios13010040

Rana, K., Mittal, J., Narang, J., Mishra, A., & Pudake, R. N. (2021). Graphene based electrochemical DNA biosensor for detection of false smut of rice (*Ustilaginoidea virens*). *Plant Pathology Journal*, *37*(3), 291–298. https://doi.org/10.5423/PPJ.OA.11.2020.0207

Regiart, M., Rinaldi-Tosi, M., Aranda, P. R., Bertolino, F. A., Villarroel-Rocha, J., Sapag, K., Messina, G. A., Raba, J., & Fernández-Baldo, M. A. (2017). Development of a nanostructured immunosensor for early and in situ detection of Xanthomonas arboricola in agricultural food production. *Talanta*, *175*, 535–541. https://doi.org/10.1016/j.talanta.2017.07.086

Sadanandom, A., & Napier, R. M. (2010). Biosensors in plants. *Current Opinion in Plant Biology*, *13*(6), 736–743. https://doi.org/10.1016/j.pbi.2010.08.010

Schackart, K. E., & Yoon, J.-Y. (2021). Machine learning enhances the performance of bioreceptor-free biosensors. *Sensors*, *21*(16), 5519. https://doi.org/10.3390/s21165519

Schroeder, V., Savagatrup, S., He, M., Lin, S., & Swager, T. M. (2019). Carbon nanotube chemical sensors. *Chemical Reviews*, *119*(1), 599–663. https://doi.org/10.1021/acs.chemrev.8b00340

Sharma, A., Kaushal, A., & Kulshrestha, S. (2017). A Nano-Au/C-MWCNT based label free amperometric immunosensor for the detection of capsicum chlorosis virus in bell pepper. *Archives of Virology*, *162*(7), 2047–2052. https://doi.org/10.1007/s00705-017-3293-5

Tahir, M. A., Bajwa, S. Z., Mansoor, S., Briddon, R. W., Khan, W. S., Scheffler, B. E., & Amin, I. (2018). Evaluation of carbon nanotube-based copper nanoparticle composite for the efficient detection of agroviruses. *Journal of Hazardous Materials*, *346*, 27–35. https://doi.org/10.1016/j.jhazmat.2017.12.007

Tahir, M. A., Hameed, S., Munawar, A., Amin, I., Mansoor, S., Khan, W. S., & Bajwa, S. Z. (2017). Investigating the potential of multiwalled carbon nanotubes-based zinc nanocomposite as a recognition interface towards plant pathogen detection. *Journal of Virological Methods*, *249*, 130–136. https://doi.org/10.1016/j.jviromet.2017.09.004

Tang, Y., Xing, D., Zhu, D., & Liu, J. (2007). An improved electrochemiluminescence polymerase chain reaction method for highly sensitive detection of plant viruses. *Analytica Chimica Acta*, *582*(2), 275–280. https://doi.org/10.1016/j.aca.2006.09.021

Tereshchenko, A., Bechelany, M., Viter, R., Khranovskyy, V., Smyntyna, V., Starodub, N., & Yakimova, R. (2016). Optical biosensors based on ZnO nanostructures: advantages and perspectives. A review. *Sensors and Actuators B: Chemical*, *229*, 664–677. https://doi.org/10.1016/j.snb.2016.01.099

Wei, J., Liu, H., Liu, F., Zhu, M., Zhou, X., & Xing, D. (2014). Miniaturized paper-based gene sensor for rapid and sensitive identification of contagious plant virus. *ACS Applied Materials & Interfaces*, *6*(24), 22577–22584. https://doi.org/10.1021/am506695g

Wei, S., Sun, Y., Xi, G., Zhang, H., Xiao, M., & Yin, R. (2018). Development of a single-tube nested PCR-lateral flow biosensor assay for rapid and accurate detection of Alternaria panax Whetz. *PLOS ONE*, *13*(11), e0206462. https://doi.org/10.1371/journal.pone.0206462

Zhang, Y., Jiang, H., Ye, T., & Juhas, M. (2021). Deep learning for imaging and detection of microorganisms. *Trends in Microbiology*, *29*(7), 569–572. https://doi.org/10.1016/j.tim.2021.01.006

Zhao, Y., Liu, L., Kong, D., Kuang, H., Wang, L., & Xu, C. (2014). Dual amplified electrochemical immunosensor for highly sensitive detection of *Pantoea stewartii* sub sp. *stewartii*. *ACS Applied Materials & Interfaces*, *6*(23), 21178–21183. https://doi.org/10.1021/am506104r

12 Applications of Versatile CRISPR Technology for the Detection of Targeted Plant Pathogens

Rashmi ER, Reena Prusty, Jyoti priya, Kavitha R, and Supreetha B G

12.1 INTRODUCTION

The world population goes on escalating like blinking an eye and it is expected to reach at least 9.8 billion by 2050 and 10.4 billion by 2100 (Adam, 2021). With the increasing pace of the world's human vehicle, the necessity for increased food supply and supplements requires attention. Among the various factors affecting the yield and quality of the crops, disease incidence is a major factor limiting the world crop yield approximately 20-40% (Savary, 2019). It is therefore necessary to develop sustainable management of plant diseases to supply sufficient food to the escalating population without affecting the yield and quality of the crops.

Crop yield and quality are arrested by various pathogens like fungus, bacteria, and viruses. Therefore, there is a need to understand the interplay between pathogens and the plants, as well as the components of the plant defense response for successful plant disease management strategies (Veillet, 2020). A multilayer protective system is activated in the event of a disease attack to combat alien organisms. This system comprises a two-tier immune sensing system with molecular receptors that identify possible pathogens inside plant cells as well as at the cell's periphery (Dodds and Rathjen, 2010). Pattern recognition receptors (PRRs) recognize pathogen proteins released in the apoplast and conserved microbial patterns, sometimes referred to as pathogen-associated molecular patterns (PAMPs), at the cell surface. This detection triggers a defense response known as PRR-triggered immunity (PTI) (Thomma, 2011). Within the cytoplasm, pathogen effectors that have been translocated to the host and generally modify plant functions to promote pathogen growth are recognized by intracellular nucleotide-binding domain and leucine-rich repeat(LRR)containing, or NOD-like receptors (NLRs), which are expressed by disease resistance genes (R genes) (Nabi *et al.*, 2024). This triggers off an NLR-mediated reaction, also referred

 DOI: 10.1201/9781003474036-12

to as NLR- or effector-triggered immunity (NTI/ETI), which activates the plant immune response and halts the growth of the pathogen (Win, 2012).

Genome editing (GE), as a new breeding technology (NBT), gained paramount popularity. Most genome editing-derived crop varieties, in contrast to genetically modified crops, are not governed by further laws in several nations and do not contain any foreign DNA (Menz, 2020). Genome editing (GE) involves nucleases that are programmed to target DNA sequences for targeted genome modification. Examples of these nucleases are Mega nucleases (MNs), Zinc finger nucleases (ZFNs), Transcription activator-like effector nucleases (TALENs), and Clustered regularly interspaced short palindromic repeats (CRISPR)-associated nucleases (Cas). Because of its wide range of applications, suited to multiplexing, and simplistic use, CRISPR-Cas technology has outperformed MNs, ZFNs, and TALENs (Gao, 2021).

12.2 OVERVIEW OF CRISPR/CAS

CRISPR is a natural immune system that shields against invading viruses. Nakata and associates in 1987 discovered a group of 29 bp repeats downstream of the *IAP* gene locus in *Escherichia coli*, which marked the beginning of the CRISPR era (Ishino, 1987). These DNA sequences were identified by Jansen and Mojica in 2002, as a distinct type of clustered repeats that present in more than 40% of bacterial species. These repeats were subsequently dubbed as Clustered Regularly Interspaced Short Palindromic Repeats (CRISPR) (Jansen, 2002). CRISPR-associated genes Cas), such as Cas9, which was demonstrated to be an RNA-directed endonuclease linked to these repeats was reported by Moineau and colleagues in 2010 (Garneau, 2010).

Upon infection with a virus, bacteria sense the presence of foreign material and integrate a short segment of viral DNA into the bacterium chromosome at a site called the CRISPR locus. The bacterium then uses this sequence to transcribe a short RNA called crRNA (CRISPR RNA), which combines with another RNA called tracrRNA (trans-activating CRISPR RNA) to form a complex. This complex is called "guide RNA" (gRNA) and it directs the recognition of viral DNA by a protein called Cas9 (Javed, 2018). Cas9 is endonucleases that derived from *Streptococcus pyogenes* and produces a double-stand break in the genome. The gRNA directs Cas9 to a particular DNA site that has a sequence complementary to the first 17–20 gRNA nucleotides and is followed by an NGG PAM (protospacer adjacent motif). At precisely -3 positions upstream of PAM, the HNH and RuvC-like nuclease domains of Cas9 cleave both strands of the DNA, producing a blunt-ended double-strand break (DSB) (Gosavi, 2020).

12.3 MECHANISM OF CRISPR/CAS

The two main channels for repairing genomic DSBs produced by site-specific nucleases are homology-directed repair (HDR) and non-homologous end joining (NHEJ). Higher plants are the primary hosts of NHEJ, which is prone to error and results in tiny insertions or deletions (indels). Indels-induced frameshift mutations offer efficient gene knockout. Alternatively, gene modification can be accomplished with ease using HR (homologous recombination) if a DNA template with homology to the sequence surrounding the DSB is available. However, HDR is less frequent in higher plants and is used for fine-tuning with an extra source of donor templates that carry the desired modification (Karkute, 2017).

The removal and elimination of the DNA between two DSBs can occur when two DSBs are induced within the same chromosomal region. To achieve complete gene deletion or even the elimination of chromosomal fragments containing gene sets or recently copied genes, two or more guide RNAs directed at certain loci can be used (Belhaj, 2015). Using this method, several versions of a specific pathway or gene family that are close to one another on the same chromosome and have possibly redundant tasks during the invasion of pathogens can be detected (Langner, 2018).

12.4 TYPES OF CRISPR/CAS

CRISPR/Cas systems are classified into two classes, Types I, III, and IV (Class 1) and Types II, V, and VI (Class 2), based on the genetic, structural, and functional characteristics of Cas proteins. These classes have different mechanisms of guide RNA biogenesis and target disruption (Garcia-Doval and Jinek, 2017). The Class 2 CRISPR-associated nucleases have been widely modified and used for nucleic acid manipulation (Ji, 2019).

12.5 CRISPR/CAS-BASED DIAGNOSTIC TOOLS (CRISPR-DX)

Detecting diseases using CRISPR-based technologies offers a remarkable advantage in effortlessly pinpointing various pathogens. Through the manipulation of guide RNA (gRNA) to target specific primers or unique nucleic acid fragments associated with a particular pathogen, this method boasts versatility in identification. The entire process, completing in approximately one hour, stands out for its cost-effectiveness and minimal reagent requirements, setting it apart from more traditional laboratory approaches. Notably, the use of recombinase polymerase amplification (RPA) rather than loop-mediated isothermal amplification (LAMP) enables target sequence amplification at room temperature, eliminating the need for intricate heating apparatus (Akbar M, 2022). Consequently, this innovative approach significantly streamlines both the time and cost involved in pathogen identification.

CRISPR/Cas technology has transformed molecular biology and biotechnology by providing a diverse set of tools for a wide range of applications. CRISPR-Cas systems were created for the first time in 2016 to detect nucleic acids for molecular diagnostics (Pardee, 2016). The single-base specificity and point-of-care compatibility of CRISPR-derived RNA-guided engineered proteins (like dCas9, Cas13, and Cas12a) have led to the report of unpredictable potential in the field of pathogen detection in recent years. The most extensively utilized pathogen detection technologies are nucleic acid amplification detection tests (NAATs) and next-generation sequencing (NGS)-based pathogen diagnostics. These approaches have established the gold standard because they can recognize and distinguish between even a few nucleic acid copies (Mukhopadhyay, 2021; Patel, 2014; Rahbari, 2021). However, constraints of

NAATs such as PCR's complicated heat cycle process high cost, and extensive personnel and instrumentation needs (Li, 2021; Zhao, 2021; Sachse, 2004; Lee, 2019; Euliano, 2019; Wu, 2020; Adams, 2017), difficult primer design (Zhang, 2019), and unstable single base resolution (Higgins, 2020) restrict their utility. In the case of NGS, sequence read length constraints and the need for a profound understanding of biology while interpreting the analyzed data render this technique unsuitable for point-of-care testing (POCT) (Boers, 2019; Lecuit, 2015). Gene chip technology is of particular importance for pathogen detection since it may considerably improve the number of tests that can be run concurrently (Ragoussis, 2006), but its specificity and sensitivity are dependent on well-designed probes that complement the target (Call, 2005). While there are benefits to both electrical and optical biosensors, such as high selectivity, sensitivity, and ease of integration, they are costly and require specialized equipment (Lee, 2019; Kampeera, 2019; Ki, 2000; Xiang, 2015). Here lies the importance of CRISPR-based diagnostic tools in overcoming all of the limits encountered during onsite and laboratory diagnostics of plant pathogens.

Plant pathogen identification and disease management should benefit greatly from Cas nuclease-based diagnostics in the near future. The specificity of Cas nucleases is used by both Cas12a- and Cas13a-based detection methods to identify a defined target sequence in pathogen DNA or RNA (Gootenberg, 2017; Li, 2018c; Chen, 2018). When compared to conventional nucleic acid detection technologies like PCR, qPCR, and digital droplet PCR, Cas-based detection permits much smaller inputs, and depending on the methodology, can reliably detect at femtomolar (10^{215}) to zeptomolar (10^{221}) sensitivity (Li, 2019a). To make use of CRISPR technology easier, less expensive, and more suitable for pathogen identification in lab and field conditions, the technology is either coupled or modified with other PCR methods. Prior to the insertion of the Cas nuclease, isothermal amplification is often performed using T7-RPA (T7 transcription-recombinase polymerase amplification) (Cas13a specific), RPA, or LAMP. In the case of SHERLOCK, SHERLOCK v2, HOLMES, and DETECTR techniques enable the degradation and measurement of a fluorescent reporter oligonucleotide since Cas nuclease is left active in these scenarios. Furthermore, immunostrip-based lateral flow assays are compatible with Cas-based methodologies (Gootenberg, 2018). Cas-based diagnostics, when combined with a portable fluorescence signal detector or immunostrip, might open the way for quick and large-scale in-field detection of plant pathogens, which will be useful for qualitative and quantitative measures of disease prevalence in plants, insect vectors, and soils. The field-deployable detection can create quarantine zones for invasive diseases or aid in integrated disease management measures for reducing disease transmission, such as informed and accurate pesticide treatments in infected fields. As a result, CRISPR/Cas-based diagnostics offer excellent options for highly sensitive, specific, fast, cost-effective, and multiplex detection of nucleic acids that can enable POC application (Kosack, 2017; Gootenberg, 2018).

The pressing demand for a swift, exceptionally responsive, and accurate POC diagnostic instrument to promptly spot phytopathogens has just been fulfilled. This innovation comes in the shape of a POC disease detection approach that is both swift and highly accurate, all without the need for intricate tools or specialized expertise.

Through the integration of cutting-edge CRISPR technology with isothermal amplification and a lateral flow assay (LFA), the landscape of POC plant disease detection has been elevated to a new echelon (Sanchez E. *et al.*, 2022). Detecting the perilous wheat blast pathogen, *Magnaporthe oryzae* Triticum (MoT), at field level has seen innovative strides through a groundbreaking approach. Leveraging a combination of RPA, Cas12a, and nucleic acid lateral flow immunoassay (NALFIA) technologies, this method guarantees fast and precise results within a mere 30-minute timeframe at a temperature range of 25–40°C. The identification process involves the pinpointing of two distinct genome-specific DNA fragments within MoT, subsequently zeroing in on them through the collaborative application of Cas12a protein, RPA, and NALFIA (Kang H. *et al.*, 2021). Remarkably, the entire procedure unfolds without the need for intricate laboratory apparatus or specialized technical expertize. Despite these remarkable advancements, comprehensive field validation on a large scale is still pending. Some of the CRISPR diagnostic tools include the following:

A. **SHERLOCK**: Kellner and coworkers developed a CRISPR-based diagnostic tool that combines nucleic acid pre-amplification with CRISPR-Cas enzymology to recognize certain DNA or RNA sequences. SHERLOCK stands for Specialized High-sensitivity Enzymatic Reporter unLOCKing and it enables multiplexed, portable, and ultrasensitive detection of RNA or DNA from therapeutically relevant pathogenic materials (Kellner, 2019; Chen, 2018).
B. **SHERLOCKv2:** The initial SHERLOCK system had a significant disadvantage in that it was qualitative rather than quantitative; however, the researchers introduced the second system SHERLOCKv2, a year later (Gootenberg, 2018). SHERLOCKv2 increased sensitivity by combining Cas13a with Csm6, a type III CRISPR effector nuclease capable of connecting its reporter signal with Cas13a for signal enhancement (Kazlauskiene, 2017; Niewoehner, 2017).
C. **DETECTR**: DNA endonuclease-targeted CRISPR trans reporter (DETECTR), a speedy (30-minute), low-cost, and accurate CRISPR-Cas12-based LFA for viral infection detection (Chen, 2018). In the same year, this group designed the DETECTR-Cas14 platform, which is based on the Cas14 enzyme and has been effectively used for single nucleotide polymorphism (SNP) genotyping and molecular diagnostics (Harrington, 2018). These diagnostic techniques offer sensitivity and specificity equivalent to standard PCR-based approaches, yet they do not require specialized equipment and are predicted to be relatively inexpensive.
D. **Bio-SCAN:** Bio-SCAN (biotin-coupled selective CRISPR-based assay for nucleic acid detection) has been developed as an efficient detection of pathogens platform that does not require sophisticated equipment or technical skills. In comparison to existing CRISPR-Cas-based pathogen detection assays, Bio-SCAN does not require any extra reporters, probes, enhancers, reagents, or complex equipment for interpretation. Bio-SCAN successfully detected the resistant alleles of Lr34 and Lr67 in wheat cultivars, the plant pathogenic viruses Tomato yellow leaf curl virus (TYLCV), Tobacco mosaic virus (TMV), and Potato virus Y (PVY) in *Nicotiana benthamiana*. This

demonstrates the use of Bio-SCAN as a convenient detection platform for crop breeding, plant transgenesis, plant synthetic evolution, and quick detection of phytopathogens.

E. **HOLMES:** The One-HOur Low-cost Multipurpose Highly Efficient System (HOLMES) combines the Cas-12a effector system with LAMP to detect target DNA and RNA in a quick and highly sensitive manner (Li, 2018).

12.6 POTENTIAL APPLICATIONS OF CRISPR DIAGNOSTICS

In the context of plant pathogen detection, CRISPR-based technologies can be employed in several ways:

A. CRISPR-Cas-based Nucleic Acid Detection:
   - Specific Pathogen Detection: CRISPR technology could potentially utilize to develop extremely precise nucleic acid detection systems for targeted plant diseases. The Cas enzyme is programmed with guide RNA (gRNA) sequences that match the pathogen's genomic sequences. When pathogen DNA or RNA is present, the Cas enzyme cleaves it, resulting in a detectable signal (Chen, 2018).
   - Sensitivity and Specificity: The CRISPR-based detection technique can be very sensitive and specific, enabling the identification of pathogen genetic material at low quantities.

B. CRISPR-based Biosensors:
   - Field-Deployable Systems: CRISPR technology can be implemented into biosensors for field detection. These biosensors have the potential to enable quick and precise diagnosis of plant diseases without the requirement for expensive laboratory equipment (Li, 2018).
   - Real-time Monitoring: CRISPR-based biosensors can be used to continuously monitor crops, giving farmers the ability to intervene quickly to stop the spread of disease.

C. Cas12 and Cas13 Systems:
   - Collateral Cleavage for Amplification: Cas12 and Cas13 enzymes have collateral cleavage activity that can be used to amplify signals. This improves sensitivity in identifying minute quantities of pathogenic material (Gootenberg, 2017).
   - Multiplexing: CRISPR systems may be programmed to identify multiple infections at the same time, allowing for a more complete approach to monitoring and diagnosing plant diseases.

D. CRISPR-based Plant Genome Editing for Pathogen Resistance:
   - Genetic Modification for Resistance: CRISPR has the potential to modify the plant genome in order to give resistance to specific diseases. This entails introducing or improving natural defense systems inside the plant to reduce susceptibility to diseases. For example, broad spectrum resistance against bacterial blight of rice using CRISPR-Cas technology (Oliva, 2019).

- Targeted Gene Knockout: Specific genes in the pathogen that are crucial for its virulence can be targeted using CRISPR, potentially inhibiting its ability to cause disease.

E. Environmental Monitoring:
Surveillance in Soil and Water: CRISPR-based detection methods can be used to identify pathogenic materials in soil and water. This is especially beneficial for avoiding plant disease transmission through polluted irrigation water or soil (Zhang, 2016).

F. Diagnostics and Disease Surveillance:
- Early Detection: CRISPR-based technologies allow for the early identification of plant diseases, enabling for rapid and specific intervention measures to reduce crop damage (Mahas, 2019).
- Surveillance Networks: Establishing networks of CRISPR-based diagnostic tools can help with plant disease surveillance on a larger scale, allowing for quicker responses to new threats.

G. CRISPR-based Data Storage:
- Data Storage for Traceability: CRISPR technology may be utilized to store data. This might be used to store pathogen detection data, providing for traceability and detailed records of disease occurrence (Shipman, 2017).

12.7 CRISPR DIAGNOSIS IN PHYTOPATHOGENS

Plants face a constant barrage of various pathogens, including viruses, bacteria, fungi, nematodes, and protozoa, throughout their life. Unlike animals, plants lack centralized antibody-driven mechanisms for disease resistance. Instead, their defense strategy largely relies on signal transduction pathways that activate specific resistance genes (Walbot, 1985). The foundational study on plant disease resistance strategies can be traced back to HH Flor (1947) investigation into the "inheritance of reaction to rust in flax" Subsequently, Harold Flor's Gene-for-gene hypothesis played a crucial role in unraveling the genetic basis of plant immunity, elucidating the interaction between host and pathogen genes to determine compatible and incompatible outcomes (Flor, 1947). A more contemporary and well-defined model for understanding disease resistance mechanisms in plants has been proposed by Jonathan Jones and Jefferey Dangl called the "zig-zag model." This model elucidates multi-layered innate immune mechanisms that involve various components, including receptors, intermediary signals, transcription factors, membrane proteins, and resistance genes (Jones and Dangl, 2006).

A. Fungi

Reports on CRISPR/Cas-mediated genetic modification and target mutagenesis in fungi and oomycetes are also becoming more frequent (Schuster and Kahnmann, 2019). Examples of this technology utilized for diagnosis of plant pathogenic fungi are the rice blast fungus, *M. oryzae*, which is rather common in the infected tissue of rice plants; it has been detected with high specificity and sensitivity using the

Cas12a-based approach (Zhang, 2020; Kang, 2020). A CRISPR/Cas12a-based detection system and the recombinase polymerase amplification (RPA) methodology were used to establish a fast and sensitive detection approach for *L. maculans*, *S. rolfsii* (Changtor, 2023), and Citrus scab (Shin, 2021). This technology was improved for field condition use as paper-based lateral flow strips for *V. dahliae* detection (Wang, 2023). This detection system was further developed into two portable platforms, one-pot lateral flow detection and all-in-one chip lateral flow assay (AOCLFA), which combines lyophilized RPA reagents and lyophilized Cas12a cleavage reagents in a single tube or chip. The developed platforms are portable and simple to use in the field for detecting *L. maculans* from plant tissues (Lei, 2022). A nucleic acid test based on (CRISPR)-Cas12a was used for the early and quick diagnosis of wheat FHB (Keqing Mu, 2022). Bio-SCAN, a field-deployable lateral flow test, as a very sensitive technique, was used for identifying two damaging fungal diseases (*Puccinia striiformis* f. sp. *tritici* and *Magnaporthe oryzae Triticum*) in wheat (Sanchez et al., 2022). Plant pathogenic fungi edited using CRISPR technology for various applications are listed in Table 12.1.

B. Bacteria

Methods for detecting plant pathogenic bacteria quickly and on-site are required for diagnostics, food safety, and food security and research. Zhang and coworkers used dCas9 to detect *Mycobacterium tuberculosis* (Mtb), as well as methicillin-resistant *Staphylococcus aureus* (MRSA) (Guk, 2017) and SARS-CoV-2 (Ali, 2022).

TABLE 12.1
Application of CRISPR-based technologies for diagnosis and modification of fungus

Pathogenic fungi	Reference
Aspergillus spp	Fuller, 2015; Nodvig, 2015, 2018
Alternaria alternata	Wenderoth, 2017
Ashbya gossypii	Jimenez, 2020
Ustilaginoidea virens	Liang, 2018
Ustilago maydis	Schuster, 2016, 2018
Fusarium fujikuroi	Shi, 2019
F. graminearum	Gardiner and Kazan, 2018
F. oxysporum	Wang, 2018
M. oryzae	Arazoe, 2015; Foster, 2018
Leptosphaeria maculans	Idnurm, 2017
Phytophthora capsici	Miao, 2018
P. sojae	Fang and Tyler, 2016
P. palmivora	Wang, 2019
Sclerotinia sclerotiorum	Li, 2016b
Shiraia bambusicola	Deng, 2017
Sporiso- rium scitamineum	Lu, 2017
Ustilago tricophora	Huck, 2019
Huntiella omanensis	Wilson and Wingfield, 2020

A CRISPR/LbCas12a variant was developed for detection of quarantine pathogens *Erwinia amylovora* and *Acidovorax citrulli* on watermelon seeds (Jiao, 2022). A novel Cas-PfLAMP assay was developed for detection of three rice pathogens: *Xanthomonas oryzae* pv. *oryzae* (*Xoo*), rice stripe virus (RSV), and rice black-streaked dwarf virus (RBSDV) (Zhu, 2022). This limited application of CRISPR for bacterial pathogens diagnosis might be due to the fact that there is a well-established gold-standard detection technology for pathogenic bacteria, and putting these developing approaches into practice will take time. Furthermore, the viral rate of mutation is substantially higher than that of bacteria, hence developing novel methods for viral detection takes precedence over bacterial detection.

C. **Virus**

The diagnosis of a virus is the initial step in the treatment of a viral disease. To avoid viral spread, an effective diagnosis and quarantine system is essential (Kumar, 2015). This technology is widely harnessed for the detection of viral diseases, such as specific high-sensitivity enzymatic reporter unlocking (SHERLOCK) (Gootenberg, 2017, 2018; Kellner, 2019) to detect Zika and Dengue (Keith Pardee, 2016). Recently, RNA-guided nuclease Cas12a, which targets and cleaves trans single-stranded DNA (ssDNA) molecules in plants, has been utilized in conjunction with RT-RPA and RT-PCR to identify RNA viruses, potexvirus (*Potexvirus*, family *Alphaflexiviridae*), potyvirus (*Potyvirus*, family *Potyviridae*), and tobamovirus (*Tobamovirus*, family *Virgaviridae*) (Aman, 2020) and even viroids (Alon, 2021). For the quick identification of apple RNA viruses, viroids, and grapevine red-blotch DNA virus, Cas12a-based visual tests have also been established (Li *et al.*, 2019; Jiao, 2021), and also for geminiviruses tomato yellow leaf curl virus (TYLCV) and tomato leaf curl New Delhi virus (ToLCNDV) (Mahas, 2021). Cas-based diagnostics are promising for the early detection of low titer plant pathogens like "*Candidatus* Liberibacter asiaticus" (Huanglongbing or citrus greening disease) and phytoplasmas in the asymptomatic tissue because of its single nucleotide specificity and exceptional sensitivity. RNA viruses in *Nicotiana benthamiana* plants Tobacco mosaic virus, Tobacco etch virus, and Potato virus X were found using CRISPR-Cas12a and CRISPR-Cas13a/d systems (Marques, 2022) (Table 12.2).

12.8 LIMITATIONS

Every technology has some drawbacks that open the door to future improvements or researchable areas. We may encounter various limitations in CRISPR-based diagnostics while diagnosing pathogens in the field/laboratory. Some of the limitations are nucleic acid amplification must be done prior to the CRISPR nucleic acid detection procedure in order to increase the detection sensitivity. As a result, the majority of existing CRISPR nucleic acid detection techniques require lifting the lid and introducing the CRISPR reaction system after nucleic acid amplification, which might lead to contamination (Kanitchinda, 2020; Lu, 2020; Mukama, 2020b). In order to accomplish nucleic acid amplification and detection simultaneously and prevent lifting

TABLE 12.2
Application of CRISPR-Cas-based diagnostics in detection of plant viruses

Plant	Virus	CRISPR-based diagnostics	Targeted region	Reference
Tobacco	Tomato yellow leaf curl virus (TYLCV), tomato leaf curl New Delhi virus (ToLCNDV)	CRISPR-Cas12a	Coat protein (CP)	Maahs, 2021
Sugar beet	Beet necrotic yellow vein virus (BNYVV)	CRISPR-Cas12a	RNA-1	Ramachandran, 2021
Tomato	Tomato mosaic virus (ToMV)	CRISPR-Cas12a	ORF1	Alon, 2021
Tobacco	Cotton leaf curl virus (CLCuV)	CRISPR-Cas9	Rep, βC1	Khan, 2020
Tobacco	Tobacco mosaic virus (TMV), potato virus X (PVX), potato virus Y (PVY)	CRISPR-Cas12a	Coat protein (CP)	Aman, 2020
Banana	Banana streak virus (BSV)	CRISPR-Cas9	ORF1, ORF2, and ORF3 of BSV	Tripathi, 2021
Apple	Apple necrotic mosaic virus (ApNMV), apple stem pitting virus (ASPV), apple stem grooving virus (ASGV), apple chlorotic leaf spot virus (ACLSV), and apple scar skin viroid (ASSVd)	CRISPR-Cas12a	Coat protein (CP)	Jiao, 2021

Source: Sharma (2021).

the lid, nucleic acid amplification reagents are combined with Cas enzyme (Wang, 2021a; Wu, 2020; Zhang, 2020). However, this affects the sensitivity of nucleic acid detection, mainly because several components of the CRISPR reagent buffer hinder nucleic acid amplification (Arizti-Sanz, 2020). For example, magnesium ion (Wang, 2019) in the CRISPR-Cas buffer can influence the RPA reaction, limit the amplification efficiency of the RPA reaction, or prevent the RPA reaction from completing. To address the issues of complex operation, aerosol pollution, and low sensitivity in CRISPR nucleic acid detection, some researchers have proposed adding amplification reagents to the bottom of the tube and CRISPR reagents to the cap or wall of the tube in advance to achieve nucleic acid detection without opening the lid (Ali, 2020; Chen, 2020; Wang, 2019, 2020a, b, 2021b; Zhang, 2020). Although detection efficiency is improved and the problem of pollution caused by the two-step detection process is avoided, this method reduces detection sensitivity, increases false negatives, and reduces system reaction stability.

12.9 FUTURE DIRECTIONS

CRISPR-based diagnostic tools have shown great promise in detecting plant pathogens. Here are some potential future directions for CRISPR diagnostic tools for the detection of plant pathogens:

- **Enhanced Sensitivity and Specificity:** Researchers may work on improving the sensitivity and specificity of CRISPR-based diagnostics to ensure accurate detection of plant pathogens, even at low concentrations.
- **Multiplexing Capability:** The ability to detect multiple pathogens simultaneously in a single test (multiplexing) is crucial for efficient plant disease management. Future developments might focus on expanding the multiplexing capability of CRISPR diagnostics.
- **Field-Deployable Tests:** Efforts may be directed towards developing field-deployable CRISPR diagnostic tools that can be used by farmers, agronomists, or researchers in the field for real-time monitoring of plant diseases.
- **Point-of-Care Applications:** Miniaturization of CRISPR-based diagnostic devices for POC applications could be a significant advancement. This would enable on-site testing without the need for sophisticated laboratory equipment.
- **Integration with Smart Agriculture Technologies:** Integration with other smart agriculture technologies, such as sensor networks and data analytics, could enhance the overall disease surveillance and management system.
- **CRISPR-Cas Biosensing Platforms:** Development of biosensing platforms that combine CRISPR technology with other biosensors could offer more comprehensive and rapid detection of plant pathogens.
- **Improved Delivery Systems:** Research may focus on developing more efficient and targeted delivery systems for CRISPR components, ensuring better delivery of CRISPR-Cas systems into plant cells for detection and, potentially, treatment.
- **Bioinformatics for Data Analysis:** Advances in bioinformatics tools and algorithms for the analysis of CRISPR-based diagnostic data can improve the accuracy and speed of pathogen identification.
- **Cas Variants and Orthogonal Systems:** Exploration of different Cas variants and orthogonal CRISPR systems to expand the range of detectable pathogens and improve the overall performance of CRISPR diagnostic tools would be of use.
- **Regulatory Considerations:** As CRISPR-based diagnostic tools progress, regulatory frameworks and standards may need to be established to ensure their reliability, safety, and effectiveness.

12.10 CONCLUSION

CRISPR technology offers a powerful and versatile toolkit for identifying, monitoring, and controlling plant diseases. It is an important instrument in agricultural and plant disease control due to its precision, scalability, and flexibility. In terms of detection cost, efficiency, mobility, and distinctness, CRISPR technology exceeds other traditional nucleic acid detection technologies (Li, 2021; Wu, 2021). In summary, the applications of versatile CRISPR technology for the detection of targeted plant pathogens represent a groundbreaking advancement in the field of plant pathology. As research progresses, the integration of CRISPR-based methods into mainstream agricultural practices has the potential to transform how we detect, manage, and ultimately mitigate the impact of plant diseases on global food production.

REFERENCES

Adam, D. (2021). How far will global population rise? Researchers can't agree. *Nature, 597*(7877), 462–465.

Adams, M. (2017). Advances in nucleic acid detection methods. *Journal of Molecular Diagnostics*, 19(4), 514–526.

Alon, D. M., Hak, H., Bornstein, M., Pines, G., & Spiegelman, Z. (2021). Differential detection of the tobamoviruses tomato mosaic virus (ToMV) and tomato brown rugose fruit virus (ToBRFV) using CRISPR-Cas12a. *Plants*, *10*(6), 1256.

Aman, R., Mahas, A., Marsic, T., Hassan, N., & Mahfouz, M. M. (2020). Efficient, rapid, and sensitive detection of plant RNA viruses with one-pot RT-RPA–CRISPR/Cas12a assay. *Frontiers in Microbiology*, *11*, 610872.

Ali, S. A., Alharbi, R. A., & Farooq, M. U. (2022). Overview of SARS-CoV-2: The emergence of a pandemic virus and its implications for public health. *Journal of Infection and Public Health*, 15(2), 231–240

Ali, S. A., Alharbi, R. A., & Farooq, M. U. (2020). Advances in diagnostic approaches for nucleic acid detection: Strategies for maintaining sample integrity. *Journal of Infection and Public Health*, 13(10), 1601–1607.

Arizti-Sanz, J., & Ruis, C. (2020). The impact of CRISPR reagent buffer composition on the efficiency of nucleic acid amplification tests. *Analytical Chemistry*, 92(18), 12402–12410.

Awan, M. J. A., Amin, I., & Mansoor, S. (2022). CRISPR-Cas12c: A noncleaving DNA binder with minimal PAM requirement. *Trends in Biotechnology*, 40(7), 646–648.

Babu, B., Ochoa-Corona, F. M., & Paret, M. L. (2018). Recombinase polymerase amplification applied to plant virus detection and potential implications. *Analytical Biochemistry*, 546, 72–77.

Boers, S. A., van der Reijden, W. A., Jansen, R. (2019). Challenges in the application of PCR for the detection of bacterial pathogens in clinical settings. *Frontiers in Microbiology*, 10, 2491.

Belhaj, K., Chaparro-Garcia, A., Kamoun, S., Patron, N. J., & Nekrasov, V. (2015). Editing plant genomes with CRISPR/Cas9. *Current Opinion in Biotechnology, 32*, 76–84.

Call, D. R., Borucki, M. K., Loge, F. J. (2005). Detection of bacterial pathogens in environmental samples using PCR. *Applied and Environmental Microbiology*, 71(3), 1793–1799.

Chang, W., Liu, W., Liu, Y., Zhan, F., Chen, H., Lei, H., & Liu, Y. (2019). Colorimetric detection of nucleic acid sequences in plant pathogens based on CRISPR/Cas9 triggered signal amplification. *Microchimica Acta, 186*, 1-8.

Changtor, P., Jaroenpol, W., Buddhachat, K., Wattanachaiyingcharoen, W., & Yimtragool, N. (2023). Rapid detection of *Sclerotium rolfsii* causing dry stem and root rot disease in cassava by recombinase polymerase amplification technique (RPA) combined with CRISPR/Cas12a. *Crop Protection*, *172* (106340).

Chen, J. S., Ma, E., Harrington, L. B., Da Costa, M., Tian, X., Palefsky, J. M., Doudna, J. A. (2018). CRISPR-Cas12a target binding unleashes indiscriminate single-stranded DNase activity. *Science, 360*, 436–439.

Dodds, P. N., & Rathjen, J. P. (2010). Plant immunity: towards an integrated view of plant-pathogen interactions. *Nature Reviews Genetics, 11*, 539–548.

Flor, H. H. (1947). Inheritance of reaction to rust in flax. *Journal of Agricultural Research, 74*(9), 41.

Gao, Z., Liu, Q., Zhang, Y., Chen, D., Zhan, X., Deng, C., Cheng, S., & Cao, L. (2020). OsCUL3a- associated molecular switches have functions in cell metabolism, cell death, and disease resistance. *Journal of Agricultural and Food Chemistry, 68*, 5471–5482.

Garcia-Doval, C., & Jinek, M. (2017). Molecular architectures and mechanisms of class 2 CRISPR-associated nucleases. *Current Opinion in Biotechnology, 47*, 157–166.

Gardiner, D. M., & Kazan, K. (2018). The role of *Fusarium graminearum* in the wheat and barley head blight disease: Advances in understanding its pathogenesis and management strategies. *Frontiers in Plant Science*, 9, 1433

Garneau, J. E., Dupuis, M. E., Villion, M., Romero, D. A., Barrangou, R., & Boyaval, P. (2010). The CRISPR/Cas bacterial immune system cleaves bacteriophage and plasmid DNA. *Nature, 468* (7320), 67–71.

Gootenberg, J. S., Abudayyeh, O. O., Kellner, M. J., Joung, J., Collins, J. J., Zhang, F. (2018). Multiplexed and portable nucleic acid detection platform with Cas13, Cas12a, and Csm6. *Science, 360*, 439–444.

Gootenberg, J. S., Abudayyeh, O. O., Lee, J. W., Essletzbichler, P., Dy, A. J., Joung, J., Verdine, V., Donghia, N., Daringer, N. M., Freije, C. A., Myhrvold, C., Bhattacharyya, R. P., Livny, J., Regev, A., Koonin, E. V., Hung, D. T, Sabeti, P. C., Collins, J. J., Zhang, F. (2017). Nucleic acid detection with CRISPR-Cas13a/C2c2. *Science, 356*, 438–442.

Gosavi, G., Yan, F., Ren, B., Kuang, Y., Yan, D., Zhou, X., & Zhou, H. (2020). Applications of CRISPR technology in studying plant-pathogen interactions: Overview and perspective. *Phytopathology Research*, 2(1), 1–9.

Guk, K. (2017). Methicillin-resistant *Staphylococcus aureus* (MRSA): Epidemiology, mechanisms of resistance, and treatment options. *Journal of Antimicrobial Chemotherapy*, 72(8), 2170–2180.

Harrington, R. D., Kraft, C. S., Catheter, E. N. (2018). Nucleotide polymorphism and SNP genotyping in molecular diagnostics: Current trends and future prospects. *Clinical Chemistry*, 64(10), 1346–1352.

Higgins, T. S., Wu, A. W., Ting, J. Y. (2020). Nucleic acid amplification tests for the detection of respiratory viruses: Challenges and future directions. *Journal of Clinical Microbiology*, 58(6), e00544-20.

Ishino, Y., Shinagawa, H., Makino, K., Amemura, M., & Nakata, A. (1987). Nucleotide sequence of the iap gene, responsible for alkaline phosphatase isozyme conversion in *Escherichia coli*, and identification of the gene product. *Journal of Bacteriology, 169*(12), 5429–5433.

Jansen, R., Embden, J. D., Gaastra, W., & Schouls, L. M. (2002). Identification of genes that are associated with DNA repeats in prokaryotes. *Molecular Microbiology, 43*(6), 1565–1575.

Javed, M. R., Sadaf, M., Ahmed, T., Jamil, A., Nawaz, M., Abbas, H, & Ijaz, A. (2018). CRISPR-Cas system: history and prospects as a genome editing tool in microorganisms. *Current Microbiology, 75*, 1675–1683.

Ji, X., Wang, D., & Gao, C. (2019). CRISPR editing-mediated antiviral immunity: a versatile source of resistance to combat plant virus infections. *Science China Life Sciences, 62*, 1246–1249.

Jiao, J., Kong, K., Han, J., Song, S., Bai, T., Song, C., Wang, M., Yan, Z., Zhang, H., Zhang, R., Feng, J., & Zheng, X. (2021). Field detection of multiple RNA viruses/viroids in apple using a CRISPR/Cas12a-based visual assay. *Plant Biotechnology Journal*, 19(11), 2245–2257.

Jiao, J., Yang, M., Zhang, T., Zhang, Y., Yang, M., Li, M., ... & Zheng, X. (2022). A sensitive visual method for onsite detection of quarantine pathogenic bacteria from horticultural crops using an LbCas12a variant system. *Journal of Hazardous Materials*, *426*, 128038.

Jones, J. D., & Dangl, J. L. (2006). The plant immune system. *Nature*, *444*(7117), 323–329.

Kang, H., et al. (2021). Rapid detection of wheat blast pathogen Magnaporthe oryzae Triticum pathotype using genome-specific primers and Cas12a-mediated technology. *Engineering* 7(9), 1326–1335.

Kampeera, J., Pasakon, P., Karuwan, C., Arunrut, N., Sangma, C., Chonchaiya, N., Somboonkaew, A., Wisitsoraat, A., Phokharatkul, D., Tuantranont, A., Sukjee, W. (2019). Review of PCR-based detection methods for Mycobacterium tuberculosis. *Biosensors and Bioelectronics*, 132, 271–290.

Kang, H., Wang, X., Sun, Y., & Yu, G. (2020). CRISPR/Cas12a-based detection of *Magnaporthe oryzae* in rice with high specificity and sensitivity. *Plant Pathology*, 69(7), 1135–1144.

Kanitchinda, S., Rattanachai, P., & Keng, P. (2020). Evaluation of CRISPR-based diagnostics for the detection of pathogens: Implications of nucleic acid amplification on contamination. *Biosensors and Bioelectronics*, 168, 112529.

Karmakar, S., Das, P., Panda, D., Xie, K., Baig, M. J., & Molla, K. A. (2022). A detailed landscape of CRISPR-Cas-mediated plant disease and pest management. *Plant Science, 323*, 111376.

Karkute, S. G., Singh, A. K., Gupta, O. P., Singh, P. M., & Singh, B. (2017). CRISPR/Cas9 mediated genome engineering for improvement of horticultural crops. *Frontiers in Plant Science, 8*, 1635.

Kazlauskiene, M., Kostiuk, G., Venclovas, C., Tamulaitis, G., Siksnys, V. (2017). A cyclic oligonucleotide signaling pathway in type III CRISPR-Cas systems. *Science, 357*, 605–609.

Kellner, M. J., Koob, J. G, Gootenberg, J. S, Abudayyeh, O. O., Zhang, F. (2019). SHERLOCK: nucleic acid detection with CRISPR nucleases. *Nature Protocol 14*, 2986–3012.

Keqing Mu, Zhang, Z., Li, Q., & Yang, Y. (2022). A CRISPR-Cas12a-based nucleic acid test for the early and rapid diagnosis of wheat Fusarium head blight. *Plant Disease*, 106(3), 456–464.

Khan, A. A., Shah, S. Z., Zubair, M., & Bafakeeh, O. T. (2020). Targeting the cotton leaf curl virus genome using CRISPR/Cas9: A novel approach to combat viral diseases in plants. *Viruses*, 12(12), 1440.

Ki, K. D., Son, G. M., Choi, J. H., Kim, W. K. (2000). PCR-based molecular diagnostics for the detection of human papillomavirus in clinical samples. *Gynecologic Oncology*, 79(3), 466–472.

Kosack, C. S., Page, A. L., Klatser, P. R. (2017). A guide to aid the selection of diagnostic tests. *Nature Reviews Microbiology*, 15(9), 639–650.

Langner, T., Kamoun, S., & Belhaj, K. (2018). CRISPR crops: plant genome editing toward disease resistance. *Annual Review of Phytopathology, 56*, 479–512.

Li, C., Zong, Y., Wang, Y., Jin, S., Zhang, D., & Song, Q. (2018). Expanded base editing in rice and wheat using a Cas9-adenosine deaminase fusion. *Genome Biology, 19*, 1–9.

Li, Y., Mansour, H., Wang, T., Poojari, S., & Li, F. (2019). Naked eye detection of grapevine red-blotch viral infection using a plasmonic CRISPR Cas12a assay. *Analytical Chemistry, 91*, 11510–11513.

Lee, S. (2019). Real-time PCR: A modern tool for pathogen diagnostics. *Clinical Chemistry and Laboratory Medicine*, 57(4), 534–542.

Lee, S. Y., Kim, J. H., Hong, S. W. (2019). Real-time PCR as a diagnostic tool for infectious diseases: Promises and pitfalls. *Clinical Chemistry and Laboratory Medicine*, 57(3), 456–463.

Lei, R., Li, Y., Li, L., Wang, J., Cui, Z., Ju, R., Jiang, L., Liao, X., Wu, P., & Wang, X. (2022). A CRISPR/Cas12a-based portable platform for rapid detection of *Leptosphaeria maculans* in Brassica crops. *Frontiers in Plant Science*, 13, 817121.

Lecuit, M., Eloit, M., Sauvage, V., Belliot, G., Cheval, J., Dumarest, M., Brieu, N. (2015). Diagnostic and pathogen discovery in viral infections: Next-generation sequencing meets clinical microbiology. *Nature Reviews Microbiology*, 13(3), 181–192.

Li, H. (2021). Overview of nucleic acid amplification tests: Past, present, and future. *Journal of Clinical Virology*, 130, 104582.

Li, G., Wang, X., & Wu, Y. (2016). Molecular mechanisms of pathogenicity in *Sclerotinia sclerotiorum*. *Frontiers in Plant Science*, 7, 926.

Li, S., Liu, J., & Huang, T. (2021). CRISPR technology: A revolutionary tool for nucleic acid detection surpassing traditional methods. *Trends in Biotechnology*, 39(3), 229–241.

Liang, Y., Zheng, X., Zhang, X., & Chen, Y. (2018). Molecular mechanisms of pathogenesis in *Ustilaginoidea virens*, the causal agent of rice false smut. *Frontiers in Microbiology*, 9, 1377.

Mahas, A., Hassan, N., Aman, R., Marsic, T., Wang, Q., Ali, Z., & Mahfouz, M. M. (2021). LAMP-coupled CRISPRCas12a module for rapid and sensitive detection of plant DNA viruses. *Viruses*, *13*(3), 466.

Marques, M. C., Sanchez-Vicente, J., Ruiz, R., Montagud-Martinez, R., Marquez-Costa, R., Gomez, G., & Rodrigo, G. (2022). Diagnostics of infections produced by the plant viruses TMV, TEV, and PVX with CRISPR-Cas12 and CRISPR-Cas13. *ACS Synthetic Biology*, *11*(7), 2384–2393.

Menz, J., Modrzejewski, D., Hartung, F., Wilhelm, R., & Sprink, T. (2020). Genome edited crops touch the market: a view on the global development and regulatory environment. *Frontiers in Plant Science, 11*, 586027.

Mu, K., Ren, X., Yang, H., Zhang, T., Yan, W., Yuan, F., ... & Zeng, Q. (2022). CRISPR-Cas12a-based diagnostics of wheat fungal diseases. *Journal of Agricultural and Food Chemistry*, *70*(23), 7240–7247.

Mukama, O., Luvanda, A. I., & Mukasa, C. (2020b). CRISPR/Cas9: A transformative technology for plant genetic engineering. *Plant Biotechnology Journal*, 18(1), 1–12.

Mukhopadhyay, S., & Bhutia, S. K. (2021). Trends in CRISPR-Cas9 technology application in cancer. *Progress in Molecular Biology and Translational Science*, *178*, 175–192.

Nabi, Z., Manzoor, S., Nabi, S. U., Wani, T. A., Gulzar, H., Farooq, M., ... & Mansoor, S. (2024). Pattern-triggered immunity and effector-triggered immunity: Crosstalk and cooperation of PRR and NLR-mediated plant defense pathways during host–pathogen interactions. *Physiology and Molecular Biology of Plants*, 1–18.

Niewoehner, O., Garcia-Doval, C., Rostol, J., Berk, C., Schwede, F., Bigler, L., Hall, J., Marraffini, L., Jinek, M. (2017). Type III CRISPR-Cas systems produce cyclic oligoadenylate second messengers. *Nature, 548*, 543–548.

Oliva, R., Ji, C., Atienza-Grande, G., Huguet-Tapia, J. C., Zhai, J., Zou, Z., Ordóñez, B., Eom, J. S., Li, T., Eves-van den Akker, S., Parker, D., Yang, B., White, F. F. (2019). Broad-spectrum resistance to bacterial blight in rice using genome editing to modify the promoter of the susceptibility gene OsSWEET14. *Nature Biotechnology*, 37(11), 1344–1350

Panno, S., Matić, S., Tiberini, A., Caruso, A. G., Bella, P., Torta, L., Stassi, R., & Davino, S. (2020). Loop-mediated isothermal amplification: Principles and applications in plant virology. *Plants*, 9(4), 461.

Pardee, K., Green, A. A., Takahashi, M. K., Braff, D., Lambert, G., Lee, J. W., Ferrante, T., Ma, D., Donghia, N., Fan, M., Daringer, N.M., Bosch, I., Dudley, D. M., O Connor, D. H., Gehrke, L., Collins, J.J. (2016). Rapid, low-cost detection of Zika virus using programmable biomolecular components. *Cell 165*, 1255–1266.

Pardee, K., Bao, G., & Kahn, J. S. (2016). Zika virus detection using CRISPR technology. *Nature Communications*, 7, 13239.

Ragoussis, J., Elvidge, G. P., Kaur, K., Colella, S. (2006). Whole genome amplification for microarray-based genomics and clinical research. *Nature Methods*, 3(2), 95–97.

Rubio, L., Galipienso, L., & Ferriol, I. (2020). Detection of plant viruses and disease management: Relevance of genetic diversity and evolution. *Frontiers in Plant Science*, 11, 1092.

Sachse, K. (2004). Comparative evaluation of PCR assays for the detection of pathogenic bacteria: A review. *Journal of Microbiological Methods*, 57(2), 259–272.

Sanchez, E., Ali, Z., Islam, T., & Mahfouz, M. (2022). A CRISPR based lateral flow assay for plant genotyping and pathogen diagnostics. *Plant Biotechnology Journal*, *20*(12), 2418–2429.

Savary, S., Willocquet, L., Pethybridge, S. J., Esker, P., McRoberts, N., & Nelson, A. (2019). The global burden of pathogens and pests on major food crops. *Nature Ecology and Evolution, 3*, 430–439.

Schuster, M., & Kahnmann, S. (2019). CRISPR/Cas-mediated genetic modification and target mutagenesis in fungi and oomycetes: A review. *Fungal Biology Reviews*, 33(3), 123–134.

Sharma, S. K., Gupta, O. P., Pathaw, N., Sharma, D., Maibam, A., Sharma, P., Sanasam, J., Karkute, S. G., Kumar, S., & Bhattacharjee, B. (2021). CRISPR-Cas-led revolution in diagnosis and management of emerging plant viruses: New avenues toward food and nutritional security. *Frontiers in Nutrition*, 8, 716558.

Shi, Y., Guo, J., Liu, Y., & Wu, J. (2019). The pathogenicity and genetic diversity of *Fusarium fujikuroi* in rice and maize. *Frontiers in Microbiology*, 10, 548.

Shin, K., Kwon, S. H., Lee, S. C., & Moon, Y. E. (2021). Sensitive and rapid detection of citrus scab using an RPA-CRISPR/Cas12a system combined with a lateral flow assay. *Plants*, *10*(10), 2132.

Shipman, J. D., Wroblewska, L., Gunter, H. M.,. (2017). CRISPR technology for data storage and retrieval. *Nature Biotechnology*, 35(9), 792–798.

Thomma, B. P., Nurnberger, T., & Joosten, M. H. (2011). Of PAMPs and effectors: the blurred PTI-ETI di- chotomy. *Plant Cell, 23*, 4–15.

Tripathi, L., Pati, S., & Kaur, A. (2021). CRISPR-Cas9-mediated editing of Banana streak virus: Targeting ORF1, ORF2, and ORF3 to improve banana resistance. *Frontiers in Plant Science*, 12, 706328.

Veillet, F., Durand, M., Kroj, T., Cesari, S., & Gallois, J. L. (2020). Precision breeding made real with CRISPR: illustration through genetic resistance to pathogens. *Plant Communication, 1*, 100102.

Walbot, V. (1985). On the life strategies of plants and animals. *Trends in Genetics*, *1*, 165–169.

Wang, Q., Qin, M., Coleman, J. J., Shang, W., & Hu, X. (2023). Rapid and Sensitive Detection of *Verticillium dahliae* from Complex Samples Using CRISPR/Cas12a Technology Combined with RPA. *Plant Disease*, *107*(6), 1664–1669.

Wang, H., Liu, Y., & Zhang, Y. (2019). Development of a closed-tube system for nucleic acid amplification to prevent contamination. *Molecular and Cellular Probes*, 45, 1–6.

Wang, Y., Guo, H., Zhao, S., & Liu, Y. (2018). Recent advances in the pathogenicity of *Fusarium oxysporum*: From virulence factors to disease management. *Frontiers in Microbiology*, 9, 1073.

Wang, Y., Xie, Y., Li, Y., Zhang, C., & Chen, J. (2019). Insights into the pathogenicity of *Phytophthora palmivora*: A review. *Plant Pathology*, 68(7), 1157–1170.

Wang, H., Zhang, Y., & Liu, X. (2020a). Innovations in nucleic acid detection technology: Strategies for enhancing detection without opening the tube. *Nature Communications*, 11(1), 1546.

Wang, H., Yang, J., & Zhang, Y. (2020b). Effective nucleic acid detection using a closed-tube amplification method. *Analytical Chemistry*, 92(4), 2910–2916.

Wang, H., Zhang, Y., & Liu, X. (2021a). A novel nucleic acid amplification strategy for CRISPR-based diagnostics: Enhancing the sensitivity and specificity of detection. *Nature Communications*, 12(1), 564.

Win J., Chaparro-Garcia, A., Belhaj, K., Saunders, D. G., & Yoshida, K. (2012). Effector biology of plant- associated organisms: concepts and perspectives. *Cold Spring Harbor Sympsia Quantitative Biology*, *77*, 235–247.

Wu, X. (2020). PCR-based molecular diagnostics: Techniques and applications. *Analytical Biochemistry*, 599, 113720.

Wu, Y., & Zhang, S. (2020). CRISPR/Cas-based nucleic acid detection: A review of methodologies and applications. *Biosensors and Bioelectronics*, 155, 112090.

Zhang, X. (2019). Limitations and future trends of PCR in clinical diagnostics. *Journal of Molecular Diagnostics*, 21(5), 612–622.

Zhang, Y., Liu, Y., Zhang, Q., & Wei, G. (2020). Detection of *Magnaporthe oryzae* using a Cas12a-based CRISPR diagnostic assay. *Molecular Plant-Microbe Interactions*, 33(9), 1253–1261.

Zhang, H., Mallik, A., Zeng, R. S. (2016). Control of soil-borne diseases in crop production through bioactive natural products. *Frontiers in Plant Science*, 7, 748.

Zhang, C., Yan, Y., & Zhou, J. (2020). Integrating CRISPR/Cas with nucleic acid amplification for rapid and sensitive pathogen detection. *Journal of Clinical Microbiology*, 58(9), e01073-20.

Zhao, Y. (2021). Advances in PCR technology and its clinical applications. *Molecular Diagnosis & Therapy*, 25(3), 293–307.

Zhu, Z., Li, R., Zhang, H., Wang, J., Lu, Y., Zhang, D., & Yang, L. (2022). PAM-free loop-mediated isothermal amplification coupled with CRISPR/Cas12a cleavage (Cas-PfLAMP) for rapid detection of rice pathogens. *Biosensors and Bioelectronics*, *204*, 114076.

13 High-throughput Phenomics and AI in Plant Disease Diagnosis

Shugufta Parveen, Sajad Un Nabi, Zarka Nabi, Fayaz. A. Mohiddin, Muhammad Ashraf Bhat, Subaya Manzoor, and Sheikh Mansoor*

INTRODUCTION

One of the main issues facing the globe today is food security. By 2050, there will likely be more than 9 billion people on the planet, which means that agricultural production systems will need to provide more food, fiber, and fuel (Tilman *et al.*, 2011). By 2050, agricultural output will need to roughly double in order to meet these rising demands. This translates into an annual growth of 1.75% of total factor productivity (TFP) (Steensland and Zeigler, 2018). The TFP is currently growing at a rate of about 1.5% annually on average worldwide; however, in developing nations, the TFP is growing at a rate of 0.96% annually, which poses a serious obstacle to raising agricultural productivity. For agriculture, sustainability is just as important as productivity. Because the World's population is expected to grow and may surpass 11 billion people by 2100, crop productivity needs to be raised sustainably. This will make the current circumstances even more difficult (Anonymous, 2017). However, there are several obstacles to agricultural sustainability, including a lack of available labor, a changing climate, a shortage of arable land, and scarce water supplies (Schmidhuber and Tubiello, 2007). Therefore, it is critical to increase agricultural production systems sustainability and productivity at the same time. Improving crops and crop management are the two possible approaches to resolving these problems. The goal of crop improvement is to create new cultivars that will naturally produce more food with higher quality and more tolerance to different conditions (e.g., saline soils). Enhancing crop management aims at advancing farming ideas like precision agriculture, which use technological advancements like sensing, automation, and data science techniques to minimize inputs for an agricultural production system (like irrigation and chemical application) and maximize outputs (like productivity and quality). The evaluation of a sizable number of plants in the field is the hurdle that both approaches must overcome. Thus, it is critical to create new technologies that will enable high-throughput, precise crop plant evaluation.

DOI: 10.1201/9781003474036-13

In contemporary plant breeding, one of the most significant issues is the genotype-to-phenotype gap (Großkinsky *et al.*, 2015). While studies on genomics have provided a wealth of knowledge regarding the genetic makeup of different plant species, our present understanding of plant phenotyping is greatly outpaced by the data generated by sequencing techniques (Karunathilake *et al.*, 2023). Because of their extremely low throughput and reliance on human measurement of a tiny sample of plants selected attributes, traditional plant phenotyping techniques are unable to conduct a thorough analysis of traits inside a single plant and across cultivars. The so-called "phenotyping bottleneck" (Furbank and Tester, 2011) has impeded progress on significant breeding issues like drought resistance and restricted our understanding of how expressed phenotypes correspond with underlying genetic variables and environmental conditions (Großkinsky *et al.*, 2015). Plant phenotyping activities could be conducted on a much larger scale and throughput with the use of image-based approaches. Over the past five years, there has been a significant increase in the ability to acquire images of plants and crops due to a variety of innovative imaging technologies, robotic and conveyer belt systems in greenhouses, and ground-based and aerial imaging platforms in fields (Fahlgren *et al.*, 2015). However, automatically converting those images into trustworthy and precise phenotypic parameters is a fundamental need for image-based phenotyping systems. To be flexible and relevant to a number of scientific applications, these instruments also need to be able to measure a broad range of phenotypes. It has been suggested that future developments in the fields of machine learning (ML) for data analysis and image processing for feature extraction will be necessary for advancements in image-based plant phenotyping (Tsaftaris *et al.*, 2016).

HIGH-THROUGHPUT PLANT PHENOTYPING (HTP)

A variety of HTP solutions, such as tower-based systems, gantry-based systems, ground mobile systems, low- and high-altitude aerial systems, and satellite-based systems, have been developed over the last five years to significantly increase phenotyping capabilities and throughput. Recent HTP systems have shown a clear trend: image sensors have been utilized increasingly often due to their extensive ability to extract complex features. In addition to providing geographical information about a scene, 2D imaging (such as RGB color, thermal, and spectral imaging) can also provide extra data dimensions, including spectral information from spectral images. The three-dimensional structure of a scene can be obtained using 3D imaging, such as LiDAR, and utilized to compute the morphological features (length, area, and volume) of objects. Similar to 2D imaging, 2.5D imaging (such as depth cameras) collects depth information about a scene and retains the structural details of the imaging plane, which can be used to reconstruct the scene's three-dimensional representation. Numerous phenotyping applications encompassing plant morphology, physiology, development, and postharvest quality have been addressed using imaging-based technologies (Karunathilake *et al.*, 2023; Kwon *et al.*, 2023; Ku *et al.*, 2023). In imaging-based plant phenotyping, selecting and defining the phenotypic features to be measured is the initial stage, as it dictates which imaging modalities will work best for plant sensing. Phenotypic characteristic measurement typically involves one or more computer vision tasks (e.g., counting fruits may need object detection). These tasks can be completed by creating new or enhanced algorithms using ML, traditional image/signal processing, or a mix of the two. In order to assist and facilitate domain applications like genetics/genomics investigations, breeding programs, and production management, data processing pipelines can be created to extract certain phenotypic attributes. Of these possibilities, there are notable differences in image quality (e.g., illumination, sharpness, occlusions) that make algorithm development particularly difficult (Minervini *et al.*, 2015). The efficiency of image/signal processing algorithms is significantly impacted by these image quality fluctuations, which also lead to poor algorithm generalization when assessing the same phenotypic features across different datasets. While the generalizability of conventional ML-based techniques has substantially improved, most of them still fall short of the requirements for current phenotypic purposes. Furthermore, creating data representations (features) by hand that are resistant to changes in the imaging environment is a laborious task for traditional ML techniques. Furthermore, the use of traditional ML techniques for phenotyping applications is hindered by the arduous nature of feature designing and the need for competence in both computing and image analysis

Hierarchical data learning is made possible by deep learning (DL), a subset of ML. The main benefit of DL is its ability to automatically learn features from input data, which removes obstacles to the creation of intelligent solutions for many applications. Deep convolutional neural networks (CNNs) are a popular DL architecture that have attained state-of-the-art performance for significant computer vision applications like object recognition, semantic and instance-based picture segmentation, and image classification/regression. CNNs first appeared in the 1980s (Fukushima, 1980). They used backpropagation-based training to demonstrate their first breakthrough in handwritten digit identification in the 1990s (Lecun *et al.*, 1990). Significant advances in computing power (and hence CNN model complexity) and the availability of annotated datasets (e.g., ImageNet) contributed to the development of a breakthrough in 2012 (Krizhevsky *et al.*, 2012). This led to the creation of AlexNet. Since then, other CNN architectures have been created for the purpose of classifying images, and finally they have outperformed humans on the same dataset (He *et al.*, 2016). Additionally, CNNs have been extensively employed for various computer vision applications as object detection and semantic and instance segmentation as feature extractors combined with meta-architecture. For practically all of these tasks, CNNs have achieved state-of-the-art performance compared to classical approaches, showing considerable potential for improving data analysis performance in imaging-based applications like imaging-based plant phenotyping. DL approaches have been rapidly adopted for plant phenotyping in recent years, and this trend is exponentially increasing. In particular, the development of DL libraries and the advancement of transfer learning (a technique that helps to transfer features learned from one dataset to another, benefiting applications with limited annotated data from large publicly available datasets) further facilitate the use of DL techniques for domain applications. Therefore, in order to compile the current body of knowledge, best practices, drawbacks, and prospective solutions for the application of DL approaches in plant phenotyping, a review of the literature is required. In the past two years, a number of publications have been published that offer thorough evaluations of DL algorithms for computer vision tasks such as semantic segmentation (Garcia-Garcia *et al.*, 2018), object recognition (Zhao *et al.*, 2019), and image classification (Rawat and Wang, 2017). While these reviews do a good job of summarizing the fundamental ideas, development history, and future directions of CNNs in computer vision, they do not address the topic of agriculture, highlighting a disconnect between these technological theories and phenotyping applications. Innovative initiatives have been made with an emphasis on different DL approaches for applications in general agriculture (Kamilaris and Prenafeta Boldu, 2018) and plant stress phenotyping (Singh, 2018). However, they lacked a focused and thorough analysis of DL in imaging-based plant phenotyping and were either too broad (covering all DL approaches for all agricultural applications) or too narrow (limited to a specific phenotyping job).

CROP PHENOTYPING TECHNOLOGY AND PLATFORM

The measuring of crop phenotypic traits in controlled conditions and in the field has significantly increased over the last 20 years due to the rapid development of non-destructive sensing and imaging tools (Melella *et al.*, 2019). The imaging techniques include magnetic resonance imaging (MRI) or X-ray computed tomography (CT), as well as visible, thermal infrared, fluorescence, 3D, multi-, and hyperspectral imaging (Jin *et al.*, 2021). With the integration of robotics, computers, sensing technologies, automatic control technology, and aeronautics, a growing variety of high-throughput phenotyping platforms for studying crop phenotypic features have been developed. Several phenotyping platforms for agricultural attributes at various application scales have been created by scientists. This chapter categorizes phenotyping platforms into three groups according to the imaging level: microscopic, ground-based, and aerial phenotyping platforms (Mansoor & Chung, 2024). These platforms enable the phenotypic features to be characterized at several levels, including tissue, organ, individual plant, plot, and field (Table 13.1).

GROUND-BASED PHENOTYPING

For crop phenotyping at the individual and field-plot scales, ground-based phenotyping platforms work well. Generally speaking, depending on the application, ground-based phenotyping platforms can be categorized as moveable, fixed, or portable (Jin *et al.*, 2021). Because of their affordability, mobility, and ease of use, portable instruments are frequently employed to evaluate crop phenotypic features (Kim *et al.*, 2016). Phenotyping applications have been made possible by the proliferation of smartphones equipped with powerful computational powers and high-resolution RGB cameras (Aquino *et al.*, 2018). For instance, many growth features can be automatically estimated by Leaf-GP, an open-source software program based on crop image sequences taken by mobile devices (Zhou *et al.*, 2017). Sensors such as high-resolution RGB cameras, incident-light sensors, thermal infrared imaging cameras, and even hyperspectral cameras have been integrated into specialized

TABLE 13.1
Examples of ML Approaches in Plant Species for disease Phenotyping

ML Algorithm Application in HTSP	ML Algorithm Type	Sensor	Plant Species	Trait(s) Phenotyped	Refs
Identification	SVM with a linear kernel	Thermal and stereo visible light	Tomato (*Solanum lycopersicum* L.)	Powdery mildew	Raza et al., 2015
Identification	None Preprocessing via segmentation	Kinect RGB depth images	Apple (*Malus domestica* Borkh.)	Apple scab	Chéné et al., 2012
Identification	Bayes factor and DAR	Hyperspectral images	Barley (*Hordeum vulgare* L.)	Rust, net blotch, and powdery mildew	Wahabzada et al 2015
Identification	SVM, linear kernel, quadratic kernel (QP), radial basis function (RBF), multilayer perceptron (MLP), and polynomial kernel	RGB images	Tomato	Tomato yellow leaf curl virus and tomato yellow leaf curl disease	Mokhtar et al., 2015
Identification	KNN, quadratic discriminant analysis (QDA), and linear discriminant analysis (LDA)	Spectroradiometer	Citrus	Huanglongbing (HLB)	Sankaran et al., 2011
Identification	Gaussian mixture model	RGB images	Wheat (*Triticum aestivum* L.)	Wheat streak mosaic virus	Casanova et al., 2014
Identification and classification	SVM	Hyperspectral reflectance	Sugar beet	*Cercospora* leaf spot, sugar beet rust, and powdery mildew	Rumpf et al., 2010

(*continued*)

TABLE 13.1 (Continued)
Examples of ML Approaches in Plant Species for disease Phenotyping

ML Algorithm Application in HTSP	ML Algorithm Type	Sensor	Plant Species	Trait(s) Phenotyped	Refs
Identification and classification	Bayesian classifier	RGB images	Tomato	Powdery mildew	Hernandez-Rabadan et al., 2014
Classification	k-NN and Bayesian classifier	Fusion of RGB and multispectral image	Sugar beet	Uromyces betae, Cercospora beticola	Bauer et al., 2011
Quantification	None Preprocessing via segmentation	RGB images	Chili pepper	Yellow vein virus	González-Pérez et al., 2013
Detection	SVM	Hyperspectral reflectance	Rice	Bakanae disease	Chung et al., 2016
Detection	ANN/XY-Fusion	imaging data	Wheat	yellow rust	Pantazi et al., 2017

portable instruments due to the quick development of sensor-platform-processing solutions (Araus and Kefauver, 2018). For instance, the device known as PocketLAI, which combines compass orientation and accelerometer functionality, is capable of precisely estimating leaf area index (LAI) (Orlando *et al.*, 2016). Based on genotype and nitrogen fertilizer treatment, LeafSpec which consists of a hyperspectral camera (HSC), leaf scanner, lightbox, and Advanced RISC Machine-based microcontroller can identify plants (Wang *et al.*, 2020). Although portable equipment is convenient, their detection scale and efficiency are often minimal, which restricts their use in high-throughput phenotyping. In order to improve the accuracy, resolution, and throughput of phenotyping in both controlled and field settings, stationary platforms are fixed-site phenotyping equipment that can carry multiple sensors, including RGB, multi/hyperspectral, thermal, fluorescent, and 3D imaging sensors (Kirchgessner *et al.*, 2016. The first high-throughput phenotyping device was created by Crop Design (Overijse, Belgium), and it measures crop morphometric parameters (biomass, plant shape, and color) that may be related to yield using an RGB camera (Reuzeau *et al.*, 2005). Throughout the whole crop growth phase, stationary phenotyping devices can offer simultaneous monitoring at high resolution. However, because of their fixed-site characteristic, they can only operate in a small region, and their upkeep and development are costly. Multiple sets of sensors can be added to an existing agricultural vehicle, such as a powered tractor, manual cart, or phenotyping robot platforms, to create a moving phenotyping platform. A person traveling behind a cart can maneuver it with an electric wheel and use light/laser radar (LiDAR) to estimate aboveground biomass, ground cover, and canopy height quickly and non-destructively (Jimenez-Berni *et al.*,2018).

AERIAL PHENOTYPING

Aerial phenotyping platforms, primarily comprising satellites and unmanned aerial vehicles (UAVs), have been extensively employed for field crop phenotyping in recent times (Jang et al., 2020). Large-scale crop breeding typically uses airborne phenotyping platforms as opposed to ground-based platforms. The primary applications of manned aircraft images are in the measurement of thousands of field plots biomass, plant height, chlorophyll and nitrogen contents, canopy temperature and structure, and other parameters (Sankaran *et al.*, 2015). Large-scale genetic studies of agricultural cultivars growth regulations and genetic variations can be conducted using satellite photos (Lobell *et al.*, 2020). Nevertheless, their limited use in crop experiment monitoring is due to their reduced image resolution (Matese *et al.*, 2015). The potential of UAV phenotyping systems has been demonstrated. UAV-based phenotyping platforms offer cheap hardware costs, flexibility, and ease of operation compared to manned aircraft and satellite platforms. Furthermore, they have the ability to produce images with significantly greater spatial resolutions (~1 mm per pixel) than those obtained from satellites (Yang *et al.*, 2020). These factors have led to an exponential expansion in the use of UAV systems in crop phenotyping (Barbedo, 2019). Multi-imaging sensor-equipped fixed-wing and rotating-wing UAV platforms have been deployed by researchers. Visible-light (RGB) cameras, infrared thermal imagers, LiDAR, multispectral cameras, and hyperspectral sensors are among the common sensors carried by UAVs. According to Elazab *et al.* (2016), RGB cameras are used for stress quantification, canopy cover estimation, and yield prediction since they can generate 2D images with high resolution. LiDAR, on the other hand, offers another method for creating 3D representations of plants (Gibbs *et al.*, 2017).

Massive volumes of phenotypic data are being made available for crop study by emerging phenotype acquisition technologies, such as new kinds of physical, chemical, and biological (physiological) sensors, graphics and image technology, artificial intelligence technology, and Internet of Things (IoT) technology. It is difficult to translate terabytes of point cloud, image, spectroscopic, infrared, X-ray, and other multi-scale phenotypic data covering plant, tissue, cell, and organ populations in labs and greenhouses into biologically significant information. Finkel (2009) suggested that plant scientists work in breeding will become more hands-off as a result of imaging and phenomics. The development of sophisticated image understanding and quantification algorithms for 2D or 3D mesh processing is necessary due to the increasing number of imaging-based, automated, non-invasive, and non-destructive high-throughput plant phenotyping platforms that produce raw data that must be reliably and precisely normalized, reconstructed, and analyzed. A unique 3D mesh-based method for temporal high-throughput analysis of *Gossypium hirsutum* vegetative growth was presented by Paproki *et al.* (2012). Image Harvest (IH) is an open-source, adaptable image analysis framework designed to handle photos from high-throughput plant phenotyping devices (Knecht *et al.*, 2016). Artificial intelligence (AI) technologies, including ML, are essential to data mining (DM) and processing because they offer pertinent data for decision-making that is intended to meet breeding objectives. ML techniques like random forests (RFs), support vector machines (SVMs), and artificial neural networks (ANNs) facilitate research in crop phenomics. Edlich-Muth *et al.* (2016) used 129 image-derived characteristics to conduct phenomic studies on 195 inbred and 382 hybrid maize cultivars, tracking their development from 16 days after sowing (DAS) to 48 DAS. A novel algorithmic method (Hyppo-X) was introduced by Kamruzzaman *et al.* (2019) to visualize complicated phenomics data and characterize the impact of environment on phenotypic features. Hyppo-X was assessed in two real-world plant (maize) datasets in order to gauge this methods effectiveness. A statistical framework that integrates DM and ML algorithms for processing phenomics data was proposed by Rahaman *et al.* (2019).

ARTIFICIAL INTELLIGENCE

The term artificial intelligence (AI) generally refers to the imitation of human intelligence in computers that are designed to think and behave like people. It is used in reference to devices that display characteristics similar to those of the human mind (Frankenfield, 2021). A number of scientific fields are facing both opportunities and threats as a result of AIs recent rise to prominence in academia. While some believe that the use of AI in robotics could result in technological unemployment, it has actually made scientific advances easier to achieve and allowed science to branch out into previously uncharted territory, such as medical diagnostics (Paschen *et al.*, 2020).

Large data repositories are required by AI technologies in order to program machines for specified activities. Large data sets are required for the training of the algorithms employed in AI techniques, which improve decision-making by strengthening early detection (Frey, 2019). In non-destructive phenomics, the data acquisition procedure entails incorporating the data into the AI algorithms from instruments/sensors (such as digital cameras and spectrometers), each of which is often outfitted with a unique, proprietary communication protocol. Prior to analysis, the sensor outputs frequently need to be converted into digital formats that are compatible (White *et al.*, 2012). Thus, three essential elements of phenomic data management involve the application of AI: developing models to comprehend the relationships between genotype and phenotype and environmental interactions; managing databases to facilitate information and resource sharing; and developing algorithms and programs to transform sensory data into phenotypic information (Pauli, 2015). Phenomics is one area where the primary applications of AI, ML, DL, and computer vision have been recognized to date (Mansoor *et al.*, 2024; Sheikh *et al.*, 2024).

MACHINE LEARNING

ML has been the main focus of AI research since the 1970s. Models and frameworks for statistical ML have been developed, including Bayesian networks, SVMs, and Perceptrons. But no single model is ideal for every activity. Finding the optimal model for a particular situation remains difficult (Zhuang *et al.*, 2017). Despite the difficulty of selecting the best model for a given task, Roscher *et al.* (2020) claim that the emergence and success of neural networks, together with the wealth of data and sophisticated computational and data processing infrastructure, have led to the widespread use of ML models and algorithms. HTP has been impacted by the application of several imaging techniques for nondestructive phenotyping. on HTP, many image sensors gather plant data on platforms that are almost real-time. Because of these sensors ability to gather vast amounts of data, phenomics is now a big data problem that is well-suited for the use of ML. Large datasets can be difficult to analyze and comprehend, but ML algorithms offer a way to process data more quickly, effectively, and effectively than with conventional methods. Statistics, optimization, decision theory, and probability theory are examples of traditional processing techniques. In order to enable feature identification in specific difficult tasks like these, ML systems make use of these processing approaches to extract patterns and features from these massive volumes of data. (Singh *et al.*, 2016). Rahaman *et al.* (2019) claim that one benefit of applying ML techniques to plant phenotyping is their capacity to simultaneously look at a mixture of features (as opposed to studying each feature separately) and explore vast datasets for patterns. This was previously a problem due to the great dimensionality and number of plant images, which made it hard to examine them using conventional processing techniques (Singh *et al.*, 2016). To date, ML techniques have shown effective for a variety of tasks, including plant organ segmentation and the identification and categorization of plant diseases (Wetterich *et al.*, 2013). With the use of supervised learning, which involves training algorithms with sample photos from big datasets, it is now possible to recognize sick plants. But this method stops looking for new and unexpected phenotypic features, which unsupervised ML, which is less accurate, would have found otherwise (Sommer and gerlich, 2013). A fast developing subset of ML technologies called DL has completely changed the way that image-based plant phenotyping is done. According to Lecun *et al.* (2015), this method is effective in identifying intricate patterns within high-dimensional data, making it suitable for a variety of scientific research assignments. Some ML approaches are difficult to apply due to the wide variability of the plant images obtained from the numerous sensors (Singh *et al.*, 2018). DL techniques have made it possible to create more dependable feature identification workflows, whereas conventional ML involves trial-and-error phases in the feature extraction process of images. Before making decisions, like classifying data, they use a vast bank of non-linear filters in an automated hierarchical feature extraction process. DL techniques use numerous hidden layers in the network, each of which boosts the networks capacity for discrimination and prediction by sequentially applying a basic operation to the images (Pound *et al.*, 2017). Plant phenotyping has made use of a variety of DL architectures through a procedure known as transfer learning. In order to do this, a network that was previously trained on a sizable dataset that is fairly comparable to the one being studied must be used, and its weights must be adjusted for the new dataset. Despite the recent flood of DL architectures, there are still certain obstacles to overcome in order to integrate them with agricultural applications. Annotated agricultural data is still not widely available, which limits the potential for transfer learning to produce high-performance feature extraction models. Furthermore, a lot of agricultural image data include significant degrees of occlusion, particularly from plant leaves and background noise, which increases the chance an error may be made by mistaking objects of interest for the backdrop. This is partially because of the changes in the environment (cloudy sky, windy weather, etc.) that greatly affect the photos and make them more difficult to deal with. In a similar vein, data samples are also susceptible to differences in plant genotypes, field topography, and imaging angles. Therefore, for DL models designed for agricultural applications, the criteria for resilience and adaptability are notably high (Ghosal *et al.*, 2019).

APPLICATION OF DEEP AND TRANSFER LEARNING IN DISEASE RECOGNITION

Because DL and transfer learning can reliably learn and distinguish visual information, their applications in agriculture have been quite successful over the past ten years (Geetharamani and Pandian, 2019; Abbas *et al.*, 2021). Numerous fascinating studies on the application of these potential methods for illness identification have been published (Al-Bayati and UStandag, 2020). Notably, researchers are increasingly utilizing transfer learning, which is a concept that is growing in popularity (Too *et al.*, 2019). Moreover, transfer learning is a collection of refined techniques rather than a single method, allowing extremely accurate models to be developed on more specialized datasets, such those for plant diseases. According to Mohanty *et al.* (2016), the fine-tuning method performs significantly better than a CNN model that is trained from the beginning. An additional model that is frequently used and advised for the analysis of hyperspectral data for the early diagnosis of illness is the neural network (NN). Its fundamental workings are modeled after the nervous system of humans, and its unique abilities like learning and generalization help in the identification of agricultural diseases. Because it can mix training sets more effectively than other ML techniques, it provides a more accurate diagnostic capability. Zhu *et al.* (2016) conducted a similar comparative study in which they tested back-propagation neural networks (BPNNs) for presymptomatic tobacco mosaic virus (TMV) disease detection and classification using hyperspectral imaging using the SVM, RF, latent Dirichlet allocation (LDA), extreme learning machine, LS-SVM, and partial least squares discrimination analysis. Similar to this, Zhu *et al.* (2017) investigated the viability of using ML classifiers and the variable selection strategy in conjunction with hyperspectral imaging as a non-invasive method for the early identification of TMV disease. The chemometric models reached an accuracy of 80%, while the BPNN attained 95% accuracy, according to the data. It is important to note that a novel pattern recognition method called the Artificial Intelligent Nose can be used to perform pattern identification techniques like the RF and SVM. The smart nose is a quick and non-invasive way to diagnose plant diseases, according to Cui *et al.* (2018) evaluation of various invasive and non-invasive approaches, along with their benefits and downsides. NNs essentially guarantee the purest and most intact spectrum information for hyperspectral data interpretation. The idea of DL, which has lately gained popularity in farming applications, originated from the most well-known study on ANNs. Many scholars have become increasingly interested in DL, especially around 2018. Crop image classification research has advanced significantly; CNNs, autoencoders, RNNs, and restricted Boltzmann machines are some of the most common and exemplary models. DL has been the subject of numerous interesting studies on agricultural disease categorization and detection. One of these studies, by Ma *et al.* (2018), offered a deep convolutional neural network (DCNN) model that could identify over four cucumber diseases. When compared to other established techniques like naive Bayes, AlexNet, and SVMs, the DCNN demonstrated an astounding 93.41% accuracy rate in distinguishing between the various cucumber diseases. Comparably, Tran *et al.* (2019) proposed a tomato growth and yield maximization monitoring system. This approach was able to categorize illnesses that occurred during growth and nutritional shortages. To safeguard tomato crops, agricultural specialists assess the symptoms in light of the findings. Similarly, in order to monitor apple tree growth at each stage and estimate yield, Tian *et al.* (2019) used a dense YOLOV3 model that prevents overfitting by using strategies for data augmentation. It was discovered that their method worked well and could be used in apple orchards, although their study included wavy lights, interlaced fruit, and complex backgrounds.

HYPERSPECTRAL IMAGING APPLIED TO DISEASE RECOGNITION

The hyperspectral imagery method has been strongly developed during the past two decades (Bioucas-Dias *et al.*, 2013), and used to identify abiotic and biotic stresses in cultivated plants (Mananze *et al.*, 2018). Hyperspectral imaging is a technique combining spectroscopy and imagery, making it possible to simultaneously obtain the spatial and spectral information of an object. Disease infection causes changes in plant biochemical and biophysical properties, such as transpiration rate, tissue structure, water, and pigment content. These changes can then alter plant spectral properties, intercellular space, and water content (Rumpf *et al.*, 2010). However, the hyperspectral system is able to capture these spectral features. Zhu *et al.* (2017) conducted a similar study to detect TSWV infection growth in tobacco, in which the authors reported that hyperspectral reflectance was gathered in the visible and near-infrared range to distinguish between healthy and infected TSWV tobacco leaves using statistical analysis methods. Primarily, the TSWV presence was identified at 14 days post inoculation. Moreover, Zhu *et al.* (2017) demonstrated that hyperspectral imaging is able to detect the TMV infection before showing any symptoms, while utilizing the successive projection algorithm (SPA) for the selection of the effective wavelength and, most significantly, for identifying various diseases. Due to the huge number of spectral values that are highly correlated in the hyperspectral dataset, high dimensionality and multicollinearity frequently appear in hyperspectral data (Ng *et al.*, 2019). Accordingly, the selection of EWs is crucial for hyperspectral analysis in order to lessen the computational complexity, increase the efficiency of using hyperspectral data, and reduce the computational complexity. Thus, to address this multi-collinearity issue, a variety of approaches and methods have been presented, such as the SPA (Zhu *et al.*, 2017), partial least squares regression models (Ng *et al.*, 2019) and genetic algorithms.

APPLICATION OF IOT IN THE FIELD OF LEAF DISEASE RECOGNITION

The IoT has improved agricultural capabilities. IoT applications can help farmers at any time during their farming activities and keep them updated with the latest crop and weather information to remotely monitor their fields. By means of IoT applications (Mishra *et al.*, 2021), farmers can make plans for the next seasons harvest. Furthermore, they can detect crop diseases at an early stage to curb the spread of disease and save their yield. Agricultural IoT apps clearly play a major role in increasing agricultural production and decreasing crop losses due to diseases. In this context, a large amount of research has been conducted to identify diseases, as shown in Table 13.3. Truong *et al.* (2017) devised an IoT-based system made up of various devices that is able to deliver real-time environmental information and send it to the cloud to be stored. These environmental data are processed and scrutinized to predict weather conditions by means of the SVM algorithm deployed in the cloud in order to detect crop fungal diseases. In addition, better results have been achieved when the IoT and image processing have been combined and implemented in the area of disease recognition. Krishna *et al.* (2019) implemented an IoT system featuring SMS alerts that enables automatic disease detection and pesticide spraying using the NodeMCU.

PHENOTYPING DATA AND ML

The substantial volume, diversity, speed, and reliability of imaging and remote-sensing data generated by real-time platforms pose a significant "big data" challenge. Managing the data produced by these near real-time platforms requires efficient archiving and retrieval mechanisms for subsequent analysis. While the analysis and interpretation of such image-based big data present inherent challenges, the potential impact on agricultural production makes it a promising approach for HTP and High-Throughput Stress Phenotyping (HTSP). ML approaches offer a scalable and modular strategy for data analysis, particularly in the emerging field of "plant stress analytics." Recent studies on HTSP have utilized images from UAV-based platforms to detect weeds in wheat (*Triticum aestivum* L.), maize (Pena *et al.*, 2015), and sunflower (*Helianthus annuus* L.) (Torres-Sanchez *et al.*, 2013). These studies leverage ML algorithms, opening new avenues for improved stress management practices on both spatial and temporal scales. The application of ML in handling and interpreting large-scale imaging and remote-sensing data showcases its potential to revolutionize agriculture by providing advanced tools for efficient and targeted stress management.

THE LEARNING PROCESS SUPERVISED VS. UNSUPERVISED

The first perspective revolves around whether the ML model is furnished with labeled data during the training process. In this context, a label refers to a characteristic (such as the health status of a plant or the type of crop plant) associated with an image. For instance, consider a scenario where the objective of an ML model is to differentiate between maize, soybean (*Glycine max* (L.) Merr.), and sunflower plants based on a collection of images. If the model is trained with labeled images indicating whether each is soybean, maize, or sunflower, the learning process is termed "supervised." On the other hand, if training images are provided without any labels, the learning process is deemed "unsupervised." It is worth noting that while some ML tools can be trained in either a supervised or an unsupervised manner, others are versatile and can be trained in both ways. In supervised learning, a model aims to learn a mapping between the input dataset and corresponding output labels. Using the earlier example, a model trained in a supervised manner learns to map all soybean examples to the soybean label, and similarly for maize and sunflower. Prominent examples of supervised schemes include SVM and regular ANNs. In contrast, an unsupervised method lacks specific output labels associated with input images. Instead, it identifies structures or features present in the images, such as the presence of tassels in maize or the presence of trifoliate leaves in soybean. The features identified in the unsupervised process may not always be immediately interpretable by a human observer or researcher. By contrast, an unsupervised method does not have specific output labels associated with input images. It identifies structures or features present in the images, such as the presence of tassels in maize, or the presence of trifoliate leaves in soybean. In many cases, the features identified in the unsupervised process may not be meaningful to a human user. Various clustering, mixture models, and dimension-reduction techniques fall under this category. Often ML models can be developed by using partly labeled data. This training scheme is known as "semi-supervised" learning.

THE MODELING OBJECTIVE – GENERATIVE VS. DISCRIMINATIVE

Another perspective for categorizing ML methods revolves around the modeling objective: whether the model is intended to differentiate between distinct data patterns or to learn and generate similar patterns synthetically. In the context of the earlier example, distinguishing between a maize and soybean plant could be based on the presence or absence of tassels or pods. A model developed to recognize this difference is referred to as a "discriminative model." Many supervised methods, such as SVMs and ANNs, naturally fall into this category. However, a discriminative model may not capture other features of the objects. For instance, a discriminative model distinguishing soybean and maize based on tassels might fail to differentiate between an image of a maize plant and that of a sorghum (*Sorghum bicolor* (L.) Moench) plant. Consequently, discriminative models are typically designed for specific predetermined tasks. In contrast, a model that comprehensively captures the overall data pattern, enabling it to generate synthetic images, is termed a "generative model." A single generative model can be versatile, proving useful for multiple decision tasks simultaneously. Prominent examples of generative models include mixture models, hidden Markov models, and Boltzmann machines.

Given a substantial volume of data, discriminative models often outperform generative models, especially in classification tasks like distinguishing images of soybean plants from maize plants. However, generative models may exhibit slightly better performance than discriminative models when dealing with limited training data. Additionally, generative models tend to be more robust against overfitting issues, where a model becomes overly specialized to the training data and performs poorly on unseen test data.

THE ROLE OF PREPROCESSING IN ML-HTSP

Preprocessing plays a pivotal role in the successful implementation of ML methods, particularly in the context of HTSP using image data. Thoughtful preprocessing of collected image datasets has consistently demonstrated significant enhancements in ML performance. The preprocessing spectrum ranges from basic operations like image cropping, contrast enhancement, and background removal to more intricate procedures such as clustering and dimensionality reduction through techniques like Principal Component Analysis (PCA). The overarching goal of preprocessing is to concentrate relevant information, as original datasets often contain excessive or conflicting data that can hinder ML performance. By improving the signal-to-noise ratio (the ratio of useful to useless information), preprocessing empowers ML models to more effectively identify patterns, discern trends, and categorize data into appropriate classes. The preprocessing stage is where domain knowledge becomes crucial. Domain experts play a pivotal role in identifying features within images that are pertinent for training the model. For instance, removing irrelevant elements like soil, dirt, and tags to isolate the plant canopy is often a critical step. Subsequently, various image processing tools are employed to transform raw datasets into more relevant ones that encapsulate the extracted features. Examples of preprocessing operations include image segmentation, contrast enhancement for edge detection, thresholding for binary data, conversion between different image formats (e.g., RGB to grayscale or RGB to hue saturation value), denoising using filters (band-pass, low-pass, Fast Fourier Transform), extracting features at different scales using transforms (Fast Fourier transform, wavelet, Haar, Hough, Radon), pixel-based classification, clustering into classes, and dimensionality reduction. Many ML tools are applicable to these preprocessing stages, showcasing their versatility. Several instances highlight the efficacy of preprocessing in real-world applications. In early site-specific weed management for wheat, a UAV-based platform with a visible-range camera utilized Otsus thresholding method to distinguish between wheat crops and weeds (Torres-Sanchez *et al.*, 2014. In sunflower fields, a quadrocopter equipped with RGB and multispectral sensors detected weeds for optimized herbicide application using a pixel-based classification approach. Additionally, in sunflower and maize, image segmentation followed by automatic object-based image analysis was employed to identify spatially and spectrally consistent objects. These examples underscore the utility and impact of preprocessing in improving the performance and accuracy of ML models in diverse applications (Pena *et al.*, 2015).

ML APPROACHES FOR STRESS IDENTIFICATION

Identification methods play a crucial role in detecting specific stresses amidst various potential stress factors in the field. In these scenarios, preprocessing of image data is vital, particularly in applications involving high-throughput imaging where automatic identification of plant stressors (e.g., weeds, nutrient deficiencies,

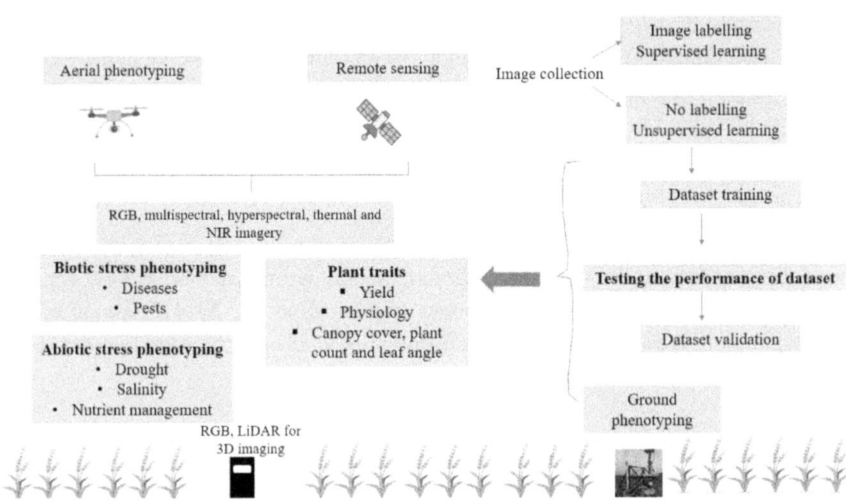

FIGURE 13.1 A thorough analysis of machine learning and high-throughput phenotyping for plant stress.

diseases, and insects) is essential. Various ML methods, including SVMs, ANNs, kernel methods, and instance-based approaches, has been employed for stress identification (Fig. 13.1).

SVM METHODS

SVM methods have found application in disease and stress identification across diverse plant species. In the citrus industry, SVMs were deployed to contain citrus greening, also known as Huanglongbing (HLB), caused by phloem-limiting bacteria (Liu et al., 2020). ML, including SVM, was utilized for the identification and estimation of HLB based on fluorescence imaging spectroscopy data of citrus leaf samples (Wetterich *et al.*, 2013). Similarly, SVMs were employed in sugar beet for the early identification of diseases such as Cercospora leaf spot, leaf rust, and powdery mildew (Rumpf *et al.*, 2010). The application extended to cotton for identifying damage by green stink bug, bacterial angular blight, and Ascochyta blight, as well as in tomatoes for identifying viruses like tomato yellow leaf curl virus and tomato yellow leaf curl disease (Camargo and Smith, 2009; Mokhtar *et al.*, 2015).

ARTIFICIAL NEURAL NETWORKS (ANNS) AND VARIANTS

Simple image processing and ANN have been effective in detecting and identifying various orchid seedling diseases, including bacterial soft rot, bacterial brown spot, and Phytopthora black rot (Huang, 2007).

DISCRIMINANT ANALYSIS

Spectral reflectance data from healthy and HLB-infected citrus leaves have been analyzed using discriminative and supervised methods such as quadratic discriminant analysis (QDA), LDA, k-nearest neighbor (k-NN), and soft independent modeling of class analogy (Sankaran *et al.*, 2011). In sugar beets, hyperspectral canopy reflectance was leveraged to identify stress symptoms induced by beet cyst nematode and Rhizoctonia crown and root rot (Hillnhutter *et al.*, 2011).

GAUSSIAN MIXTURE MODELS

Preprocessing followed by the application of Gaussian mixture models has been used to identify both biotic and abiotic stresses (Casanova *et al.*, 2014). Canopy images were taken in a stressed wheat field, at differential levels of irrigation (abiotic stress) and inoculated with wheat streak mosaic virus (biotic stress).

SELF-ORGANIZING MAP (SOM) AND BAYES CLASSIFIER

Both supervised (Bayesian classifier) and unsupervised (SOM) ML approaches were used for segmenting diseased tomato plants. Diseased regions on the tomato canopy were identified using the preprocessed tomato images by a SOM model

(Hernandez-Rabadan *et al.*, 2014). Examples from various crops are as follows illustrating the use of classification algorithms in various stresses.

ANN APPLICATIONS

In one application, a 3D model of leaves was developed to distinguish between two sugar beet diseases, *Cercospora beticola* and *Uromyces betae*, utilizing a combination of K-means clustering and ANNs. This approach incorporated fused information from RGB and multispectral camera imaging (Bauer et al., 2011). Another instance involved the analysis of thermal and hyperspectral imaging to identify and classify symptoms caused by three *Alternaria* species in oilseed brassica. Among various classifiers tested using the Weka software, the BPNN model exhibited the highest prediction accuracy for classifying *Alternaria* species (Baranowski *et al.*, 2015).

LDA, K-MEANS, AND COUPLED METHODS

ML methods were employed to classify the presence of aflatoxins, toxic compounds produced by fungi (Camardo et al., 2021) Aspergillus flavus and Aspergillus parasiticus. Input data included fluorescence (UV illumination) and reflectance (halogen excitations). Preprocessing involved feature extraction using Guyon's SVM-RFE, classical Fisher discriminant power, and PCA. Extracted features were utilized in classifiers such as LDA and Multi-Layer Perceptron (MLP). Another study focused on classifying RGB images of clover plants into "healthy" and "injured" classes due to varying ozone levels. Various pixel-classifying methods, including LDA, K-means clustering, FPM-T2, and FPM-RSS, were compared (Kruse *et al.*, 2014).

PREDICTION OF STRESS

Predicting plant stress at an early stage, imperceptible to the human eye, holds significant implications for timely and cost-effective stress control. While there are limited reported activities in using ML for stress prediction, this emerging field represents the next frontier in research efforts. The potential applications extend to prescription farming and precision agriculture, promising transformative impacts on agricultural practices.

SUPPORT VECTOR MACHINES

SVMs have proven instrumental in predicting plant diseases, as demonstrated in the case of rice blast disease. A weather-based prediction model utilizing SVM, compared to alternative methods such as BPNN, generalized regression neural network, and conventional multiple regression, exhibited superior performance in cross-location and cross-year models. This underscores the efficacy of SVM in early forecasting of plant diseases (Kaundal *et al.*, 2006). Moreover, SVMs versatility extends to predicting drought-induced stress in barley through the analysis of hyperspectral images, employing an SVM variant and an ordinal classification approach (Behmann *et al.*, 2014). In another instance, SVM was employed in barley for the presymptomatic detection of water stress. The Differential Absorption Ratio (DAR) algorithm, applied to images captured by a hyperspectral camera, demonstrated effectiveness in predicting stress conditions before visible symptoms emerged, showcasing SVMs role in proactive plant stress detection (Kersting *et al.,* 2012).

CONCLUSION

Crop diseases are one of the main challenges in the farming sector. Thus, there is a need to identify crop diseases at the earliest stage to reduce disease severity and to curb disease propagation on farms. Accordingly, prominent and advanced research has been conducted in recent years on several kinds of disease identification techniques, as presented in this chapter. This chapter provided readers with insights into the automatic crop disease detection process. By applying ML to sensor data, farm management systems are evolving into real AI systems, providing richer recommendations and insights for the subsequent decisions and actions with the ultimate scope of production improvement. For this scope, in the future, it is expected that the use of ML models will be even more widespread, allowing for the possibility of integrated and applicable tools. At the moment, all of the approaches regard individual approaches and solutions and are not adequately connected with the decision-making process, as seen in other application domains. This integration of automated data recording, data analysis, ML implementation, and decision-making or support will provide practical tools in line with the so-called knowledge-based agriculture for increasing production levels and bioproduct quality.

REFERENCES

Abbas, A., Jain, S., Gour, M. & Vankudothu, S. (2021). Tomato plant disease detection using transfer learning with C-GAN synthetic images. *Computers and Electronics in Agriculture, 187*, 106279.

Al-bayati, J. S. H. & Üstünda g, B. B. (2020). Evolutionary feature optimization for plant leaf disease detection by deep neural networks. *International Journal of Computational Intelligence Systems, 13*, 12–23.

Annonymous. (2017). United Nations Department of Economic and Social Affairs, World Population Prospects: The 2017 Revision, United Nations.

Aquino, A., Barrio, I., Diago, M. P., Millan, B., & Tardaguila, J. (2018). vitisBerry: An Android-smartphone application to early evaluate the number of grapevine berries by means of image analysis. *Computers and Electronics in Agriculture*, *148*, 19–28.

Araus, J. L., & Kefauver, S. C. (2018). Breeding to adapt agriculture to climate change: Affordable phenotyping solutions. *Current Opinion in Plant Biology*, *45*, 237–247.

Ataş, M., Yardimci, Y., & Temizel, A. (2012). A new approach to aflatoxin detection in chili pepper by machine vision. *Computers and Electronics in Agriculture*, *87*, 129–141.

Baranowski, P., Jedryczka, M., Mazurek, W., Babula-Skowronska, D., Siedliska, A., & Kaczmarek, J. (2015). Hyperspectral and thermal imaging of oilseed rape (Brassica napus) response to fungal species of the genus Alternaria. *PloS One*, *10*(3), e0122913.

Barbedo, J. G. A. (2019). A review on the use of unmanned aerial vehicles and imaging sensors for monitoring and assessing plant stresses. *Drones*, *3*(2), 40.

Bauer, S. D., Korč, F., & Förstner, W. (2011). The potential of automatic methods of classification to identify leaf diseases from multispectral images. *Precision Agriculture*, *12*, 361–377.

Behmann, J., Schmitter, P., Steinrücken, J., & Plümer, L. (2014). Ordinal classification for efficient plant stress prediction in hyperspectral data. *The International Archives of the Photogrammetry. Remote Sensing and Spatial Information Sciences*, *40*, 29–36.

Bioucas-Dias, J. M., Plaza, A., Camps-Valls, G., Scheunders, P., Nasrabadi, N. & Chanussot, J. (2013). Hyperspectral remote sensing data analysis and future challenges. *IEEE Geoscience Remote Sensing Magazine, 1*, 6–36.

Camargo, A. & Smith, J. S. (2009). An image-processing based algorithm to automatically identify plant disease visual symptoms. *Biosystems Engineering,* 102, 9–21.

Camardo Leggieri, M., Mazzoni, M., & Battilani, P. (2021). Machine learning for predicting mycotoxin occurrence in maize. *Frontiers in Microbiology*, *12*, 661132.

Casanova, J. J., O'Shaughnessy, S. A., Evett, S. R., & Rush, C. M. (2014). Development of a wireless computer vision instrument to detect biotic stress in wheat. *Sensors*, *14*(9), 17753–17769.

Chéné, Y., Rousseau, D., Lucidarme, P., Bertheloot, J., Caffier, V., Morel, P., ... & Chapeau-Blondeau, F. (2012). On the use of depth camera for 3D phenotyping of entire plants. *Computers and Electronics in Agriculture*, *82*, 122–127.

Chung, C. L., Huang, K. J., Chen, S. Y., Lai, M. H., Chen, Y. C., & Kuo, Y.F. (2016). Detecting Bakanae disease in rice seedlings by machine vision. *Computers and Electronics in Agriculture*, 121, 404–411.

Cui, S., Ling, P., Zhu, H., & Keener, H. M. (2018). Plant pest detection using an artificial nose system: A review. *Sensors, 18*, 378.

Edlich-Muth, C., Muraya, M. M., Altmann, T., & Selbig, J. (2016). Phenomic prediction of maize hybrids. *Biosystems*, *146*, 102–109.

Elazab, A., Ordóñez, R. A., Savin, R., Slafer, G. A., & Araus, J. L. (2016). Detecting interactive effects of N fertilization and heat stress on maize productivity by remote sensing techniques. *European Journal of Agronomy*, *73*, 11–24.

Fahlgren, N., Gehan, M. A., & Baxter, I. (2015). Lights, camera, action: High-throughput plant phenotyping is ready for a close-up. *Current Opinion in Plant Biology*, *24*, 93–99.

Finkel E., With 'Phenomics,' plant scientists hope to shift breeding into overdrive. *Science* 325,380-381(2009). DOI: 10.1126/science.325_380

Frankenfield, J. (2021). Artificial Intelligence (AI). Available online: www.investopedia.com/terms/a/artificial-intelligence-ai.asp

Frey, L. J. (2019). Artificial intelligence and integrated genotype–phenotype identification. *Genes*, *10*(1), 18.

Fukushima, K. (1980). Neocognitron: A self-organizing neural network model for a mechanism of pattern recognition unaffected by shift in position. *Biological Cybernetics*, *36*(4), 193–202.

Furbank, R. T., & Tester, M. (2011). Phenomics–technologies to relieve the phenotyping bottleneck. *Trends in Plant Science*, *16*(12), 635–644.

Garcia-Garcia, A., Orts-Escolano, S., Oprea, S., Villena-Martinez, V., Martinez-Gonzalez, P., & Garcia-Rodriguez, J. (2018). A survey on deep learning techniques for image and video semantic segmentation. *Applied Soft Computing*, *70*, 41–65.

Geetharamani, G. & Pandian, A. (2019). Identification of plant leaf diseases using a nine-layer deep convolutional neural network. *Computers and Electrical Engineering, 76*, 323–338.

Ghosal, S., Zheng, B., Chapman, S. C., Potgieter, A. B., Jordan, D. R., Wang, X., ... & Guo, W. (2019). A weakly supervised deep learning framework for sorghum head detection and counting. *Plant Phenomics*.

Gibbs, J. A., Pound, M., French, A. P., Wells, D. M., Murchie, E., & Pridmore, T. (2017). Approaches to three-dimensional reconstruction of plant shoot topology and geometry. *Functional Plant Biology*, *44*(1), 62–75.

González-Pérez, J. L., Espino-Gudiño, M. C., Gudiño-Bazaldúa, J., Rojas-Rentería, J. L., Rodríguez-Hernández, V., & Castaño, V. M. (2013). Color image segmentation using perceptual spaces through applets for determining and preventing diseases in chili peppers. *African Journal of Biotechnology*, *12*(7), 679–688.

Großkinsky, D. K., Svensgaard, J., Christensen, S., & Roitsch, T. (2015). Plant phenomics and the need for physiological phenotyping across scales to narrow the genotype-to-phenotype knowledge gap. *Journal of Experimental Botany*, *66*(18), 5429–5440.

He, K., Zhang, X., Ren, S., & Sun, J. (2016). Deep residual learning for image recognition. In *Proceedings of the IEEE Conference on Computer Vision and Pattern Recognition* (pp. 770–778).

Hernández-Rabadán, D. L., Ramos-Quintana, F., & Guerrero Juk, J. (2014). Integrating SOMs and a Bayesian classifier for segmenting diseased plants in uncontrolled environments. *Scientific World Journal*, 2014, 214674. doi: 10.1155/2014/214674.

Hillnhütter, C., Mahlein, A. K., Sikora, R. A., & Oerke, E. C. (2011). Remote sensing to detect plant stress induced by Heterodera schachtii and Rhizoctonia solani in sugar beet fields. *Field Crops Research*, *122*(1), 70–77.

Huang, K. Y. (2007). Application of artificial neural network for detecting Phalaenopsis seedling diseases using color and texture features. *Computers and Electronics in Agriculture, 57*, 3–11

Jang, G., Kim, J., Yu, J. K., Kim, H. J., Kim, Y., Kim, D. W., ... & Chung, Y. S. (2020). Cost-effective unmanned aerial vehicle (UAV) platform for field plant breeding application. *Remote Sensing*, *12*(6), 998.

Jimenez-Berni, J. A., Deery, D. M., Rozas-Larraondo, P., Condon, A. T. G., Rebetzke, G. J., James, R. A., ... & Sirault, X. R. (2018). High throughput determination of plant height, ground cover, and above-ground biomass in wheat with LiDAR. *Frontiers in Plant Science*, *9*, 237.

Jin, X., Zarco-Tejada, P. J., Schmidhalter, U., Reynolds, M. P., Hawkesford, M. J., Varshney, R. K., ... & Li, S. (2020). High-throughput estimation of crop traits: A review of ground and aerial phenotyping platforms. *IEEE Geoscience and Remote Sensing Magazine*, *9*(1), 200–231.

Kamilaris, A., & Prenafeta-Boldú, F. X. (2018). Deep learning in agriculture: A survey. *Computers and Electronics in Agriculture*, *147*, 70–90.

Kamruzzaman, M., Kalyanaraman, A., Krishnamoorthy, B., Hey, S., & Schnable, P. S. (2019). Hyppo-X: A scalable exploratory framework for analyzing complex phenomics data. *IEEE/ACM Transactions on Computational Biology and Bioinformatics*, *18*(4), 1535–1548.

Karunathilake, E. M. B. M., Le, A. T., Heo, S., Chung, Y. S., & Mansoor, S. (2023). The path to smart farming: Innovations and opportunities in precision agriculture. *Agriculture*, *13*(8), 1593.

Kaundal, R., Kapoor, A. S., & Raghava, G. P. (2006). Machine learning techniques in disease forecasting: A case study on rice blast prediction. *BMC Bioinformatics*, *7*, 1–16.

Kersting, K., Xu, Z., Wahabzada, M., Bauckhage, C., Thurau, C., Roemer, C., ... & Pluemer, L. (2012). Pre-symptomatic prediction of plant drought stress using dirichlet-aggregation regression on hyperspectral images. In *Proceedings of the AAAI Conference on Artificial Intelligence* (Vol. 26, No. 1, pp. 302–308).

Kim, T., Kim, J. I., Visbal-Onufrak, M. A., Chapple, C., & Kim, Y. L. (2016). Nonspectroscopic imaging for quantitative chlorophyll sensing. *Journal of Biomedical Optics*, *21*(1), 016008–016008.

Kirchgessner, N., Liebisch, F., Yu, K., Pfeifer, J., Friedli, M., Hund, A., & Walter, A. (2016). The ETH field phenotyping platform FIP: a cable-suspended multi-sensor system. *Functional Plant Biology*, *44*(1), 154–168.

Knecht, A. C., Campbell, M. T., Caprez, A., Swanson, D. R., & Walia, H. (2016). Image Harvest: an open-source platform for high-throughput plant image processing and analysis. *Journal of Experimental Botany*, *67*(11), 3587–3599.

Krishna, M., Sulthana, S. F., Sireesha, V., Prasanna, Y., & Sucharitha, V. (2019). Plant disease detection and pesticide spraying using dip and IoT. *Journal of Emerging Technologies and Innovative Research, 6*, 54–58.

Krizhevsky, A., Sutskever, I., & Hinton, G. E. (2012). Imagenet classification with deep convolutional neural networks. *Advances in Neural Information Processing Systems*, 25.

Kruse, O. M. O., Prats-Montalbán, J. M., Indahl, U. G., Kvaal, K., Ferrer, A., & Futsaether, C. M. (2014). Pixel classification methods for identifying and quantifying leaf surface injury from digital images. *Computers and electronics in Agriculture*, *108*, 155–165.

Ku, K. B., Mansoor, S., Han, G. D., Chung, Y. S., & Tuan, T. T. (2023). Identification of new cold tolerant Zoysia grass species using high-resolution RGB and multi-spectral imaging. *Scientific Reports, 13*(1), 13209.

Kwon, S. H., Ku, K. B., Le, A. T., Han, G. D., Park, Y., Kim, J., ... & Mansoor, S. (2024). Author Correction: Enhancing citrus fruit yield investigations through flight height optimization with UAV imaging. *Scientific Reports*, *14*.

LeCun, Y., Bengio, Y., & Hinton, G. (2015). Deep learning. *Nature*, *521*(7553), 436–444.

LeCun, Y., Boser, B., Denker, J., Henderson, D., Howard, R., Hubbard, W., & Jackel, L. (1990). Handwritten digit recognition with a back-propagation network. In *Advances in Neural Information Processing Systems*, Morgan Kaufmann, San Mateo, Vol. 2, 96–404.

Liu, Y., Xiao, H., Hao, Y., Ye, L., Jiang, X., Wang, H., & Sun, X. (2020). Diagnosis of citrus greening using raman spectroscopy-based pattern recognition. *Journal of Applied Spectroscopy*, *87*, 150-158.

Lobell, D. B., Di Tommaso, S., You, C., Djima, I. Y., Burke, M., & Kilic, T. (2020). Sight for sorghums: Comparisons of satellite-and ground-based sorghum yield estimates in Mali. *Remote Sensing*, *12*(1), 100.

Ma, J., Du, K., Zheng, F., Zhang, L., Gong, Z. & Sun, Z. (2018). A recognition method for cucumber diseases using leaf symptom images based on deep convolutional neural network. *Computers and Electronics in Agriculture, 154*, 18–24.

Mananze, S., Pôças, I., & Cunha, M. (2018). Retrieval of maize leaf area index using hyperspectral and multispectral data. *Remote Sensing, 10*, 1942.

Mansoor, S., & Chung, Y. S. (2024). Functional phenotyping: understanding the dynamic response of plants to drought stress. *Current Plant Biology*, 100331.

Mansoor, S., Karunathilake, E. M. B. M., Tuan, T. T., & Chung, Y. S. (2024). Genomics, Phenomics, and Machine Learning in Transforming Plant Research: Advancements and Challenges. *Horticultural Plant Journal.*

Matese, A., Toscano, P., Di Gennaro, S. F., Genesio, L., Vaccari, F. P., Primicerio, J., ... & Gioli, B. (2015). Intercomparison of UAV, aircraft and satellite remote sensing platforms for precision viticulture. *Remote Sensing*, *7*(3), 2971–2990.

Milella, A., Marani, R., Petitti, A., & Reina, G. (2019). In-field high throughput grapevine phenotyping with a consumer-grade depth camera. *Computers and Electronics in Agriculture*, *156*, 293–306.

Minervini, M., Scharr, H., & Tsaftaris, S. A. (2015). Image analysis: The new bottleneck in plant phenotyping [applications corner]. *IEEE Signal Processing Magazine*, *32*(4), 126–131.

Mishra, M., Choudhury, P., & Pati, B. (2021). Modified ride-NN optimizer for the IoT based plant disease detection. *Journal of Ambient Intelligence and Humanized Computing, 12*, 691–703.

Mohanty, S. P., Hughes, D. P., & Salathé, M. (2016). Using deep learning for image-based plant disease detection. *Frontiers of Plant Science, 7*, 1419.

Mokhtar, U., Ali, M. A., Hassanien, A. E., & Hefny, H. (2015). Identifying two of tomatoes leaf viruses using support vector machine. In *Information Systems Design and Intelligent Applications: Proceedings of Second International Conference INDIA 2015, Volume 1* (pp. 771–782). Springer India.

Ng, W., Minasny, B., Malone, B. P., Sarathjith, M. & Das, B. S. (2019). Optimizing wavelength selection by using informative vectors for parsimonious infrared spectra modelling. *Computers and Electronics in Agriculture, 158*, 201–210.

Orlando, F., Movedi, E., Coduto, D., Parisi, S., Brancadoro, L., Pagani, V., ... & Confalonieri, R. (2016). Estimating leaf area index (LAI) in vineyards using the PocketLAI smart-app. *Sensors*, *16*(12), 2004, 1–12.

Pantazi, X. E., Moshou, D., Oberti, R., West, J., Mouazen, A. M., & Bochtis, D. (2017). Detection of biotic and abiotic stresses in crops by using hierarchical self organizing classifiers. *Precision Agriculture*, *18*, 383–393.

Paproki, A., Sirault, X., Berry, S., Furbank, R., & Fripp, J. (2012). A novel mesh processing based technique for 3D plant analysis. *BMC Plant Biology*, *12*, 1–13.

Paschen, U., Pitt, C., & Kietzmann, J. (2020). Artificial intelligence: Building blocks and an innovation typology. *Business Horizons*, *63*(2), 147–155.

Pauli, D. (2015). High-throughput phenotyping technologies in cotton and beyond. In *Proceedings of the Advances in Field-Based High-Throughput Phenotyping and Data Management: Grains and Specialty Crops*, Spokane, WA, USA. pp. 1–11.

Pena, J. M., Torres-Sánchez, J., Serrano-Pérez, A., De Castro, A. I., & López-Granados, F. (2015). Quantifying efficacy and limits of unmanned aerial vehicle (UAV) technology for weed seedling detection as affected by sensor resolution. *Sensors*, *15*(3), 5609–5626.

Pound, M. P., Atkinson, J. A., Townsend, A. J., Wilson, M. H., Griffiths, M., Jackson, A. S., ... & French, A. P. (2017). Deep machine learning provides state-of-the-art performance in image-based plant phenotyping. *Gigascience*, *6*(10), gix083.

Rahaman, M. M., Ahsan, M. A., & Chen, M. (2019). Data-mining techniques for image-based plant phenotypic traits identification and classification. *Scientific Reports*, *9*(1), 19526.

Rawat, W., & Wang, Z. (2017). Deep convolutional neural networks for image classification: A comprehensive review. *Neural Computation*, *29*(9), 2352–2449.

Raza, S. E. A., Prince, G., Clarkson, J. P., & Rajpoot, N. M. (2015). Automatic detection of diseased tomato plants using thermal and stereo visible light images. *PloS One*, *10*(4), e0123262.

Reuzeau, C., Pen, J., Frankard, V., de Wolf, J., Peerbolte, R., Broekaert, W. & van Camp, W. (2005). TraitMill: A discovery engine for identifying yield-enhancement genes in cereals. *Molecular Plant Breeding, 3*, 753–759.

Roscher, R., Bohn, B., Duarte, M. F., & Garcke, J. (2020). Explainable machine learning for scientific insights and discoveries. *IEEE Access, 8*, 42200–42216.

Rumpf, T., Mahlein, A. K., Steiner, U., Oerke, E. C., Dehne, H. W., & Plümer, L. (2010). Early detection and classification of plant diseases with support vector machines based on hyperspectral reflectance. *Computers and Electronics in Agriculture, 74*(1), 91–99.

Sankaran, S. et al. (2011) Visible-near infrared spectroscopy for detection of Huanglongbing in citrus orchards. *Computers and Electronics in Agriculture, 77*, 127–134.

Sankaran, S., Khot, L. R., Espinoza, C. Z., Jarolmasjed, S., Sathuvalli, V. R., Vandemark, G. J., ... & Pavek, M. J. (2015). Low-altitude, high-resolution aerial imaging systems for row and field crop phenotyping: A review. *European Journal of Agronomy, 70*, 112–123.

Sankaran, S., Mishra, A., Maja, J. M., & Ehsani, R. (2011). Visible-near infrared spectroscopy for detection of Huanglongbing in citrus orchards. *Computers and Electronics in Agriculture, 77*(2), 127–134.

Schmidhuber, J. & Tubiello, F. N. (2007). Global food security under climate change. In *Proceedings of the National Academy of Sciences of the United States of America* 104:19703–19708.

Sheikh, M., Iqra, F., Ambreen, H., Pravin, K. A., Ikra, M., & Chung, Y. S. (2024). Integrating artificial intelligence and high-throughput phenotyping for crop improvement. *Journal of Integrative Agriculture, 23*(6), 1787–1802.

Singh, A. K., Ganapathysubramanian, B., Sarkar, S., & Singh, A. (2018). Deep learning for plant stress phenotyping: trends and future perspectives. *Trends in Plant Science, 23*(10), 883–898.

Singh, A., Ganapathysubramanian, B., Singh, A. K., & Sarkar, S. (2016). Machine learning for high-throughput stress phenotyping in plants. *Trends in Plant Science, 21*(2), 110–124.

Sommer, C., & Gerlich, D. W. (2013). Machine learning in cell biology–teaching computers to recognize phenotypes. *Journal of Cell Science, 126*(24), 5529–5539.

Steensland, A. & Zeigler, M. (2018). Global Agricultural Productivity Report https://globalagriculturalproductivity.org/.

Tian, Y., Yang, G., Wang, Z., Wang, H., Li, E. & Liang, Z. (2019). Apple detection during different growth stages in orchards using the improved YOLO-V3 model. *Computers and Electronics in Agriculture, 157*, 417–426.

Tilman, D., Balzer, C., Hill, J., & Befort, B. L. (2011). "Global food demand and the sustainable intensification of agriculture," *Proceedings of the National Academy of Sciences of the United States of America 108*: 20260–20264.

Too, E. C., Yujian, L., Njuki, S. & Yingchun, L. (2019). A comparative study of fine-tuning deep learning models for plant disease identification. *Computers and Electronics in Agriculture, 161*, 272–279.

Torres-Sánchez, J., López-Granados, F., De Castro, A. I., & Peña-Barragán, J. M. (2013). Configuration and specifications of an Unmanned Aerial Vehicle (UAV) for early site specific weed management. *PloS One, 8*. e58210. 10.1371/journal.pone.005.

Torres-Sánchez, J., Peña, J. M., de Castro, A. I., & López-Granados, F. (2014). Multi-temporal mapping of the vegetation fraction in early-season wheat fields using images from UAV. *Computers and Electronics in Agriculture, 103*, 104–113.

Tran, T. T., Choi, J. W., Le, T. T. H. & Kim, J. W. (2019). A comparative study of deep CNN in forecasting and classifying the macronutrient deficiencies on development of tomato plant. *Applied Sciences, 9*, 1601.

Truong, T., Dinh, A. & Wahid, K. (2017). An IoT environmental data collection system for fungal detection in crop fields. In *Proceedings of the 2017 IEEE 30th Canadian Conference on Electrical and Computer Engineering (CCECE)*, Windsor, ON, Canada, pp. 1–4.

Tsaftaris, S. A., Minervini, M., & Scharr, H. (2016). Machine learning for plant phenotyping needs image processing. *Trends in Plant Science*, *21*(12), 989–991.

Wahabzada, M. et al. (2015). Metro maps of plant disease dynam- ics-automated mining of differences using hyperspectral images. *PLoS One*, *10*, e0116902.

Wang, L., Jin, J., Song, Z., Wang, J., Zhang, L., Rehman, T. U., ... & Tuinstra, M. R. (2020). LeafSpec: An accurate and portable hyperspectral corn leaf imager. *Computers and Electronics in Agriculture*, *169*, 105209.

Wetterich, C. B., Kumar, R., Sankaran, S., Belasque Junior, J., Ehsani, R., & Marcassa, L. G. (2013). A comparative study on application of computer vision and fluorescence imaging spectroscopy for detection of Huanglongbing citrus disease in the USA and Brazil. *Journal of Spectroscopy*, *2013,* 1–6.

White, J. W., Andrade-Sanchez, P., Gore, M. A., Bronson, K. F., Coffelt, T. A., Conley, M. M., ... & Wang, G. (2012). Field-based phenomics for plant genetics research. *Field Crops Research*, *133*, 101–112.

Yang, W., Feng, H., Zhang, X., Zhang, J., Doonan, J. H., Batchelor, W. D., ... & Yan, J. (2020). Crop phenomics and high-throughput phenotyping: past decades, current challenges, and future perspectives. *Molecular Plant*, *13*(2), 187–214.

Zhao, Z. Q., Zheng, P., Xu, S. T., & Wu, X. (2019). Object detection with deep learning: A review. *IEEE Transactions on Neural Networks and Learning Systems*, *30*(11), 3212–3232.

Zhou, J., Applegate, C., Alonso, A. D., Reynolds, D., Orford, S., Mackiewicz, M., ... & Pullen, N. (2017). Leaf-GP: an open and automated software application for measuring growth phenotypes for arabidopsis and wheat. *Plant Methods*, *13*(1), 1–17.

Zhu, H., Cen, H., Zhang, C. & He, Y. (2016). Early detection and classification of tobacco leaves inoculated with tobacco mosaic virus based on hyperspectral imaging technique. In *Proceedings of the 2016 ASABE Annual International Meeting*, Orlando, FL, USA, pp. 1.

Zhu, H., Chu, B., Zhang, C., Liu, F., Jiang, L. & He, Y. (2017). Hyperspectral imaging for presymptomatic detection of tobacco disease with successive projections algorithm and machine-learning classifiers. *Scientific Reports*, *7*, 4125.

Zhuang, Y. T., Wu, F., Chen, C., & Pan, Y. H. (2017). Challenges and opportunities: From big data to knowledge in AI 2.0. *Frontiers of Information Technology & Electronic Engineering*, *18*, 3–14.

14 Artificial Intelligence Techniques in Disease Diagnosis

Ronit Jaiswal, Sajad Un Nabi, Girish Kumar Jha, Mahendra K Verma, and Rajeev Ranjan Kumar

14.1 INTRODUCTION

The term artificial intelligence (AI) was initially coined by John McCarthy in 1955 when he provided the definition as "the science and engineering of creating intelligent machines." AI involves methodologies that replicate human cognitive functions and has seamlessly integrated into numerous household devices. As an illustration, a smartphone utilizes AI-based applications to transcribe audio messages into text or translate them into different languages. These applications streamline tasks through machine learning (ML) algorithms, which autonomously learn and discern patterns from data without the need for explicit programming. Prominent ML algorithms encompass artificial neural networks (ANNs), support vector machines (SVMs), decision trees, and genetic algorithms. Nevertheless, the efficacy of traditional ML, also referred to as shallow learning (SL) techniques, is heavily contingent on the quality of data features supplied by domain experts. Recognizing this constraint, a subset of ML known as deep learning (DL) emerged. DL enhances conventional ML by introducing multiple processing layers into models, facilitating a hierarchical representation of raw data through several layers of abstraction (Schmidhuber, 2015). Essentially, DL constitutes an ANN with numerous hidden layers, whereas the traditional ANN typically incorporates one or two hidden layers (Jha and Sinha, 2014). Deep neural networks (NNs) possess the capability to independently learn and adapt by filtering information through multiple hidden layers. This approach allows them to tackle intricate problems with heightened performance and precision. DL has demonstrated successful applications across various domains, including agriculture (Kamilaris *et al.*, 2017).

14.2 ARTIFICIAL INTELLIGENCE IN AGRICULTURE

In recent years, the agricultural landscape has undergone a transformative shift with the integration of AI technologies. This convergence has given rise to the concept of "smart farming," (Tyagi, 2016) marking a departure from traditional agricultural methods. As global population growth continues to strain food production

DOI: 10.1201/9781003474036-14

systems, the need for innovative solutions becomes imperative. AI, with its diverse applications and capabilities, emerges as a key player in addressing the challenges faced by the agricultural sector. Smart farming represents a paradigm shift in agricultural practices, where information technologies and AI converge to optimize complex farming systems. The integration of information and communication technologies facilitates the creation of intelligent systems that enhance the efficiency and productivity of agriculture. The agricultural sector, encompassing crops, forestry, livestock, and aquaculture, stands to benefit significantly from the incorporation of AI. There are many applications of AI in agriculture based on different activities such as:

- **Crop and Soil Monitoring:** One of the primary applications of AI in agriculture is the real-time monitoring of crops and soil health. ML algorithms analyze data collected from various sources, including sensors and satellite imagery, providing farmers with valuable insights into crop conditions. This enables timely interventions, allowing for precise irrigation, fertilization, and pest control strategies.
- **Disease Diagnosis:** Early detection of plant diseases is crucial for effective disease management. AI-powered systems, equipped with machine vision and sensor technologies, aid in the early classification of plant diseases. By analyzing visual cues and data patterns, these systems assist farmers in implementing targeted strategies to mitigate the impact of diseases, reducing crop losses.
- **Agricultural Robots:** The integration of sensors, machine vision, and AI models has paved the way for the use of agricultural robots. These robots can perform tasks such as harvesting with remarkable precision and speed, addressing the challenges posed by labor shortages in the agricultural sector. The deployment of robots not only enhances efficiency but also minimizes crop wastage compared to traditional harvesting methods.
- **Predictive Insights:** AI facilitates the generation of predictive insights that empower farmers to make informed decisions. Predictive analytics, coupled with ML algorithms, enables the forecasting of crop yields and optimal harvesting times. This foresight allows farmers to plan and manage their operations more effectively, maximizing overall agricultural productivity.
- **Intelligent Spraying:** The application of AI in intelligent spraying systems is instrumental in optimizing the use of chemicals for pest control. By analyzing data on pest prevalence and environmental conditions, AI-driven systems can determine the precise amount and timing of chemical sprays. This not only reduces the environmental impact but also leads to cost savings for farmers.

These applications collectively bring resource efficiency, precision farming, and increased profitability to agriculture, transforming it into a more sustainable and productive sector. While challenges like the initial cost and data privacy concerns exist, ongoing research aims to overcome these hurdles, promising a future where AI plays a central role in ensuring global food security through efficient and sustainable agricultural practices.

14.3 PLANT DISEASE DIAGNOSIS

Diseases affecting plants can manifest in various parts, with leaves being a particularly significant and prominent component due to their role in photosynthesis, producing chlorophyll essential for plant growth (Chouhan *et al.*, 2020). Leaf diseases have a direct impact on plant yield and survival, causing drops or withering, consequently leading to decreased crop productivity and increased production costs. Traditionally, farmers relied on manual inspections by labor and experts, presenting drawbacks such as high manpower requirements, training needs, fatigue in manual tasks, and challenges in timely and large-scale disease detection, often subjective due to human errors and bias. ML approaches have emerged as more effective solutions, with image analysis of plant leaves demonstrating their efficacy and simplicity in plant leaf pathology classification. Plant leaf images, easily obtainable using cameras, provide advantages in terms of dataset collection, analysis, and reproducibility, enabling the extraction of valuable features for improved agricultural production quality and quantity. Plant disease identification and classification must be efficient and timely in order to reduce production losses. In the era of precision agriculture (PA) or smart agriculture (SA), ML technologies have been researched and implemented, particularly in plant leaf pathology classification, enabling automatic detection of plant leaf diseases at an early stage (Ahmed and Yadav, 2023; Bondre and Patil, 2023; Upadhye *et al.*, 2023; Yao *et al.*, 2023)

DL and conventional (shallow) ML are the two methods currently used to classify leaf diseases (DL). A branch of ML, deep learning (DL) is a field of study that focuses on developing new ways to enhance the effectiveness and usefulness of ML techniques. In shallow learning (SL), the pivotal role of feature extraction often necessitates expert involvement for engineering valuable features. Conversely, DL has the potential to alleviate the cost of feature engineering by enabling effective learning across extensive datasets. While DL may encounter challenges related to data intensity, the ease of collecting leaf images and farmer participation in disease annotation mitigate this issue. To diminish reliance on labeled data, data augmentation methods are employed to generate additional training data and fortify model robustness. Transfer learning emerges as a potential strategy for decreasing the demand for tagged leaf data by leveraging pre-trained models from other tasks. The effectiveness of ML techniques depends on the quality and amount of data. Therefore, this survey differs from previous ones in that it discusses the suitability of public datasets for ML model evaluation, as well as their availability and quality. The process of disease identification involves several key steps (Bondre and Patil, 2023):

a) **Image Collection:** Photos of the crops can be captured using various devices such as a digital camera, digital printer, or handheld camera. Both diseased and healthy crop images are obtained to construct a comprehensive dataset for analysis.
b) **Image Preprocessing and Segmentation:** Techniques like image resizing, smoothing, contrast enhancement, and image enhancement are employed to eliminate noise and unwanted artifacts from the images. In the segmentation step, methods like K-means, Otsu process, RGB (red, green, blue) to HSI

(hue, saturation, lightness) conversion, and RGB to HSV (hue, saturation, and value) conversion are used to divide the images into distinct functional parts.

c) **Feature Extraction:** Features such as shape, color, texture, and edges are extracted from the images. Various techniques, including global color-histogram, local binary pattern, color coherence vector, and spatial gray-level dependency matrices, are applied during this stage.
d) **Classification:** Image detection algorithms such as SVMs, ANNs, backpropagation, and CNNs are employed to identify specific diseases on plant leaves. This step involves extending the analysis to the dataset after feature extraction.
e) **Testing:** The model's evaluation is carried out using a subset of features not utilized in the training process. Alternatively, a picture of a diseased plant can be processed through all the aforementioned steps, checking if the classifier correctly identifies the disease. This phase serves to verify the accuracy of the model.

Through these systematic steps, the identification of plant diseases becomes a structured process, leveraging image analysis and ML techniques to enhance the efficiency and accuracy of disease detection in agriculture.

14.4 SYMPTOMS AND DISEASES OF LEAVES

Based on the origin of the principal agent, plant diseases are classified as either contagious or non-infectious. When plant diseases impede essential processes including photosynthesis, translocation, water absorption, nitrogen uptake, fruit and flower production, and overall plant development and growth, they can seriously damage plants. Plant diseases can be caused by a variety of organisms, including as bacteria, fungus, viruses, viroids, nematodes, and phytoplasma. Bacteria, for instance, can cause rot, wilt, scab, bacterial blot, and bacterial blight. Greymold, aphids, powdery mildew, spotted wilt, mealy bugs, downy mildew, Cylindrocladium, curly-top, and mosaic are examples of fungal-induced illnesses.

Each plant exhibits distinctive characteristics that aid in identifying the type of disease, whether caused by bacteria, fungi, or viruses. The symptoms of a disease are crucial in determining its true nature and cause. The images in Figure 14.1 illustrate bacterial, fungal, and viral diseases, discussed as follows:

a) Bacterial diseases manifest in distinct ways (spots, blight, rots, scab, etc.) depending on the specific type. Bacterial spots are characterized by small black spots and water-soaked areas on infected leaves, visible from both sides. Contributing factors include weak soil drainage, a moist climate, humid temperatures, low air circulation, and heavy rainfall. In bacterial blight disease, untreated leaf blades develop yellow water-soaked lesions along the margins, with bacterial discharge occurring in the morning. This disease thrives in conditions of heavy rain, warm temperatures, high humidity, weak air circulation, poor soil drainage, and limited sunshine. Scab disease is

aBacterial disease: Fire blight in apple

bFungal disease: Apple Alternaria leaf spot

cViral disease: Apple mosaic

FIGURE 14.1 Diseased leaf images.

influenced by soil pH and moisture content, spreading due to warm and dry soil conditions. In rot disease, infection begins on outer leaves in direct soil contact, spreading to the head and resulting in mushy, fuzzy, and dark areas. Hot and humid weather with abundant rainfall contributes to the development of this disease.

b) Fungal diseases exhibit diverse symptoms depending on the specific type. Gray-mold disease easily spreads in gardens, especially during moist, mild, or cold weather conditions. Tree stress is exacerbated by aphids, which are frequently seen on sap-dripping tree leaves. These insects have a pear form and come in different colours. When downy mildew first appears, top leaf surfaces have yellow patches that eventually turn dark yellow. The center of the leaf spot then turns brown with a yellow border. The incidence of diseases is associated with both variable damp weather and inadequate plant aeration. Powdery mildew first appears on leaves as spherical powdery-white patches that eventually turn yellow-brown and finally black. It then appears as a dusty-white to grey coating. Mealy bugs are soft-bodied, wingless insects that appear as white, cottony masses on plant stems, leaves, and fruit in colder climates.

c) Viral infections have distinct symptoms, such as mosaic sickness. The leaves that are infected may seem bent, stunted, or exhibit white, light yellow, or pale green color spots. Plant stress, contaminated planting materials, and weedy fields are a few of the factors that lead to the disease's spread. The virus is spread by spotted-wilt insects, which prick leaves to remove plant material and feed on a variety of crops. Other signs include leaves with black or golden spots, dark streaks in the terminal branches of the plant, slowed development, and perhaps dying back of the growing tips. When a plant has curly-top disease, its young leaves and shoots grow more slowly, taking longer to mature and eventually becoming dwarfed and bushy. The spread of this disease is facilitated by weedy land and contaminated transplant conditions.

14.5 DATASETS

The role of data in modern AI is considered crucial, particularly with the recent prominence of DL techniques. An important factor influencing the effectiveness of big models used in DL is the amount and calibre of training data (Goodfellow *et al.*, 2016). Image databases are considered essential for computational vision problems in both research and real-world applications. In the field of leaf disease categorization, several researchers are working to compile information on plant diseases for the general audience. Plant Pathology 2021 FGVC8 dataset, Plant Village dataset, Plant Leaves dataset, Plantae K dataset, PlantDoc dataset, Maize Leaf (NLB) dataset, Citrus dataset, Rice Diseases Image dataset, UCI Rice Leaf Diseases dataset, and other noteworthy publicly available datasets are only a few examples.

- Plant Pathology 2021-FGVC8 is a Kaggle challenge dataset for apple leaf diseases, and is part of the FGVC8 workshop at CVPR 2021. With approximately 23,000 images, each apple leaf is labeled with one or more of six health categories: "healthy," "complex," "rust," "frog eye leaf spot," "powdery mildew," and "scab." The "complex" category is used to denote an unhealthy leaf with an unidentified cause, contributing to its value for multi-class apple leaf disease classification.
- The Plant Village Dataset, widely utilized for leaf disease classification, includes two versions, an original one and an augmented one. The original dataset comprises 54,305 images of leaf diseases or healthy from 14 plant species, each with 1–10 classes of diseases or healthy (Hughes and Salathe, 2015). In 2019, Hughes and Salathe released a data augmentation version, employing six methods like image rotation, flipping, scaling, noise injection, PCA color augmentation, and gamma correction leading to increasing the dataset to 61,486 images.
- The Plant Leaves dataset comprises 4,502 high-resolution JPG images categorizing healthy and diseased leaves into 22 categories based on species and health state. The dataset includes tree types like Alstonia Scholaris, Arjun, Bael, Basil, Mango, Guava, Lemon, Jamun, Pomegranate, Jatropha, Chinar, and Pongamia Pinnata.
- The Plantae_K dataset consists of 2,153 high-resolution JPG images categorizing healthy and diseased plant leaves into 16 categories based on species and health state, such as "apple healthy" and "apple diseased." This dataset encompasses eight fruit types, including Apple, Grapes, Cherry, Pear, Apricot, Walnut, Cranberry, and Peach.
- The PlantDoc dataset aims to provide a real-world images collection. Singh *et al.* (2020) raised concerns about the Plant Village images being captured in laboratory setups, potentially impacting the efficacy of trained models in real cultivation field conditions. The PlantDoc dataset, which offers a sizable non-laboratory dataset for leaf disease classification, was created in order to address this problem. Images from PlantDoc provide a more realistic picture because they are unformatted and have messy backgrounds. Featuring 2,598 leaf photos from 13 different plant species, PlantDoc offers 17 distinct disease categories

as well as 38 classes that show combinations of illnesses and species (e.g., Apple Scab Leaf). Expert annotations played a role in creating this dataset.

- The Maize Leaf (NLB) Dataset, collected by Wiesner-Hanks *et al.* (2018), makes use of a variety of camera setups, such as hand cameras, cameras placed on drones, and cameras on a 5 m boom. The maize leaf dataset, the largest publicly available dataset for a single plant species to date, has 105,735 annotated Northern leaf blight (NLB) lesions spread across over 18,222 photos.
- The Citrus dataset is split into two folders: 150 photos of citrus fruits and 609 photos of citrus leaves are found in one folder. Five categories—canker, black spot, melanose, greening, and healthy—are included in each folder. Every photograph within both files has been expertly annotated.
- The Rice Diseases Image Dataset includes four categories of rice leaves: Leaf Blast (523 photos), Hispa (565 images), Brown Spot (523 images), and Healthy (1,488 images) (779 images). This dataset has been the subject of study in various works, such as those conducted by Kathiresan *et al.* (2021) and Bari *et al.* (2021), focusing on leaf disease classification.
- The UCI Rice Leaf Diseases dataset is designed for the identification and categorization of rice plant diseases as described by Prajapati *et al.* (2017), consists of three disease categories: Bacterial leaf blight, Brown spot, and Leaf smut, with 40 images in each category, totalling 120 images. Although its tiny size makes it useful for rapid testing and ML prototyping, it might not be as applicable to DL techniques, which usually require larger datasets.

A comprehensive dataset for crop disease identification using cameras involves several critical considerations. The process begins by selecting the specific crop species and categorizing the various diseases affecting it. Image annotation with corresponding disease labels is essential for training ML models. Choosing suitable cameras, such as handheld devices or drones, is crucial for capturing images under diverse field conditions. The images should be well-composed, capturing the entire plant or specific parts, and taken from multiple angles and viewpoints to enhance model adaptability.

14.6 ARTIFICIAL INTELLIGENCE TECHNIQUES

Presently, AI techniques for leaf disease classification can be broadly categorized into two directions: SL and DL as illustrated in Table 14.1. SL methods typically involve leaf localization as a preliminary step, followed by disease classification based on the identified leaves. Feature extraction is crucial in this approach to capture leaf characteristics before classification. DL, on the other hand, has gained significance for leaf disease classification because to its ability to do end-to-end learning and prediction, eliminating the requirement for explicit feature engineering. However, the advantages and drawbacks of SL and DL approaches are still a subject of ongoing discussion. SL is criticized for its limitations in handling large datasets, complex processing pipelines, and the necessity for feature extraction. On the other hand, DL is computationally expensive and demands substantial amounts of data. Recent efforts to address data scarcity involve augmented learning techniques, where artificial data generation and reuse of pre-trained models from other domains/tasks are employed to enhance performance. The computational expenses of DL have also seen mitigation with advancements in hardware and computing systems.

14.6.1 Shallow Learning

SL, also known as traditional ML or shallow models, refers to a class of algorithms that have a limited number of layers (typically one or two) in their architecture. These models are characterized by their simplicity and reliance on manually engineered features. Common examples of SL models include linear regression, logistic regression, decision trees, support vector machines, k-nearest neighbours, and shallow neural networks. Major differences between SL and DL are presented Table 14.1.

14.6.1.1 Feature Detection Techniques

K-means clustering: This is a popular unsupervised ML algorithm used for data clustering and segmentation (Javidan *et al.*, 2023). It aims to partition a dataset into K clusters, where each data point belongs to the cluster with the nearest mean. In the context of disease identification, this means grouping pixels or regions with similar characteristics, potentially corresponding to healthy or diseased portions of a plant. K-means is computationally efficient and relatively easy to implement. It provides a straightforward way to identify patterns within images, aiding in the localization of diseased regions. The algorithm is versatile and can be adapted to various image characteristics, including color and texture.

It can be applied to cluster pixels based on color information in images. For instance, it can distinguish between healthy and diseased areas by grouping pixels with similar color properties. K-means can be extended to cluster pixels based on texture features, aiding in the identification of textural patterns associated with diseases. This is particularly useful in scenarios where color alone may not provide sufficient discrimination. By segmenting the image into clusters, K-means helps isolate regions potentially affected by diseases. This segmentation can serve as valuable input for further analysis and classification.

Grey-Level Co-occurrence Matrix (GLCM): GLCM is a feature extraction technique widely used in image processing for capturing texture information (Manik *et al.*, 2020). In the context of disease identification, GLCM can be utilized to quantify the spatial relationships between pixel intensities, offering valuable insights into the texture properties of images. GLCM is constructed by calculating the frquency of pixel pairs with specific intensity values at a certain spatial relationship within an image. The resulting matrix provides information about the texture patterns, such as coarseness or smoothness.

GLCM is well-suited for capturing intricate texture details, making it valuable for disease identification where subtle variations may be indicative of specific conditions. The extracted features from GLCM can be used as input for ML models, aiding in the training and classification process. GLCM-derived features, such as contrast, entropy, and homogeneity, can serve as discriminative features for ML models aiming

TABLE 14.1
Comparison of SL and DL

S.N.		SL	DL
1.	Feature Engineering	• Explicit feature engineering is often required. This involves selecting and extracting relevant features from the input data, such as texture, color, or shape characteristics. • Features need to be carefully crafted by domain experts, making the process reliant on human expertise and understanding of the problem.	• Eliminates the need for manual feature engineering. Deep neural networks can automatically learn hierarchical representations from raw data, extracting relevant features as part of the training process. • Advantageous in scenarios where manually crafting features might be challenging or impractical.
2.	Data Requirements	• SL models may struggle with large and complex datasets. Their performance tends to plateau as the dataset size increases, and they may not fully exploit the potential of big data.	• DL models thrive on large datasets. They are data-hungry and can leverage massive amounts of information to learn intricate patterns, resulting in improved generalization to unseen data.
3.	Model Complexity	• SL models typically have a simpler architecture with fewer layers. They are less computationally intensive and are suitable for scenarios where a less complex model is sufficient.	• DL models, with their deep neural architectures, are inherently more complex. This complexity enables them to capture intricate relationships within the data but comes at the cost of increased computational requirements.
4.	Computational Efficiency	• SL models are computationally efficient and have lower training times. They are often preferred in scenarios where quick model deployment is essential.	• DL models, especially deep neural networks, can be computationally expensive and require substantial computing resources. Advances in hardware, such as GPUs and TPUs, have mitigated some of these challenges.
5.	Interpretability	• SL models generally provide more interpretable results. The relationship between input features and the output is often more transparent, aiding in understanding the decision-making process.	• DL models, especially deep neural networks, are often considered as "black boxes" due to their complex architectures. Understanding the internal representations and decision-making mechanisms can be challenging.

(*continued*)

TABLE 14.1 (Continued)
Comparison of SL and DL

S.N.		SL	DL
6.	Performance on Large Datasets	• SL models may struggle to scale and fully exploit the potential of large datasets. They might reach a performance plateau beyond a certain dataset size.	• DL excels when provided with large datasets. The ability to learn hierarchical features allows deep neural networks to handle the complexity and variability present in extensive datasets.
7.	Robustness to Noisy Data	• SL models might be more sensitive to noisy data. Their performance can be affected by irrelevant or misleading features.	• DL models, with their ability to learn hierarchical representations, can be more robust to noisy data. They can automatically filter out irrelevant information during training.
8.	Application Areas	• SL is often applied in situations where the dataset is not massive, and the problem is not highly complex. It is suitable for tasks with well-defined features and simpler relationships.	• DL is well-suited for tasks where the data is abundant and exhibits complex patterns. It has shown remarkable success in image recognition, natural language processing, and tasks involving high-dimensional data.

to distinguish between healthy and diseased samples. GLCM considers the spatial relationships between pixel intensities, providing a comprehensive texture analysis that enhances the ability to characterize subtle differences in disease-related patterns.

14.6.1.2 Classification Techniques

Among many classifiers including multilayer perceptron (MLP), support vector machine (SVM), decision trees, random forest (RF), etc., SVM outperforms the others (Ahmed and Yadav, 2023; Sahu and Pandey, 2023; Yao *et al.*, 2023)

Support Vector Machine (SVM): This is a powerful ML technique widely employed in the field of agriculture for disease detection in plants. SVM (Liu *et al.*, 2012) is a supervised learning algorithm that can efficiently classify and analyze data, making it valuable in identifying diseases affecting plant health. SVM operates by finding the optimal hyperplane that best separates different classes in a high-dimensional space. In the context of plant disease detection, these classes represent healthy and diseased plants. SVM aims to maximize the margin between classes, enhancing its ability to generalize well to unseen data.

Effective disease detection relies on the selection of informative features that distinguish between healthy and diseased plants. SVM accommodates various feature representations, including spectral, textural, and morphological features extracted from plant images or sensor data. The choice of features significantly influences the performance of the SVM model. SVM utilizes kernel functions to transform input data into a higher-dimensional space, allowing for better separation of classes. Common kernel functions for plant disease detection include linear, polynomial, and radial basis function (RBF) kernels. The selection of an appropriate kernel is crucial for capturing complex relationships in the data.

Multilayer Perceptron (MLP): This kind of ANN consists of an input layer, one or more hidden layers, and an output layer, among other layers (Jaiswal *et al.*, 2022). It is a feedforward neural network capable of learning complex relationships in data. It comprises of neurons in one layer connected to neurons in the next layer through weighted connections. Each connection has an associated weight, and the network learns these weights during training to map input patterns to output patterns. Each neuron consists of one activation function. Non-linear activation functions, such as the sigmoid or ReLU (Rectified Linear Unit), are commonly used in MLP to introduce non-linearity and enable the network to learn from non-trivial patterns. Backpropagation is used in MLP training, where errors are transmitted backward through the network and corrected by adjusting the weights to minimise the errors. Gradient descent optimization algorithms are often employed.

Decision Trees: These are a supervised ML approach that may be applied to regression and classification problems (Resti *et al.*, 2022). They model decisions based on a series of conditions and their outcomes, forming a tree-like structure. The tree consists of nodes representing conditions and branches representing possible outcomes based on those conditions. The tree is built recursively by selecting the most informative features at each node. Decision trees split the dataset based on features that maximize information gain or minimize impurity, depending on the algorithm used (e.g., ID3, C4.5, CART). Popular impurity measures include Gini impurity and entropy. Decision trees can be trimmed to prevent overfitting by eliminating branches that do not considerably increase accuracy. Pruning aids in the model's generalisation to fresh, untested data. By analysing variables like colour, texture, and shape of plant photos, decision trees can be used to diagnose plant diseases. The decision-making process involves a series of splits, eventually leading to a disease classification.

Random Forest: This is an ensemble learning technique where a large number of decision trees are built during training. The output is based on the mean forecast of the individual trees for regression or the mode of the classes for classification (Chaudhary *et al.*, 2016). At each split, a random subset of features and a random subset of training data are used to build each tree in the forest. The introduction of this randomness instills diversity among the trees, enhancing the model's robustness and reducing susceptibility to overfitting. In classification tasks, each tree "votes" for a class, and the predicted class is determined by the class with the highest number of votes. This ensemble approach improves accuracy and generalization. RF is a form of bagging (Bootstrap Aggregating), and it can be seen as a combination of bagging and feature randomization. It is effective in reducing overfitting and increasing the model's stability. RF can enhance disease detection in plants by aggregating the predictions of multiple decision trees. Each tree contributes to the final decision, increasing accuracy and robustness.

14.6.2 Deep Learning

DL methods, as previously discussed, are subdivisions of ML that, built upon ANNs, possess increased depth either in the number of layers or in complexity. These techniques are specifically designed to autonomously extract patterns or features from raw data. The distinctive feature of DL techniques lies in their capacity for layer-wise abstraction in a hierarchical manner, progressing from raw input to a final abstract level. This capability allows them to understand and learn from even the most complex and non-linear datasets. Different DL techniques utilize various mechanisms for feature extraction; some involve additional hidden layers, others use gates and memory cells, and some embrace innovative structures such as convolutions and pooling, or encoding–decoding strategies. Regardless of the specific method or mechanism, their primary objective is to construct a model that performs multilevel abstractions from the data, yielding meaningful outcomes. DL techniques exhibit diversity, and their classification is rooted in learning algorithms, broadly categorized into unsupervised and supervised approaches. Supervised DL methods comprise recurrent neural networks (RNN) or long short-term memory (LSTM) and convolutional neural networks (CNN), which incorporate feedback. In contrast, unsupervised or semi-supervised techniques encompass autoencoders (AEs), self-organizing maps (SOMs), and restricted Boltzmann machines.

14.6.2.1 Convolution Neural Networks

CNNs (or ConvNets) (Goodfellow *et al.*, 2016; LeCun *et al.*, 2015) stand out as one of the most extensively applied DL techniques, specifically designed to handle high-dimensional data such as image data represented as a three-dimensional grid of pixels. CNNs are a subset of neural networks (NNs) (Figure 14.2) that use convolution

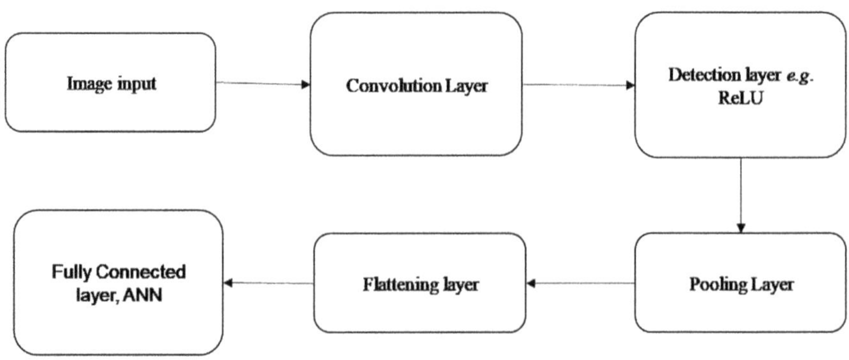

FIGURE 14.2 Flowchart of CNN.

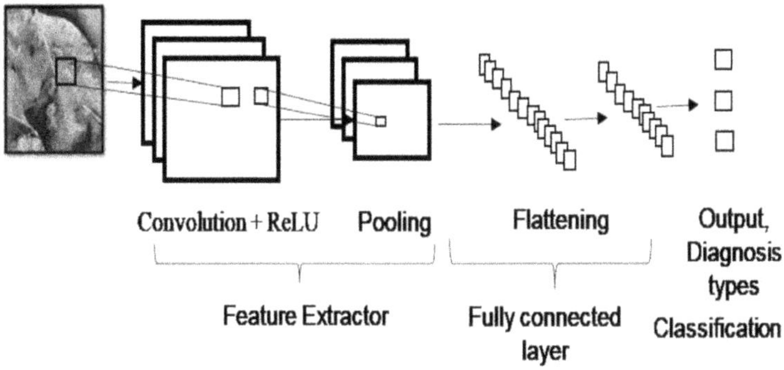

FIGURE 14.3 Typical architecture and analysis procedure of a CNN model.

operations (the combined integration of two functions) in one or more layers as opposed to the traditional matrix multiplication. The first function in this case is called the input, and the second, which adapts as it learns, is called the kernel or feature detector. The output that is produced is sometimes referred to as the feature map. Three separate processes are involved at different stages in a typical CNN layer (Figure 14.3). Multiple parallel convolutions are performed in the first step to produce a set of linear activations. Through the use of a non-linear activation function, such as the ReLU, also known as the detector stage, the second stage adds nonlinearity to each linear activation. In the third step, pooling operations are performed. A pooling function uses the mean, maximum, or minimum of nearby outputs as a summary statistic to replace the output at a given position. After pooling, the numerous pooled feature maps are flattened into a vector of input data via the flattening operation. This vector is then sent to the fully connected layer or ANN to generate classification results.

A multitude of CNNs has been developed to address various challenges in image classification. It is straightforward to select and apply an existing model to classify diseases from leaf images. Some of the available models are as follows:

- LeNet and GoogLeNet: LeNet, an early CNN, serves as the inspiration for many contemporary deep CNN models despite its relatively shallow architecture. It consists of seven layers, including three convolutional layers, two subsampling layers, and two fully connected layers. Although it achieved lower accuracy (94.0%) in Plant Village compared to other approaches, its evolution, GoogLeNet (Inception V1), introduced improvements such as batch normalization, image distortions, and additional layers. In various studies, GoogLeNet showcased enhanced performance, achieving 95.69% accuracy in Apple disease classification and an impressive 98.9% accuracy for Maize leaf disease classification.
- AlexNet, among the earliest deep CNN models, has been extensively utilized in various leaf disease classification studies. It consists of eight layers, including five convolutional layers, three fully connected layers, and the use of ReLU activation functions. In assessments for tomato diseases, it demonstrated promising results with 95.75% accuracy in one study and 90.1% accuracy in another. For

grape diseases, AlexNet achieved an accuracy of 86.5%. However, compared to other deep CNNs, its standalone performance was often inferior.

- VGG, which stands for Very Deep Convolutional Networks or VGGNet, presents an effective strategy for increasing the depth of CNNs. VGG-16 has been implemented in studies involving tomato leaf datasets, with notable applications reported. It is characterized by its simplicity, with 16 or 19 weight layers, and the extensive use of 3x3 convolutional filters. In one instance, a pre-trained model achieved 77.2% accuracy, while another study proposed an improved training approach leading to significantly higher performance, reaching 90.1% accuracy. Furthermore, a deeper version of VGG, VGG-19, demonstrated success in classifying tomato leaf diseases, achieving an accuracy of 96.86%.
- ResNet, among the various deep CNN models, stands out for its powerful structure, enabling the training of models with numerous layers for superior performance. In the context of tomato leaves, ResNet-50 achieved an accuracy of 98.40%. Another study by Guan (2021) applied ResNet to achieve an accuracy of 82.78% on a modified Plant Village dataset. For Betelvine leaf disease, Kumar *et al.* (2020) demonstrated that ResNet-34 outperformed other models, achieving an impressive 99.40% accuracy and a 0.9651 F1-score. These results surpass the performance of SVM (50.69% & 50.57%), Decision Tree (72.23% & 72.02%), Logistic Regression (80.99% & 80.88%), and K-NN (87.86% & 88.06%).

14.7 DISCUSSION

The exploration of DL techniques for plant disease diagnosis has unveiled promising avenues for revolutionizing agriculture and crop management. The deployment of CNNs and other DL models has exhibited substantial potential in automating and enhancing the accuracy of plant disease identification. This discussion delves into key aspects, challenges, and future prospects associated with the application of DL techniques in the realm of plant disease diagnosis.

Accuracy and Reliability: The studies presented in this chapter underscore the commendable accuracy achieved by DL models in distinguishing and classifying various plant diseases. The robustness of these models, especially in image-based diagnoses, showcases their capability to outperform traditional methods. However, it is essential to consider factors influencing model performance, such as dataset quality, model architecture, and generalization across diverse plant species and environmental conditions.

Transferability and Generalization: An important facet of the discussion involves the transferability of DL models. While these models exhibit proficiency within specific datasets, challenges arise when applying them to new or unseen datasets. Enhancing the generalization capabilities of models across different crops, geographical regions, and environmental variations is crucial for their widespread adoption in real-world agricultural scenarios.

Data Challenges and Augmentation: The availability of comprehensive and diverse datasets remains a critical factor in the efficacy of DL models. Addressing challenges related to data collection, annotation, and imbalance is imperative. Furthermore, the discussion explores the role of data augmentation techniques in mitigating data scarcity issues, thereby contributing to the robustness of the trained models.

Interpretability and Explainability: The inherent complexity of DL models poses challenges in understanding the decision-making processes. Discussing techniques for model interpretability and explainability becomes pertinent, especially in applications where the rationale behind disease predictions is crucial for gaining the trust of end-users, such as farmers and agricultural practitioners.

Integration with Precision Agriculture: The potential integration of DL techniques with precision agriculture practices is a key focal point. Discussing how these models can contribute to real-time monitoring, early detection, and precise intervention strategies will pave the way for sustainable agriculture and resource optimization.

Future Directions: Looking forward, potential avenues for future research and development include investigating advanced model architectures, leveraging multimodal data sources, and embracing emerging technologies like edge computing for on-device inference. Additionally, collaborative efforts among researchers, agricultural experts, and technology developers are crucial for addressing interdisciplinary challenges in the field.

REFERENCES

Ahmed, I., & Yadav, P. K. (2023). A systematic analysis of machine learning and deep learning based approaches for identifying and diagnosing plant diseases. *Sustainable Operations and Computers*, *4*(March), 96–104. https://doi.org/10.1016/j.susoc.2023.03.001

Bari, B. S., Islam, M. N., Rashid, M., Hasan, M. J., Razman, M. A. M., Musa, R. M., Ab Nasir, A. F., & P.P. Abdul Majeed, A. (2021). A real-time approach of diagnosing rice leaf disease using deep learning-based faster R-CNN framework. *PeerJ Computer Science*, *7*, e432. https://doi.org/10.7717/peerj-cs.432

Bondre, S., & Patil, D. (2023). Recent advances in agricultural disease image recognition technologies: A review. *Concurrency and Computation: Practice and Experience*, *35*(9), 1–14. https://doi.org/10.1002/cpe.7644

Chaudhary, A., Kolhe, S., & Kamal, R. (2016). An improved random forest classifier for multi-class classification. *Information Processing in Agriculture*, *3*(4), 215–222. https://doi.org/10.1016/j.inpa.2016.08.002

Chouhan, S. S., Singh, U. P., & Jain, S. (2020). Applications of computer vision in plant pathology: A survey. *Archives of Computational Methods in Engineering*, *27*(2), 611–632. https://doi.org/10.1007/s11831-019-09324-0

Goodfellow, I., Bengio, Y., & Courville, A. (2016). *Deep Learning*. Cambridge, MIT Press.

Guan, X. (2021). *A novel method of plant leaf disease detection based on deep learning and convolutional neural network. In* 6th International conference on intelligent computing and signal processing (ICSP) *(pp. 816-819).* IEEE. https://doi.org/10.1109/ICSP51882.2021.9408806

Hughes, D. P., & Salathe, M. (2015). *An open access repository of images on plant health to enable the development of mobile disease diagnostics.* http://arxiv.org/abs/1511.08060

Jaiswal, R., Jha, G. K., Kumar, R. R., & Choudhary, K. (2022). Deep long short-term memory based model for agricultural price forecasting. *Neural Computing and Applications*, *34*(6), 4661–4676. https://doi.org/10.1007/s00521-021-06621-3

Javidan, S. M., Banakar, A., Vakilian, K. A., & Ampatzidis, Y. (2023). Diagnosis of grape leaf diseases using automatic K-means clustering and machine learning. *Smart Agricultural Technology*, *3*(June 2022), 100081. https://doi.org/10.1016/j.atech.2022.100081

Jha, G. K., & Sinha, K. (2014). Time-delay neural networks for time series prediction: An application to the monthly wholesale price of oilseeds in India. *Neural Computing and Applications*, *24*(3–4), 563–571. https://doi.org/10.1007/s00521-012-1264-z

Kamilaris, A., Kartakoullis, A., & Prenafeta-Boldú, F. X. (2017). A review on the practice of big data analysis in agriculture. *Computers and Electronics in Agriculture*, *143*, 23–37. https://doi.org/10.1016/J.COMPAG.2017.09.037

Kathiresan, G., Anirudh, M., Nagharjun, M., & Karthik, R. (2021). Disease detection in rice leaves using transfer learning techniques. *Journal of Physics: Conference Series*, *1911*(1), 012004. https://doi.org/10.1088/1742-6596/1911/1/012004

Kumar, V., Arora, H., & Sisodia, J. (2020). Resnet-based approach for detection and classification of plant leaf diseases. *In* 2020 international conference on electronics and sustainable communication systems (ICESC) *(pp. 495-502).* IEEE. https://doi.org/10.1109/ICESC48915.2020.9155585

LeCun, Y., Bengio, Y., & Hinton, G. (2015). Deep learning. *Nature*, *521*(7553), 436–444. https://doi.org/10.1038/nature14539

Liu, X., Gao, C., & Li, P. (2012). A comparative analysis of support vector machines and extreme learning machines. *Neural Networks*, *33*, 58–66. https://doi.org/10.1016/j.neunet.2012.04.002

Manik, F. Y., Saputra, S. K., & Ginting, D. S. B. (2020, June). Plant classification based on extraction feature gray level co-occurrence matrix using k-nearest neighbour. *In* Journal of Physics: Conference Series (Vol. 1566, No. 1, p. 012107*)*. IOP Publishing.

Prajapati, H. B., Shah, J. P., & Dabhi, V. K. (2017). Detection and classification of rice plant diseases. *Intelligent Decision Technologies*, *11*(3), 357–373. https://doi.org/10.3233/IDT-170301

Resti, Y., Irsan, C., Amini, M., Yani, I., Passarella, R., & Zayantii, D. A. (2022). Performance improvement of decision tree model using fuzzy membership function for classification of corn plant diseases and pests. *Science and Technology Indonesia*, *7*(3), 284–290. https://doi.org/10.26554/sti.2022.7.3.284-290

Sahu, S. K., & Pandey, M. (2023). An optimal hybrid multiclass SVM for plant leaf disease detection using spatial fuzzy C-means model. *Expert Systems with Applications*, *214*, 118989. https://doi.org/10.1016/j.eswa.2022.118989

Schmidhuber, J. (2015). Deep learning in neural networks: An overview. *Neural Networks*, *61*, 85–117. https://doi.org/10.1016/j.neunet.2014.09.003

Singh, D., Jain, N., Jain, P., Kayal, P., Kumawat, S., & Batra, N. (2020). PlantDoc. *Proceedings of the 7th ACM IKDD CoDS and 25th COMAD*, 249–253. https://doi.org/10.1145/3371158.3371196

Tyagi, A. C. (2016). Towards a second green revolution. *Irrigation and Drainage*, *65*(4), 388–389. https://doi.org/10.1002/ird.2076

Upadhye, S. A., Dhanvijay, M. R., & Patil, S. M. (2023). Sugarcane disease detection using CNN-Deep learning method: An Indian perspective. *Universal Journal of Agricultural Research, 11*(1), 80–97. https://doi.org/10.13189/ujar.2023.110108

Wiesner-Hanks, T., Stewart, E. L., Kaczmar, N., DeChant, C., Wu, H., Nelson, R. J., Lipson, H., & Gore, M. A. (2018). Image set for deep learning: Field images of maize annotated with disease symptoms. *BMC Research Notes, 11*(1), 440. https://doi.org/10.1186/s13104-018-3548-6

Yao, J., Tran, S. N., Sawyer, S., & Garg, S. (2023). Machine learning for leaf disease classification: Data, techniques and applications. In *Artificial Intelligence Review* (Issue 0123456789). Springer Netherlands. https://doi.org/10.1007/s10462-023-10610-4

Yustika Manik, F., Kana Saputra, S., & Sartika Br Ginting, D. (2020). Plant classification based on extraction feature gray level co-occurrence matrix using k-nearest neighbour. *Journal of Physics: Conference Series, 1566*(1), 012107. https://doi.org/10.1088/1742-6596/1566/1/012107

15 Applications of Nanotechnology for Rapid Detection of Plant Pathogens

Ravpreet Kaur, Nitika Gupta, and Deeksha Joshi

15.1 INTRODUCTION

Over the years, application of modern technology in agriculture has enabled global food production to more or less keep pace with the increasing human population; but a collateral damage of this has been environmental deterioration. Plant diseases still threaten food security with continuous emergence and spread of new pathogen lineages often facilitated by the changing climate conditions. Moreover, despite the major technological advancements in agriculture, even now it is estimated that 18–26% of the annual crop production worldwide is lost to insect pests while diseases account for 17–24% losses in various crops (Culliney, 2014; Savary *et al.*, 2019). These losses are especially higher in the developing nations of Asia and Africa, where most of the future increase in world population is expected. It has been estimated that by 2050, the food production will have to increase by 60% to meet the demands of the world population (Fedoroff, 2015; Food and Agriculture Organization of the United Nations, 2019; Ristaino *et al.*, 2021). If we look at the global crop-wise scenario; the average yield losses due to diseases and pests are around 22.6% in maize, 21.5% in wheat, 7.2% in potatoes, 30.3% in rice, and 21.4% in soybeans (Savary *et al.*, 2019). Besides these direct losses in terms of yield loss, plant diseases also impose indirect losses in the form of loss of diversity among species, mitigation costs, and downstream effects on human health (Ristaino *et al.*, 2021; Savary *et al.*, 2019).

At present, the protection of crops from insect pests and diseases relies heavily on the use of chemical pesticides. The worldwide annual pesticide application is to the tune of ~2.7 million metric tons of active ingredient (worth US $ 41 billion) with the Americas accounting for the highest pesticide consumption (FAO, 2022). Although chemical pesticides offer several advantages like high availability, fast action, and reliability, they also pose serious environmental side effects with a bulk (i.e., 90% of the applied pesticides) lost to the air during application or as run-off. This results in the contamination of soil and water due to the accumulation of pesticides besides adding to the economic burden of the farmers (Stephenson, 2003). Long-term indiscriminate use of pesticides has resulted in increased development of resistance in pathogens and insects to these pesticides, reduced pollinator population, and reduced

 DOI: 10.1201/9781003474036-15

soil biodiversity, and has affected the habitat of other non-target species like birds (Tilman *et al.*, 2002). Besides chemical pesticides, reliance on other disease management options like resistant varieties, biological control agents, etc., has also grown over the years but despite the application of these multifarious measures; losses due to diseases and insect pests continue to be high. In this scenario; rapid, early, accurate and reliable diagnosis of plant diseases becomes essential to attain an early and additional advantage in disease control thereby reducing disease-associated direct and indirect losses in crops.

Traditionally, most plant diseases are detected by visual characterization of symptoms on affected plants/ plant parts and biological identification of pathogens. Although this approach provides a reasonably accurate disease diagnosis and pathogen identification, it can be done successfully only when the disease has already progressed, therefore posing a challenge to its effective management (Tran *et al.*, 2020). In contrast, accurate monitoring of plant health along with an early diagnosis of pathogens and diseases can help in more effective and timely disease management and curb its spread, ultimately reducing losses. Precise and accurate disease diagnostics has been achieved in past through methods like enzyme-linked immunosorbent assay (ELISA), PCR-assisted techniques of DNA hybridization, DTBIA (direct dot-blot immunoassay), etc. Traditional techniques such as colony counting and fluorescence *in situ* hybridization (FISH) are also commonly utilized in laboratories globally (de La Rica & Stevens, 2012; Hijri, 2009; Oluwaseun *et al.*, 2018). However, the widespread application and adoption of these techniques has been restricted by factors like time-consuming protocols, requirement of significant sample volumes, expensive reagents and specific instruments and laboratory setups, lower sensitivity, etc., especially in developing nations. In this scenario, it becomes essential to harness novel scientific and the innovative approaches like nanotechnology for their application in disease diagnostics. The present chapter attempts to expand on the various areas where nanotechnology offers potential for exploitation in the field of phytopathogen diagnostics.

15.2 NANOTECHNOLOGY

The term "nanotechnology " was first defined by Norio Taniguchi in 1974, and it is a multi-disciplinary science defined as the self-assembly or manipulation of atoms, molecules, or clusters of molecules into structures to develop materials with different or novel properties (Kumari *et al.*, 2018). Nanoparticles (NP) have one or more dimensions in the order of 100 nm or less and exhibit a high surface-to-volume ratio that gives them increased reactivity and distinctive optical and biochemical activity distinguishing them from bulk materials of the same composition, enabling them to overcome some of the limitations of traditional therapeutic and diagnostic agents (Mocan *et al.*, 2017). The unique characteristics of nanomaterials and nanotechnology make them promising tools in various fields including biology. Application of nanotechnology in biological systems is also referred to as nano-biotechnology, "an aspect of nanotechnology and biology where nanotechnology provides the tools and techniques to work from nanoscale principles to investigate, understand, and transform biological systems." Nanotechnology has emerged as a paradigm shifter especially in agriculture, offering innovative solutions to address challenges related to crop productivity, resource utilization, and environmental sustainability. In the agricultural sector, nanotechnology has been explored to address various issues like modulating crop resistance to biotic and abiotic stresses, increasing their productivity, targeted fertilizer and pesticide delivery using nanomaterials, increased nutrient uptake, etc. (He *et al.*, 2019; Kashyap *et al.*, 2015). Some major areas where nanotechnology is being explored in agriculture include:

- Manipulation of crops
- Nanotechnology-based pathogen detection
- Nano-biosensors for pesticide residue detection
- Nanomaterials as carriers for biocontrol agents and their products
- Nano-fertilisers and nano-complexes
- Nanopesticides and herbicides
- Nanotechnology-based Smart Pesticide Delivery systems

15.2.1 Classification of Nanoparticles

NP can be categorized into different groups based on different criterion including function, chemical structure, morphology, and physiochemical properties (Figure 15.1). Based on their chemical structure, NPs can be broadly classified into inorganic (metal and metal oxide NPs) and organic NPs (ferritin, micelles, dendrimers, and liposomes). Inorganic NPs are based on metal oxides (ZnO, CeO_2, CuO, and TiO_2) and metals (Ag, Au, Ni, Au, and Pd). All NPs possess distinct characteristics

Nanoparticles Classification

- **Function:**
  Basic Nanoparticles: Zinc oxide, Copper oxide
  Functionalized NPs: Core-shell NPs, Post functionalized NPs
  Capped NPs: nanotube, nanowire, nano rod
- **Chemical Structure:**
  Organic NPs: liposomes, micelles, dendrimers, polymer
  Inorganic NPs: Composite, metal and alloy, metal oxide, semimetal, semi metal oxide
- **Surface Morphology:**
  High Aspect ratio: nanowire, nanorods, nanotube
  Low aspect ratio: Colloids, suspension, powders
- **Physiochemical Properties:**
  Metallic NPs: quantum dots, alkali and noble metals
  Carbon based NPs: Single, double or multiwalled C nanotubes
  Lipid-based NPs: triglycerides
  Semiconductor NPs: CdSe-ZnS, SiO2
  Polymeric NPs: nanospheres, nanocomposites, nanocapsules
  Ceramic NPs: Zinc oxide, aluminium oxide based

FIGURE 15.1 Classification of Nanomaterials.

and can be generated by conventional or unconventional methods. NPs may further be categorized as nanocapsules (in which the material is enclosed within a polymer membrane and confined to a cavity), nanospheres (where the material for delivery is uniformly dispersed), magnetic NPs, ceramic NPs, solid lipid NPs, semiconductor NPs, polymeric NPs, polymeric micelles, metallic NPs, etc.

Figure 15.1 illustrates the systemic classification of NPs based on function, chemical structure, surface morphology and physiochemical properties.

15.2.2 Synthesis of Nanoparticles

The synthesis of NPs can be carried out through two processes, referred to as top-down and bottom-up approaches (Figure 15.2) (Kulkarni, 2015b, 2015a). Top-down approaches use methods like laser ablation, lithography, thermal decomposition, sputtering, milling, and grinding to break down bulk materials into nanoscale substances. The top-down strategy mostly relies on physical methods of NP synthesis as it requires various equipments like big tube furnaces, etc., to smash bulk material into smaller parts (Chen *et al.*, 2016). In contrast, bottom-up approaches involve the self-assembly of atoms into new nuclei, which then grow into particles at the nanoscale. Bottom-up approaches rely predominantly on chemical methods like the sol-gels process, chemical vapour deposition, spinning, and pyrolysis as well as biological synthesis methods using various microbes and plant extracts. Bottom-up approaches employing chemical methods generally require toxic starting materials, and high temperatures and may result in contaminants and have high running costs to produce the final products (Ahmad *et al.*, 2022; Kulkarni, 2015b, 2015a). However, biological methods offer a more environmentally favourable method for NP production.

Figure 15.2 delineates the synthesis of NPs through both bottom-up and top-down approaches, showcasing the versatility of methods employed for precision engineering at the nanoscale.

15.3 NANOTECHNOLOGY IN PLANT PATHOGEN DETECTION

Nanomaterials have immense potential to recognize pathogens, external stimuli, etc., and can be exploited to develop rapid, sensitive, accurate, and reliable disease diagnostic methods. Since they have submicron size (1–100 nm), these nanomaterials can offer several unique benefits including higher surface-to-volume ratios, distinctive chemical, photosensitive, and electrical properties that are not otherwise present in the bulk forms of these materials. This makes nanomaterials an excellent potential option for exploitation in disease diagnostics. The unique optical properties of the nanomaterials make them an ideal candidate for utilization in highly sensitive and accurate detection of plant pathogens and disease-specific compounds. Nanotechnology also provides a required thrust for development of novel high precision instruments for exploitation in disease diagnostics (Figure 15.3). Taking the example of nanobiosensors, these sensors outperform other sensors in selectivity, sensitivity, and detection limits besides having the potential to be miniaturized and

SYNTHESIS OF NANOPARTICLES

BOTTOM-UP APPROACHES	**PHYSICAL METHODS** (Gaseous or liquid state precursors) Abrasion Condensation Evaporation **CHEMICAL METHODS** Electrochemical Emulsification Flame pyrolysis Chemical Vapour deposition (CVD) Polyol Sol-Gel Microemulsion Aerosol Hydrothermal **BIOLOGICAL METHODS** Micro-organisms Viruses Plants Human Cells
TOP-DOWN APPROACHES	**PHYSICAL METHODS** (Solid state precursors) Laser ablation Mechanical milling Crushing Etching Grinding Sonification Photochemical reduction UV Irradiation

FIGURE 15.2 Synthesis of Nanoparticles.

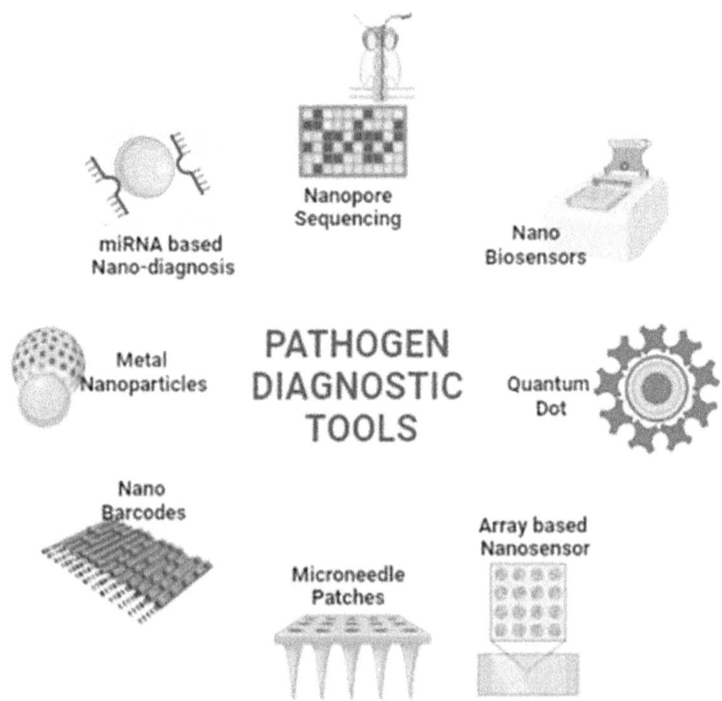

FIGURE 15.3 Various Applications of Nanotechnology in Plant Disease Diagnostics.

can be an effective tool in disease diagnostics. Similarly, techniques like nanopore DNA sequencing, nano-imaging, etc., can provide early identification of pathogens with high specificity (López *et al.*, 2009). Nano-diagnostic kits are a very promising avenue for field-level diagnosis of pathogens assisting farmers in disease management. To date different nanotechnological applications have been exploited for pathogen detection in various crops (Table 15.1).

Figure 15.3 showcases cutting-edge tools for pathogen detection, including nanopore sequencing, nano-biosensors, quantum dots, array-based nanosensors, microneedle patches, nano-barcodes, metal NPs, and miRNA-based nano-diagnosis. These technologies offer diverse and precise methods for rapid and sensitive pathogen identification across a range of applications.

15.3.1 Nanoparticle-based Bio-barcodes

NP based bio-barcodes function as identification tags in agricultural biotechnology, benefiting from their small size and having the potential to recognize enzyme-free ultra-sensitive proteins and DNA. This method is advantageous over conventional methods like ELISA, qPCR, etc., in that it is more sensitive and extensive and aids in rapid recognition of the pathogen. The nano-barcode assay involves the use of two types of probes:

i. Magnetic microbeads (MMB) serve as carriers for antibodies or DNA, functioning as biological probes designed for target recognition.
ii. Gold nanoparticles (Au-NPs) incorporate polyclonal antibodies or an oligonucleotide (bio-barcode). This method is responsive to detecting proteins and DNA without the need for enzymes or PCR.

In this method, one probe comprises magnetic beads coated with monoclonal antibodies for the protein, which can adsorb onto the target protein. The second probe is of AuNPs coated with anti-target protein antibodies and barcode DNA. A magnetic field is used to form a sandwich-like complex of the two probes with a test sample containing a target protein to form "magnetic microsphere-target protein-AuNP". The oligonucleotides are later released and the target protein content can be determined by the selected detection methods (Wang *et al.*, 2019). The bio-barcode assay is an efficient approach to identify protein complexes and nucleic acids and has been used successfully to detect several human pathogens as well as detection of pathogens and toxins in food products (Wang *et al.*, 2019). In plants, it has been used to detect *Pseudomonas aeruginosa* using fluorescence bio-barcoding technique and gold NPs. This method was found to be quick, simple, and sensitive, with a broad range of detection (5–200 ng/mL) using fluorescence spectroscopy (Amini *et al.*, 2017). It has also been used for the detection of *Aspergillus flavus* aflatoxin B1(AFB1) in food products with a sensitivity of approximately 10^{-8} ng/mL (Yu *et al.*, 2018).

TABLE 15.1
Plant pathogen detection through different NPs technological applications

Application of Nanoparticle	Nanoparticle	Disease/Causal Organism/ Compound Detected	Target Plants	Bio-Recognition Method	References
Nanoparticle-based Bio-barcodes	Au Nanoparticles	*Pseudomonas aeroginosa*	-	Fluorescence spectroscopy (the detection limit was 1.2 ng/ml)	(Amini *et al.*, 2017)
	bDNA-GNPs (bio barcode on gold nanoparticle)	*Aspergillus flavus* aflatoxin B1(AFB1)	-	Lateral flow assay	(Yu *et al.*, 2018)
	Fluorescent NPs	*Phakopsora Pachyrhizi*	Soybean	IgG antibodies fluorescence 2.2 ng mL^-1	(Peng & Chen, 2019)
Nanopore DNA sequencing platform	Au NPs/SPCE	Detection of plant pathogen DNA	-	Recombinase polymerase amplification DPV/ 214 pmol L^-1	(Chaudhary *et al.*, 2021)
		Verticillium dahliae	Pepper	MinION	(Chalupowicz *et al.*, 2019)
		Tomato brown rugose fruit virus *(ToBRFV) infection,* tobamovirus, tomato mottle mosaic virus (ToMMV), Pepino mosaic virus (PepMV)	Tomato	ONT sequencing technology	(Abou Kubaa *et al.*, 2023)
		Cucumber green mottle mosaic virus *(CGMMV)*	Cucumber	MinION sequencing	(Mackie *et al.*, 2022)
		Penicillium digitatum	Lemon	MinION	(Chalupowicz *et al.*, 2019)

		Jasmine virus *C*	*Jasminum officinale*	Oxford Nanopore Technology (ONT	(Amoia *et al.*, 2022)
		Candidatus Liberibacter asiaticus, Plum pox virus	Peaches	MinION	(Bronzato Badial *et al.*, 2018)
Array-based nanosensors	Au NPs/SPCE	CMV	Cucumber	Chronoamperometry	(Chartuprayoon *et al.*, 2013)
	Au NPs/SPCE	CTV	Citrus	Thiolated ssDNA EIS/ LOD: 100 nM	(Li *et al.*, 2021)
	Au NPs/SPCE	CTV detection	-	Thiolated primer EIS LOD: 1 pg µL^-1	(Khater, De La Escosura-Muñiz, *et al.*, 2019)
	Au NPs	*Phytophthora infestans*	Tomato	(Cys)-capped C/0.4 ppm	(H. Wang *et al.*, 2020)
	Au NPs	*Xanthomonas campestris*	*Brassica*	Colorimetric/102 CFU mL^-1	(Freitas *et al.*, 2019)
	Au NPs reverse primer (20-mer)	Yellow leaf curl virus	Tomato	Reverse primer (20-mer) LSPR/5 ng µL^-1	(Miranda *et al.*, 2013)
	N- and B-doped MWCNTs/ Interdigitated Electroless nickel immersion gold electrodes	VOCs produced by *Aspergillus* and *Rhizopus* fungi	Strawberry	E-nose	(Berto *et al.*, 2019)
	SWCNTs/ Silicon wafer covered with SiO2	p-Ethylphenol released by *Phytophthora*	Strawberry	ssDNA E-nose 0.13% of Pethylphenol	(Sahayaraj *et al.*, 2015)
Nano-biosensors	SWCNTs/Gold microelectrodes on a Si/SiO2 wafer	Sec-delivered effector 1 (SDE1)	Citrus	Anti-SDE1 polyclonal FET/LOD: 5 nM	(Vinayaka & Thakur, 2011)
	Au NPs/GCE	PSS	Stewartia	HRP LSV/7.8 × 10^3 cfu mL^-1	(Fang *et al.*, 2014)

(*continued*)

TABLE 15.1 (Continued)
Plant pathogen detection through different NPs technological applications

Application of Nanoparticle	Nanoparticle	Disease/Causal Organism/ Compound Detected	Target Plants	Bio-Recognition Method	References
	GO/ITO	GBNV	Groundnut	Anti-GBNV DPV/ LOD: 5.7 ng mL^-1	(Rana *et al.*, 2021)
	PPY nanoribbon/ Gold microelectrode	Cucumber mosaic virus (CMV)	Cucumber	Anti-CMV IgG Chemiresistive Microelectrode/LOD 10 ng mL^-1	(Chaudhary *et al.*, 2021)
	Au NPs/SPCE	Rice tungro disease (RTBV/ RTSV)	Rice	Cyclic voltametry	(Uda *et al.*, 2017)
	Au NPs/Carbon ink 8-WE SPCE	CTV	Sweet orange trees	Anti-bodies Ab1 and Ab2 Amperometry/ LOD: 0.3 fg mL^-1	(Uda *et al.*, 2013)
	rGO and Au NPs/ Kirigami-based structure with AgNW electrodes	VOCs exhaled by *Phytophthora infestans* infection	Tomato	Chemiresistive sensor array/>97% accuracy	(Greenshields *et al.*, 2016)
	Au NPs-SA	*Alternaria panax* Whetz Ginseng	Mouse anti-Fam antibody and BSA-Biotin	LFA/0.01 pg µL^-1	(Razmi *et al.*, 2019)
	Au NPs	*Phytophthora infestans*	Potato	Streptavidin-biotinylated T and C LFA/0.01 pg µL^-1	(Wei *et al.*, 2018)
	Au NPs and silver NPs	Leafroll virus	Potato	Anti-PLRV antibodies LFA 0.2 ng mL^-1	(Zhan *et al.*, 2018)
	Carbon NPs	*X. arboricola* pv. *Pruni*	Stone fruits and almond	Polyclonal antibodies 2626.1-WC LFA 10^4 CFU mL^-1	(Panferov *et al.*, 2018)

	TiO2 and SnO2 NPs /SPCE	p-ethylguaiacol, volatile compound due to *Phytophthora cactorum* fungus infection	-	CV and DPV/LOD: 35–62 nmol L^-1	(Khater, Escosura-Muñiz *et al.*, 2019)
	GO/Paper electrodes	False smut caused by *Ustilaginoidea virens*	Rice	ssDNA CV and LSV/10 fmol L^-1	(Zhao *et al.*, 2014)
Nano-sensors for microRNA-based disease diagnosis	AuNPs	-	-	graphene FET (field-effect transistor) biosensor	(Cai *et al.*, 2015)
	Au NPs/GCE	*Xanthomonas axonopodis*	Citrus	Anti-PthA FET-SWV/ LOD: 0.01 nM	(Huang *et al.*, 2014)
Quantum dots	1 CdTe-Qds	*Phytoplasma aurantifolia*	Lime	(FRET)-Based DNA Biosensor	(Rad *et al.*, 2012)
	CdSe QD -ssDNA Conjugate	*Ganoderma boninense*	Oil palm	(FRET)-Based DNA Biosensor	(Bakhori *et al.*, 2013)
	(CdSe) quantum dots	*Banana bunch top virus*	Banana	Qd-FRET	(Majumder *et al.*, 2020)
Metal and metalloid NPs	Florescent Si	*X. campestris*	Nightshade plants	with antibody doped with Ru (II)	(Yao *et al.*, 2009)
	Au-ssDNA	*R. solanacearum*	Potato	-	(Khaledian *et al.*, 2017)
Microneedle Patches	polyvinyl alcohol (PVA)	*Phytophthora infestans*	Tomato leaf	extraction of plant pathogen DNA	(Paul *et al.*, 2019, 2020)
Nano-diagnostic kits	4mycosensor	ZEA, T-2/HT-2, DON and FB1/ FB2 mycotoxins	Corn, wheat, oat and barley	tetraplex competitive antibody-based assay	(Pimentel, 2009)

15.3.2 Nanopore DNA sequencing platform

Nanopore sequencing platforms have been effective in pathogen detection especially plant viral detection and are widely employed for plant virus monitoring, new virus discovery, and other related tasks. In nanopore sequencing, the protein nanopore and enzyme are specifically crafted to manage a single strand of DNA, subjecting it to direct electrical scrutiny as it traverses the nanopore. Positioned above a microwell within a polymer bilayer membrane, the protein nanopore allows ionic current detection by a sensor chip in each microwell when a single DNA molecule passes through. Conventional sequencing technologies can only sequence short lengths of DNA, which later have to be reassembled. This limitation makes it difficult to sequence repetitive regions and ensure accurate genome assemblies using conventional methods. In contrast, nanopore sequencing can span entire repetitive regions and also resolve structural variants. Since nanopore technology can sequence large lengths without the need for amplification, it can also allow identification of base modifications. However, the rapid movement of the DNA strand through the nanopore may pose a challenge for accurate detection at times. Nanopore sequencing, particularly with the portable sequencer MinION device (Oxford Nanopore Technologies), has shown versatility in plant virus detection, exemplified by the rapid identification of viral species in peaches. The technology's efficiency is evident in diagnosing plant infections like tomato yellow leaf curl virus and watermelon chlorotic stunt virus, completing assays in under two hours. A novel nanopore based diagnostic technology for *Xylella fastidiosa* demonstrates sensitivity comparable to conventional tests with quick results and the potential for on-site monitoring of plant pathogens. This technology has also been used for the diagnosis of fungal pathogens like *Penicillium digitatum* in lemon, and *Candidatus Liberibacter asiaticus* and Plum pox virus in peach. However, challenges, including lower read accuracy compared to short-read sequencers and the need for improved bioinformatics tools, underscore the promising yet evolving nature of Nanopore sequencing in plant disease diagnostics.

15.3.3 Array-based Nanosensors

Array-based nanosensors offer the option of multiplexing and distinguishing different analytes in array-based sensor assemblies. These assemblies comprise various chromophores or synthetic nanomaterials, which are very similar to the human olfactory system. One of the most common array-based sensors is the "electronic nose" (e-nose). The term "electronic nose" was coined in 1988 by Gardner and Bartlett wherein they defined e-nose (EN) as "an instrument which comprises an array of electronic chemical sensors with partial specificity and appropriate pattern recognition system, capable of recognizing simple, or complex odour, which incorporates electronic transducers instead of chemical sensors in the array" (Ariffin *et al.*, 2013; Geng *et al.*, 2019; Röck *et al.*, 2008). The high surface area-to-volume ratio of the nanomaterials is highly advantageous for these array-based sensors since most of the material is made available for interaction with the target analytes. These sensors have the advantage of fast response time, high sensor response values, and detection of very low concentrations of target analytes. Besides e-nose devices, chemical sensor arrays also offer a rapid, easy, and sensitive method for disease detection. Chemical sensor arrays are cost-effective, easier to construct, and more resilient to environmental interferences relying on specific chemical interactions thus ensuring superior chemical specificity. Chemical sensor arrays can be readily created by incorporating various analyte-responsive dyes, like Lewis acid-base colourants, Bronsted acidic or basic colourants, solvatochromic or vapochromic dyes, etc., onto hydrophobic nanoporous substrates like modified silica sol-gels (Askim *et al.*, 2013).

Identifying and recognizing the volatile chemical profile of diseased plants can is an important aspect of plant disease detection (Cellini *et al.*, 2017). Studies have shown that affected plants release distinctive volatile organic compounds (VOCs) in response to various biotic or abiotic stresses (Cellini *et al.*, 2018; Shivashakarappa *et al.*, 2022). E-noses, coated with conductive polymers or metal oxides, serve as multi-layered gas sensors for detecting VOC biomarkers related to plant stress events such as physical injury, pathogen infection, or pest invasion (Mahmud *et al.*, 2018). An important focus area has been on the detection of ethylene, an important phytohormone, using chemiresistive or colorimetric methods. A nanoplasmonic sensor array comprising AuNPs and a molecularly imprinted sol-gel (AuNPs@MISG) has been designed for the selective detection of terpenes. This technology has been used for distinguishing tomato late blight and other fungal infections with similar symptoms on tomato leaves employing a smartphone-integrated VOC sensing technology using plasmonic NPs (Li *et al.*, 2019). In the array, gold NPs were functionalized with chemical ligands that were reactive to specific leaf VOCs, and printed on paper strips with different VOCs indicated by different colour change patterns with the final results scanned using a smartphone microscope for digital pattern recognition and differentiation. Besides these array-based sensors, other on-site tools for studying plant health include portable vapour analyzers, integrated with gas chromatography (GC), tandem mass spectroscopy (MS), or ion mobility spectroscopy (IMS). Compact MEMS devices integrated with small-scale GC-MS/IMS sensors also present effective non-optical methods for monitoring plant health, providing simplicity in device production and swift target sampling (McCartney *et al.*, 2016; Sharma *et al.*, 2019).

15.3.4 Nano-biosensors

A sensor can be defined as "a device that can detect the presence of analyte in a sample and quantify it." A sensor typically consists of a detection system called a receptor, a transducer, and a readout system. A biosensor is a device designed to detect or quantify a biochemical molecule, such as a particular DNA sequence or a particular protein. In biosensors, the receptor is a biological element and is fixed on the converter in different ways. The success of a biosensor depends largely on how the bioreceptor layer (biological component) is fixed on the transducer.

The nano-biosensor is a modified biosensor and is a compact analytical device incorporating a biologically sensitized element onto a physicochemical transducer with a miniature structure when compared to conventional biosensors and provides a potential advantage over normal biosensors in terms of enhanced detection sensitivity/

specificity (Duhan *et al.*, 2017). Nano-biosensors relay information about essential life processes like photosynthesis, respiration, growth, and nutrition transportation in plants. Conventional biosensors and biosensing technologies, like surface-enhanced Raman scattering fluorescence, chemiluminescence, and advanced electrochemical measurements, etc., are now being integrated with nanomaterials to develop nanosensors with higher performance potential. Currently, nano sensors stand out for their high sensitivity, specificity, and user-friendly nature, making them valuable devices for detecting pathogens in crop fields, which can ultimately contribute towards the prevention, treatment, and eradication of various diseases affecting plants. Furthermore, these sensors can also be placed in crop fields and connected to a Global Positioning System (GPS) for real-time monitoring of disease, soil quality, and crop health. Primarily, biosensors are formed through a ligand-receptor binding reaction to a signal transducer, comprising a bioreceptor probe and a transducer that converts biological signals to an observable/electrical signal. The interaction between the analyte and bioreceptor is designed to generate a measurable effect detected by the transducer, which then converts the information into an electrical signal. Nano-biosensors are developed on a nanoscale with immobilized biological elements to identify the presence and concentration of specific substances within a biological analyte (Vishwakarma *et al.*, 2018). NPs have been explored in different roles in the designing of biosensors. The five major categories in which NPs have been used include receiver stabilization, catalysis, improved electron transfer, labelling, and actual participation in detection reactions.

Nano-based biosensors have been developed to detect a wide range of plant pathogens (bacteria, fungi, phytoplasmas, nematodes, viroids, and viruses), abiotic stresses, phytohormones, and microRNAs (mi RNAs) (Ingle *et al.*, 2014). A nanowire-based biosensor was developed for the detection of cucumber mosaic virus and papaya ringspot virus (Ariffin *et al.*, 2013). Similarly, a nano-structured sensing electrode using gold NPs has been successfully used for the detection of *Odontoglossum* ringspot virus (ORSV) in orchid leaves (Jian *et al.*, 2018). Ag nanorods have been explored as sensor substrates for the identification of various plant pathogens and their toxins using surface-enhanced Raman spectroscopy (SERS). Pt nanosensors functionalized with IgG antibodies were introduced to matrix-assisted laser desorption/ionization mass spectrometry (MALDI MS) and exploited for detection of plant pathogenic bacteria in soils. Fluorescent silica NPs doped with a Ru (II) complex and conjugated with a secondary antibody successfully detected *Xanthomonas campestris* causing bacterial spot disease in the nightshade plant. Similarly, Au NPs functionalized with a specific single-stranded DNA were used to detect as low as 15 ng of genomic DNA of *Ralstonia solanacearum*, the potato bacterial wilt pathogen. The nano biosensor-based technology has the potential to provide a new dimension in plant diseases/insect pest diagnostic systems. However, further research on sensors and sensing techniques is essential for widespread practical application of nanosensors for disease diagnostics.

15.3.5 Nano Sensors for MicroRNA-based Disease Diagnosis

MicroRNAs (miRNA) are key regulators of gene expression and are involved in controlling various biological processes such as cellular development, differentiation, proliferation, metabolism, and apoptosis. The miRNA is an endogenous non-coding RNA, which is typically 18–23 nucleotides in length and plays an important role in post-transcriptional regulation. miRNAs are active in cytoplasm, although their synthesis and transcription take place in the nucleus. It has been observed that miRNA profile changes in a diseased state, and it has been suggested that miRNAs may be a powerful tool for controlling various pathological diseases in humans, plants, and animals. Besides exploring their role in crop protection, in recent years, the utilization of RNA molecules in nanodiagnosis has also garnered significant attention due to diversity in shapes (aptamers, ribozymes, tiny interfering RNAs, and microRNAs), and function (Nair & Kumar, 2013). RNA nanotechnology presents a novel approach for detecting and addressing tumors and viral infections in plants. Alterations in gene products induced by disease attacks are primarily regulated by miRNAs in various organisms, making their sensitive and accurate evaluation essential for early disease diagnosis. Conventionally, the detection of miRNA has been done using real-time polymerase chain reaction (qRT-PCR), microarrays, Northern blotting, and next-generation sequencing (NGS) methods. These techniques primarily show the average gene expression level but cannot indicate the heterogeneity and transient spatio-temporal variations of miRNAs in living cells. In this context, the use of nanotechnology for miRNA detection offers various advantages over conventional techniques with the main difference between nanotechnology-based and conventional methods being in the transduction mechanism. The unique physicochemical properties of nano-structured materials help to enhance signal readout. Many NPs have excellent optical properties which makes them suited for developing sensitive detection technology. Some NPs like silver nanoclusters (AgNCs) are bright and stable fluorescence emitters and can be used either directly or in fluorescence resonance energy transfer strategies. Gold NPs (AuNPs) and carbonaceous NPs can also be used as efficient fluorescence quenchers in fluorescence recovery approaches. In the presence of target miRNA, the fluorophores, released from the NP, can emit fluorescence in a concentration-dependent manner thus helping in detection. An AuNP-decorated graphene FET biosensor has demonstrated exceptional sensitivity and selectivity for miRNA detection, offering a label-free and accurate point-of-care tool for disease diagnostics (Cai *et al.*, 2015). However, to ensure an accurate diagnosis, the identification of miRNAs via expression profiling is essential. Although challenges remain in rapidly identifying plant miRNAs responsible for diseases, NP tags in miRNA assays hold promise for enhanced detection limits, paving the way for cutting-edge methods in miRNA detection and their application in diagnosing plant pathogens and crop protection.

15.3.6 Quantum Dots

Quantum dots (QDs) are semiconductor nanocrystals which can be exploited as optical nanosensors. The wavelength of light emitted by QDs is dependent on their size with the wavelength of the infrared light released increasing with size. As such, QDs have a widespread excitation spectrum with about 10–100 times greater molar extinction coefficient, distinctive adjustable fluorescence peak, and longer fluorescence lifespan. As such QDs allow for the activation of many colours of QDs from a single source without the emission signals overlapping. Due to these unique photophysical properties, QD-FRET-based systems have been successfully applied as biosensors for plant imaging and disease identification. Additionally, the ultra-small size of the QDs (1–10 nm) makes them ideal fluorescent imaging contrast agents and allows for their rapid in-planta uptake and transport. The fluorescent signals can then be readily captured to track their distribution in the living system. QDs based on different nanomaterials have been developed and explored for plant disease diagnosis. Rispail *et al.* (2014) developed a CdSe–ZnS core–shell QD with 3-mercaptopropionic acid coating, which was taken up quickly by fungal hyphae. Non-metal-based QDs like carbon dots (CDs) have also garnered wide attention due to the ease of their preparation from fruits and vegetables, low cytotoxicity and high biocompatibility. All these factors make carbon dots a valuable tool in bacterial and fungal imaging.

QDs work on the principle of fluorescence enhancement or quenching and also fluorescence resonance energy transfer (FRET). Safarnejad *et al.* (2017) reported a FRET-based complex sensor for probing *Citrus tristeza virus* (CTV). In their study, CdTe QDs were conjugated with CTV coat protein (CP) and CP-labelled rhodamine dye employed as donor–acceptor complexes. In the presence of target viruses the replacement of CP-rhodamine with free CP occured, giving a recovered fluorescence of QDs. A study on *Polymyxa betae,* the vector of Beet necrotic yellow vein virus (BNYVV), employed antibody-functionalized (QDs) in a FRET-based biosensor to target *P. betae* (Safarpour *et al.*, 2012). Additionally, DNA/RNA-based affinity sensors have also been employed for disease detection by Eun & Wong, (2000); Eun *et al.* (2002). QD-fluorescence resonance energy transfer-based sensors have been successfully used to detect witche's broom disease of lime caused by *Phytoplasma aurantifolia*. An optical FRET based DNA biosensor with an artificial QD has also been developed for the detection of specific DNA sequences of the pathogen *Ganoderma boninense* (Bakhori *et al.*, 2013). However, a potential challenge in the application of QDs is the extensive use of toxic metal elements (Pb, Zn, and Hg) as optical nanoprobes. To overcome these negative impacts, focus has now shifted towards using environmentally safe elements like carbon or silica. Another potential limitation in optical imaging of plant biomarkers using NPs is the possible interferences that can occur due to autofluorescence of host plant tissues. Host plants have common biological pigments (e.g., chlorophyll or carotenoid), which may have similar emission bands as the target pathogen. Therefore, rational selection of excitation/emission wavelengths of engineered nanosensors is critical to ensure their successful implementation for plant disease diagnostics.

15.3.7 Metal and Metalloid NPs

Various metal and metalloid NPs have been effectively explored as diagnostic tools for microorganisms causing diseases in plants. Among metal NPs, gold, platinum, and silver are the most common nanomaterials employed in agricultural diagnostics with factors like ease of production of NPs of these metals, good conductivity, and catalytic properties contributing towards their use (Darr et al, 2017). Ag, Au, Si, and metal oxide based NPs employ recognition ligands (e.g., antibodies or DNA oligos) and signal transducers to detect and quantify molecular targets. Unique sensor responses are recorded indicating distinct optical properties, based on the localized surface plasmon resonance (LSPR) of NPs. Presently, several self-assembled NPs with oligonucleotides, and proteins have been explored to construct sensitive and target-specific nano biosensors. Yüksel *et al.* (2015) developed a label-free platform for DNA hybridization using SERS to detect *Phytophthora ramorum*. The method utilized adenine on the target DNA as an endogenous marker where hybridization between adenine-free capture probes and the target DNA, facilitated on a silver NP SERS-substrate, served as an indicator of the presence of the plant pathogen (Yüksel *et al.*, 2015). For on field detection of genomic DNA of the pathogen *Ralstonia solanacearum* in soil, a probe was developed using gold NPs functionalized with single-stranded oligonucelotides that interact with the bacterium's genomic DNA. The colour change from red to purple indicated the presence or absence of *R. solanacearum* DNA, providing a simple and visible on-field detection method (Khaledian *et al.*, 2017). Silica NPs have been used for the identification of *Xanthomonas axonopodis* pv. *vesicatoria* (Xav), causing spot disease in solanaceae plants. The application involved the creation of fluorescent silica NP probes in conjunction with a secondary antibody where the silica NPs were infused with a fluorescent dye, Rubpy, facilitating the easy and rapid Xav detection through Fluorescence-Linked Immunosorbent Assay (ELISA)(Darr *et al.*, 2017).

15.3.8 Microneedle Patches

Disease and pathogen diagnostics traditionally employ either nucleic acid amplification or protein-based molecular assays. These conventional assays use techniques like polymerase chain reaction (PCR), loop-mediated isothermal amplification (LAMP), enzyme-linked immunosorbent assay (ELISA), and recombinase polymerase amplification (RPA). However, these methods have several limitations such as complicated sample preparation, long assay time, requirement of expensive instruments, etc. Microneedle patches aim to integrate nanotechnology with standard biomolecular analyses to provide an added advantage of sensitivity and portability over traditional diagnostics. Skin-penetrating microneedle (MN) patches have had considerable application in the healthcare sector due to their high efficiency in non-invasive delivery of drug molecules as well as extraction of diagnostic analytes and monitoring of various clinical parameters. A typical microneedle is of the size 5–10 μm tip diameter, but to improve sensing and sample interaction parameters, the microneedles can either be loaded with functional NPs or engineered with nanoporous structures. These microneedle patches have extensive application in human medicine especially transdermal drug delivery and subdermal biosensing. In the case of plant disease diagnostics, however, this technique is still in its infancy. In one of the very few studies conducted, a polymer-based microneedle patch in combination with real-time PCR (i.e., quantitative PCR, or qPCR) technique was developed for rapid and sensitive detection of late blight pathogen *Phytophthora infestans* (Paul *et al.*, 2019). The microneedle patches mainly assist by reducing time taken for sample preparation. Isolation of specific biomarkers from robust and rigid plant tissues in the field is a challenge. To address this, a microneedle patch made of polyvinyl alcohol (PVA) has been developed that is sufficiently strong to break the cell walls of plant leaf tissues to release plant or pathogenic DNA. This ultimately helps in reducing sample preparation time from several hours to less than 1 min. The isolated DNA using microneedle patches is ready for amplification without requirmemnt of additional sample purification steps. In this method, only a microneedle patch and an aqueous buffer solution are required. and the patch is made of an affordable polymer similar in size to a postcard stamp. Microneedle patches have the potential to be exploited for rapid amplification and diagnosis of plant pathogens in field conditions.

15.3.9 Nano-diagnostic Kits

Another potential application of nanotechnology for disease diagnostics is through the development of "nano diagnostic kits," which are also often referred to as a "lab in a box." They comprise a compact container, no larger or heavier than a briefcase, equipped with sophisticated measurement instruments, reagents, power supplies, and other essential components and can be used to monitor important plant processes in a small space (Cai *et al.*, 2015). A nano diagnostic kit, based on a tetraplex competitive antibody-based test in dipstick format, providing real-time diagnosis for mycotoxins such as ZEA, T-2/HT-2, DON, and FB1/FB2 in maize, oat, wheat, and barley samples, all below their respective European maximum residue limits (MRL) has been developed (Pimentel, 2009; Shivashakarappa *et al.*, 2022). Nevertheless, challenges persist in the advancement of nanodiagnostic kits for use in agriculture and related sectors. Overcoming issues like identifying and selecting crucial receptors such as antigens, antibodies, and nucleotides to enhance molecular diagnostics and enable strain differentiation is pivotal. Establishing global criteria for evaluating tests and detection levels is essential for comparing detection limits across studies. Additionally, streamlining the purification process and isolating key genes are vital steps for genomic target identification of specific diseases.

15.4 CONCLUSION

The escalating challenges posed by plant pathogens and their impact on global food security demand innovative solutions that transcend traditional diagnostic methods. This is further emphasized by the complex interplay of climate change and evolving pathogen lineages, both calling for a fundamental change in agricultural diagnostics. This chapter has delved into the transformative potential of nanotechnology in revolutionizing the field of phytopathogen detection and control. The early detection of pathogens and plant diseases is essential to safeguard crops from potential outbreaks. The integration of nanotechnology systems into agriculture provides the much needed impetus for pathogen diagnostics. The integration of autonomous NP sensors connected to a GPS system for real-time monitoring would be immensely beneficial, potentially offering early warnings for diseases. Such a proactive approach empowers farmers to strategically deploy insecticides, pesticides, and antimicrobials, mitigating economic losses while ensuring safety and security. The scope of nanotechnology in agriculture extends far beyond mere detection. NPs can be leveraged not only for the detection of pathogens but also for improved growth, development, and stress response mechanisms of plants. As we stand on the precipice of a global population surge and the imperative to enhance food production, embracing the possibilities offered by nanotechnology in plant pathogen detection and management becomes paramount. Nanotechnology applications in agriculture mark a pivotal step towards a resilient and sustainable agricultural future.

REFERENCES

Abou Kubaa, R., Amoia, S. S., Altamura, G., Minafra, A., Chiumenti, M., & Cillo, F. (2023). Nanopore Technology Applied to Targeted Detection of Tomato Brown Rugose Fruit Virus Allows Sequencing of Related Viruses and the Diagnosis of Mixed Infections. *Plants*, *12*(5), 999. https://doi.org/10.3390/plants12050999

Ahmad, Z., Tahseen, S., Wasi, A., Ganie, I. B., Shahzad, A., Emamverdian, A., Ramakrishnan, M., & Ding, Y. (2022). Nanotechnological Interventions in Agriculture. *Nanomaterials*, *12*(15), 2667. https://doi.org/10.3390/nano12152667

Amini, B., Kamali, M., Salouti, M., & Yaghmaei, P. (2017). Fluorescence Bio-barcode DNA Assay based on Gold and Magnetic Nanoparticles for Detection of Exotoxin A Gene Sequence. *Biosensors and Bioelectronics*, *92*, 679–686. https://doi.org/10.1016/j.bios.2016.10.030

Amoia, S. S., Minafra, A., Nicoloso, V., Loconsole, G., & Chiumenti, M. (2022). A New Jasmine Virus C Isolate Identified by Nanopore Sequencing Is Associated to Yellow Mosaic Symptoms of Jasminum officinale in Italy. *Plants*, *11*(3), 309. https://doi.org/10.3390/plants11030309

Ariffin, S. A. B., Adam, T., Hashim, U., Faridah Sfaridah, S., Zamri, I., & Uda, M. N. A. (2013). Plant Diseases Detection Using Nanowire as Biosensor Transducer. *Advanced Materials Research*, *832*, 113–117. https://doi.org/10.4028/www.scientific.net/AMR.832.113

Askim, J. R., Mahmoudi, M., & Suslick, K. S. (2013). Optical sensor arrays for chemical sensing: The optoelectronic nose. *Chemical Society Reviews*, *42*(22), 8649. https://doi.org/10.1039/c3cs60179j

Bakhori, N., Yusof, N., Abdullah, A., & Hussein, M. (2013). Development of a Fluorescence Resonance Energy Transfer (FRET)-Based DNA Biosensor for Detection of Synthetic Oligonucleotide of Ganoderma boninense. *Biosensors*, *3*(4), 419–428. https://doi.org/10.3390/bios3040419

Berto, M., Vecchi, E., Baiamonte, L., Condò, C., Sensi, M., Di Lauro, M., Sola, M., De Stradis, A., Biscarini, F., Minafra, A., & Bortolotti, C. A. (2019). Label-free Detection of Plant Viruses with Organic Transistor Biosensors. *Sensors and Actuators B: Chemical*, *281*, 150–156. https://doi.org/10.1016/j.snb.2018.10.080

Bronzato Badial, A., Sherman, D., Stone, A., Gopakumar, A., Wilson, V., Schneider, W., & King, J. (2018). Nanopore Sequencing as a Surveillance Tool for Plant Pathogens in Plant and Insect Tissues. *Plant Disease*, *102*(8), 1648–1652. https://doi.org/10.1094/PDIS-04-17-0488-RE

Cai, B., Huang, L., Zhang, H., Sun, Z., Zhang, Z., & Zhang, G.-J. (2015). Gold Nanoparticles-decorated Graphene Field-effect Transistor Biosensor for Femtomolar MicroRNA

Detection. *Biosensors and Bioelectronics*, *74*, 329–334. https://doi.org/10.1016/j.bios.2015.06.068

Cellini, A., Blasioli, S., Biondi, E., Bertaccini, A., Braschi, I., & Spinelli, F. (2017). Potential Applications and Limitations of Electronic Nose Devices for Plant Disease Diagnosis. *Sensors*, *17*(11), 2596. https://doi.org/10.3390/s17112596

Cellini, A., Buriani, G., Rocchi, L., Rondelli, E., Savioli, S., Rodriguez Estrada, M. T., Cristescu, S. M., Costa, G., & Spinelli, F. (2018). Biological Relevance of Volatile Organic Compounds Emitted during the pathogenic Interactions Between Apple Plants and *Erwinia Amylovora*. *Molecular Plant Pathology*, *19*(1), 158–168. https://doi.org/10.1111/mpp.12509

Chalupowicz, L., Dombrovsky, A., Gaba, V., Luria, N., Reuven, M., Beerman, A., Lachman, O., Dror, O., Nissan, G., & Manulis-Sasson, S. (2019). Diagnosis of Plant Diseases using the Nanopore Sequencing Platform. *Plant Pathology*, *68*(2), 229–238. https://doi.org/10.1111/ppa.12957

Chartuprayoon, N., Rheem, Y., Ng, J. C. K., Nam, J., Chen, W., & Myung, N. V. (2013). Polypyrrole Nanoribbon based Chemiresistive Immunosensors for Viral Plant Pathogen Detection. *Analytical Methods*, *5*(14), 3497. https://doi.org/10.1039/c3ay40371h

Chaudhary, M., Verma, S., Kumar, A., Basavaraj, Y. B., Tiwari, P., Singh, S., Chauhan, S. K., Kumar, P., & Singh, S. P. (2021). Graphene Oxide based Electrochemical Immunosensor for Rapid Detection of groundnut Bud Necrosis Orthotospovirus in Agricultural Crops. *Talanta*, *235*, 122717. https://doi.org/10.1016/j.talanta.2021.122717

Culliney, T. (2014). *Crop Losses to Arthropods* (pp. 201–225).

Darr, J. A., Zhang, J., Makwana, N. M., & Weng, X. (2017). Continuous Hydrothermal Synthesis of Inorganic Nanoparticles: Applications and Future Directions. *Chemical Reviews*, *117*(17), 11125–11238. https://doi.org/10.1021/acs.chemrev.6b00417

De La Rica, R., & Stevens, M. M. (2012). Plasmonic ELISA for the ultrasensitive detection of disease biomarkers with the naked eye. *Nature Nanotechnology*, *7*(12), 821–824. https://doi.org/10.1038/nnano.2012.186

Duhan, J. S., Kumar, R., Kumar, N., Kaur, P., Nehra, K., & Duhan, S. (2017). Nanotechnology: The new perspective in precision agriculture. *Biotechnology Reports*, *15*, 11–23. https://doi.org/10.1016/j.btre.2017.03.002

Eun, A. J.-C., Huang, L., Chew, F.-T., Fong-Yau Li, S., & Wong, S.-M. (2002). Detection of Two Orchid Viruses Using Quartz Crystal Microbalance-Based DNA Biosensors. *Phytopathology®*, *92*(6), 654–658. https://doi.org/10.1094/PHYTO.2002.92.6.654

Eun, A. J.-C., & Wong, S.-M. (2000). Molecular Beacons: A New Approach to Plant Virus Detection. *Phytopathology®*, *90*(3), 269–275. https://doi.org/10.1094/PHYTO.2000.90.3.269

Fang, Y., Umasankar, Y., & Ramasamy, R. P. (2014). Electrochemical Detection of p-Ethylguaiacol, a Fungi Infected Fruit Volatile using Metal Oxide Nanoparticles. *The Analyst*, *139*(15), 3804–3810. https://doi.org/10.1039/C4AN00384E

FAO. (2022). FAOSTAT: Pesticides Trade. In: FAO. Rome. www.fao.org/faostat/en/#data/RT

Fedoroff, N. V. (2015). Food in a future of 10 billion. *Agriculture & Food Security*, *4*(1), 11. https://doi.org/10.1186/s40066-015-0031-7

Food and Agriculture Organization of the United Nations. (2019). *The State of Food and Agriculture 2019: Moving Forward on Food Loss and Waste Reduction*. UN. https://doi.org/10.18356/32f21f8c-en

Freitas, T. A., Proença, C. A., Baldo, T. A., Materón, E. M., Wong, A., Magnani, R. F., & Faria, R. C. (2019). Ultrasensitive Immunoassay for Detection of Citrus Tristeza Virus in Citrus Sample using Disposable Microfluidic Electrochemical Device. *Talanta*, *205*, 120110. https://doi.org/10.1016/j.talanta.2019.07.005

Geng, Y., Peveler, W. J., & Rotello, V. M. (2019). Array-based "Chemical Nose" Sensing in Diagnostics and Drug Discovery. *Angewandte Chemie International Edition*, *58*(16), 5190–5200. https://doi.org/10.1002/anie.201809607

Greenshields, M., Cunha, B., Coville, N., Pimentel, I., Zawadneak, M., Dobrovolski, S., Souza, M., & Hümmelgen, I. (2016). Fungi Active Microbial Metabolism Detection of Rhizopus sp. and Aspergillus sp. Section Nigri on Strawberry Using a Set of Chemical Sensors Based on Carbon Nanostructures. *Chemosensors*, *4*(3), 19. https://doi.org/10.3390/chemosensors4030019

He, X., Deng, H., & Hwang, H. (2019). The Current Application of Nanotechnology in Food and Agriculture. *Journal of Food and Drug Analysis*, *27*(1), 1–21. https://doi.org/10.1016/j.jfda.2018.12.002

Hijri, M. (2009). The Use of Fluorescent in situ Hybridisation in Plant Fungal Identification and Genotyping. In R. Burns (Ed.), *Plant Pathology* (Vol. 508, pp. 131–145). Humana Press. https://doi.org/10.1007/978-1-59745-062-1_11

Huang, X., Xu, J., Ji, H.-F., Li, G., & Chen, H. (2014). Quartz Crystal Microbalance Based Biosensor for Rapid and Sensitive Detection of Maize Chlorotic Mottle Virus. *Analytical Methods*, *6*(13), 4530–4536. https://doi.org/10.1039/C4AY00292J

Ingle, A. P., Seabra, A. B., Duran, N., & Rai, M. (2014). Nanoremediation. In *Microbial Biodegradation and Bioremediation* (pp. 233–250). Elsevier. https://doi.org/10.1016/B978-0-12-800021-2.00009-1

Jian, Y.-S., Lee, C.-H., Jan, F.-J., & Wang, G.-J. (2018). Detection of Odontoglossum Ringspot Virus Infected Phalaenopsis using a Nano-Structured Biosensor. *Journal of the Electrochemical Society*, *165*(9), H449–H454. https://doi.org/10.1149/2.0351809jes

Kashyap, P. L., Xiang, X., & Heiden, P. (2015). Chitosan Nanoparticle based Delivery Systems for Sustainable Agriculture. *International Journal of Biological Macromolecules*, *77*, 36–51. https://doi.org/10.1016/j.ijbiomac.2015.02.039

Khaledian, S., Nikkhah, M., Shams-bakhsh, M., & Hoseinzadeh, S. (2017). A Sensitive Biosensor based on Gold Nanoparticles to detect Ralstonia Solanacearum in Soil. *Journal of General Plant Pathology*, *83*(4), 231–239. https://doi.org/10.1007/s10327-017-0721-z

Khater, M., De La Escosura-Muñiz, A., Quesada-González, D., & Merkoçi, A. (2019). Electrochemical Detection of Plant Virus using Gold Nanoparticle-modified Electrodes. *Analytica Chimica Acta*, *1046*, 123–131. https://doi.org/10.1016/j.aca.2018.09.031

Khater, M., Escosura-Muñiz, A. D. L., Altet, L., & Merkoçi, A. (2019). In Situ Plant Virus Nucleic Acid Isothermal Amplification Detection on Gold Nanoparticle-Modified Electrodes. *Analytical Chemistry*, *91*(7), 4790–4796. https://doi.org/10.1021/acs.analchem.9b00340

Kulkarni, S. K. (2015a). Synthesis of Nanomaterials–I (Chemical Methods). In *Nanotechnology: Principles and Practices*. Springer.

Kulkarni, S. K. (2015b). Synthesis of Nanomaterials–I (Physical Methods). In *Nanotechnology: Principles and Practices*. Springer.

Kumar, S., Tao, C., Chien, M., Hellner, B., Balijepalli, A., Robertson, J. W. F., Li, Z., Russo, J. J., Reiner, J. E., Kasianowicz, J. J., & Ju, J. (2012). PEG-Labeled Nucleotides and Nanopore Detection for Single Molecule DNASequencing by Synthesis. *Scientific Reports*, *2*(1), 684. https://doi.org/10.1038/srep00684

Kumari, P., Ekka, J. P., Tirkey, S. K., & Ahmad, E. (2018). *Agro-Nanotechnology: An Innovative Approach for Diagnosis of Plants*. *7*(4S), 352–357.

Li, Z., Liu, Y., Hossain, O., Paul, R., Yao, S., Wu, S., Ristaino, J. B., Zhu, Y., & Wei, Q. (2021). Real-time Monitoring of Plant Stresses via Chemiresistive Profiling of Leaf Volatiles by a Wearable Sensor. *Matter*, *4*(7), 2553–2570. https://doi.org/10.1016/j.matt.2021.06.009

Li, Z., Paul, R., Ba Tis, T., Saville, A. C., Hansel, J. C., Yu, T., Ristaino, J. B., & Wei, Q. (2019). Non-invasive Plant Disease Diagnostics Enabled by Smartphone-based Fingerprinting of Leaf Volatiles. *Nature Plants*, *5*(8), 856–866. https://doi.org/10.1038/s41477-019-0476-y

López, M. M., Llop, P., Olmos, A., Marco-Noales, E., Cambra, M., & Bertolini, E. (2009). Are Molecular Tools Solving the Challenges Posed by Detection of Plant Pathogenic Bacteria and Viruses? *Current Issues in Molecular Biology*, *11*(1), 13–46.

Mackie, J., Kinoti, W. M., Chahal, S. I., Lovelock, D. A., Campbell, P. R., Tran-Nguyen, L. T. T., Rodoni, B. C., & Constable, F. E. (2022). Targeted Whole Genome Sequencing (TWG-Seq) of Cucumber Green Mottle Mosaic Virus Using Tiled Amplicon Multiplex PCR and Nanopore Sequencing. *Plants*, *11*(20), 2716. https://doi.org/10.3390/plants11202716

Mahmud, M. M., Constantino, N., Seok, C., Yamaner, F. Y., Dean, R. A., & Oralkan, O. (2018). A CMUT – Based Electronic Nose for Real-Time Monitoring of Volatiles Emitted by Plants: Preliminary Results. *2018 IEEE SENSORS*, 1–4. https://doi.org/10.1109/ICSENS.2018.8589740

Majumder, S., Bhattacharya, B., Singh, P. K., Johari, S., Singh, B., & Rahman, R. (2020). Impedimetric Detection of Banana Bunchy Top Virus Using CdSe Quantum Dots for Signal Amplification. *SN Applied Sciences*, *2*(4), 638. https://doi.org/10.1007/s42452-020-2345-8

McCartney, M. M., Spitulski, S. L., Pasamontes, A., Peirano, D. J., Schirle, M. J., Cumeras, R., Simmons, J. D., Ware, J. L., Brown, J. F., Poh, A. J. Y., Dike, S. C., Foster, E. K., Godfrey, K. E., & Davis, C. E. (2016). Coupling a Branch Enclosure with Differential Mobility Spectrometry to Isolate and Measure Plant Volatiles in Contained Greenhouse Settings. *Talanta*, *146*, 148–154. https://doi.org/10.1016/j.talanta.2015.08.039

Miranda, B. S., Linares, E. M., Thalhammer, S., & Kubota, L. T. (2013). Development of a Disposable and Highly Sensitive Paper-based Immunosensor for Early Diagnosis of Asian Soybean Rust. *Biosensors and Bioelectronics*, *45*, 123–128. https://doi.org/10.1016/j.bios.2013.01.048

Mocan, T., Matea, C. T., Pop, T., Mosteanu, O., Buzoianu, A. D., Puia, C., Iancu, C., & Mocan, L. (2017). Development of Nanoparticle-based Optical Sensors for Pathogenic Bacterial Detection. *Journal of Nanobiotechnology*, *15*(1), 25. https://doi.org/10.1186/s12951-017-0260-y

Nair, R., & Kumar, D. S. (2013). Plant Diseases—Control and Remedy Through Nanotechnology. In N. Tuteja & S. S. Gill (Eds.), *Crop Improvement Under Adverse Conditions* (pp. 231–243). Springer New York. https://doi.org/10.1007/978-1-4614-4633-0_10

Oluwaseun, A. C., Phazang, P., & Sarin, N. B. (2018). Biosensing Technologies for the Detection of Pathogens—A Prospective Way for Rapid Analysis. In T. Rinken & K. Kivirand (Eds.), *Biosensing Technologies for the Detection of Pathogens—A Prospective Way for Rapid Analysis*. InTech. https://www.intechopen.com/books/6319

Panferov, V. G., Safenkova, I. V., Byzova, N. A., Varitsev, Y. A., Zherdev, A. V., & Dzantiev, B. B. (2018). Silver-enhanced Lateral Flow Immunoassay for Highly-sensitive Detection of Potato Leafroll Virus. *Food and Agricultural Immunology*, *29*(1), 445–457. https://doi.org/10.1080/09540105.2017.1401044

Paul, R., Ostermann, E., Gu, Z., Ristaino, J. B., & Wei, Q. (2020). DNA Extraction from Plant Leaves using a Microneedle Patch. *Current Protocols in Plant Biology*, *5*(1), e20104. https://doi.org/10.1002/cppb.20104

Paul, R., Saville, A. C., Hansel, J. C., Ye, Y., Ball, C., Williams, A., Chang, X., Chen, G., Gu, Z., Ristaino, J. B., & Wei, Q. (2019). Extraction of Plant DNA by Microneedle Patch for Rapid Detection of Plant Diseases. *ACS Nano*, *13*(6), 6540–6549. https://doi.org/10.1021/acsnano.9b00193

Peng, H., & Chen, I. A. (2019). Rapid Colorimetric Detection of Bacterial Species through the Capture of Gold Nanoparticles by Chimeric Phages. *ACS Nano*, acsnano.8b06395. https://doi.org/10.1021/acsnano.8b06395

Pimentel, D. (2009). Invasive Plants: Their Role in Species Extinctions and Economic Losses to Agriculture in the USA. In Inderjit (Ed.), *Management of Invasive Weeds* (Vol. 5, pp. 1–7). Springer Netherlands. https://doi.org/10.1007/978-1-4020-9202-2_1

Rad, F., Mohsenifar, A., Tabatabaei, M., Safarnejad, M. R., Shahryari, F., Safarpour, H., Foroutan, A., Mardi, M., Davoudi, D., & Fotokian, M. (2012). Detection of Candidatus Phytoplasma Aurantifolia with a Quantum Dots Fret-Based Bsiosensor. *Journal of Plant Pathology*, *94*(3), 525–534. JSTOR.

Rana, K., Mittal, J., Narang, J., Mishra, A., & Pudake, R. N. (2021). Graphene Based Electrochemical DNA Biosensor for Detection of False Smut of Rice (Ustilaginoidea Virens). *The Plant Pathology Journal*, *37*(3), 291–298. https://doi.org/10.5423/PPJ.OA.11.2020.0207

Razmi, A., Golestanipour, A., Nikkhah, M., Bagheri, A., Shamsbakhsh, M., & Malekzadeh-Shafaroudi, S. (2019). Localized Surface Plasmon Resonance Biosensing of Tomato Yellow Leaf Curl Virus. *Journal of Virological Methods*, *267*, 1–7. https://doi.org/10.1016/j.jviromet.2019.02.004

Rispail, N., Matteis, L.D., Santos, R., Miguel, A.S., Custardoy, L., Testillano, P.S., Risueno, M. C., Perez-de-Luque, A., Maycock, C., Fevereiro, P., Oliva, A., Fernandez-Pacheco, R., Ibarra, M.R., de l Fuente, J.M., Marquina, C., Rubiales, D., & Prats E. (2014). Quantum Dot and Superparamagnetic Nanoparticle Interaction with Pathogenic Fungi: Internalization and Toxicity Profile. *ACS Applied Materials and Interfaces*, 6(12), 9100–9110. https://doi.org/10.1021/am501029g

Ristaino, J. B., Anderson, P. K., Bebber, D. P., Brauman, K. A., Cunniffe, N. J., Fedoroff, N. V., Finegold, C., Garrett, K. A., Gilligan, C. A., Jones, C. M., Martin, M. D., MacDonald, G. K., Neenan, P., Records, A., Schmale, D. G., Tateosian, L., & Wei, Q. (2021). The Persistent Threat of Emerging Plant Disease Pandemics to Global Food Security. *Proceedings of the National Academy of Sciences*, *118*(23), e2022239118. https://doi.org/10.1073/pnas.2022239118

Röck, F., Barsan, N., & Weimar, U. (2008). Electronic Nose: Current Status and Future Trends. *Chemical Reviews*, *108*(2), 705–725. https://doi.org/10.1021/cr068121q

Rodrigues, E. S., Montanha, G. S., Gomes, M. H. F., Duran, N. M., Corrêa, C. G., Romeu, S. L. Z., Pereira, A. E. S., De Oliveira, J. L., De Almeida, E., Pérez-de-Luque, A., Ghoshal, S., Santaella, C., De Lima, R., Fraceto, L. F., & De Carvalho, H. W. P. (2021). Are Nanomaterials Leading to more Efficient Agriculture? Outputs from 2009-2022 Research Metadata Analysis. *agriRxiv*, *2021*, 20210198122. https://doi.org/10.31220/agriRxiv.2021.00054

Safarnejad, M. R., Samiee, F., Tabatabie, M., & Mohsenifar, A. (2017). *Development of Quantum Dot-based Nanobiosensors Against Citrus Tristeza Virus (CTV). 213*(6), 54–60.

Safarpour, H., Safarnejad, M. R., Tabatabaei, M., Mohsenifar, A., Rad, F., Basirat, M., Shahryari, F., & Hasanzadeh, F. (2012). Development of a Quantum Dots FRET-based Biosensor for Efficient Detection of Polymyxa Betae. *Canadian Journal of Plant Pathology*, *34*(4), 507–515. https://doi.org/10.1080/07060661.2012.709885

Sahayaraj, K., Roobadevi, M., Rajesh, S., & Azizi, S. (2015). Vernonia Cinerea (L.) Less. Silver Nanocomposite and its Antibacterial Activity Against a Cotton Pathogen. *Research on Chemical Intermediates*, *41*(8), 5495–5507. https://doi.org/10.1007/s11164-014-1676-8

Savary, S., Willocquet, L., Pethybridge, S. J., Esker, P., McRoberts, N., & Nelson, A. (2019). The Global Burden of Pathogens and Pests on Major Food Crops. *Nature Ecology & Evolution*, *3*(3), 430–439. https://doi.org/10.1038/s41559-018-0793-y

Sharma, R., Zhou, M., Hunter, M. D., & Fan, X. (2019). Rapid In Situ Analysis of Plant Emission for Disease Diagnosis Using a Portable Gas Chromatography Device. *Journal of Agricultural and Food Chemistry*, *67*(26), 7530–7537. https://doi.org/10.1021/acs.jafc.9b02500

Shivashakarappa, K., Reddy, V., Tupakula, V. K., Farnian, A., Vuppula, A., & Gunnaiah, R. (2022). Nanotechnology for the Detection of Plant Pathogens. *Plant Nano Biology*, *2*, 100018. https://doi.org/10.1016/j.plana.2022.100018

Shu, Y., Shu, D., Haque, F., & Guo, P. (2013). Fabrication of pRNA Nanoparticles to Deliver Therapeutic RNAs and Bioactive Compounds into Tumor Cells. *Nature Protocols*, *8*(9), 1635–1659. https://doi.org/10.1038/nprot.2013.097

Stephenson, G. R. (2003). *Environmental Fate and Effects of Pesticides* (J. R. Coats & H. Yamamoto, Eds.; Vol. 853). American Chemical Society. https://doi.org/10.1021/bk-2003-0853

Tilman, D., Cassman, K. G., Matson, P. A., Naylor, R., & Polasky, S. (2002). Agricultural Sustainability and Intensive Production Practices. *Nature*, *418*(6898), 671–677. https://doi.org/10.1038/nature01014

Tran, T.-T., Clark, K., Ma, W., & Mulchandani, A. (2020). Detection of a Secreted Protein Biomarker for Citrus Huanglongbing using a Single-walled Carbon Nanotubes-based Chemiresistive Biosensor. *Biosensors and Bioelectronics*, *147*, 111766. https://doi.org/10.1016/j.bios.2019.111766

Uda, M. N. A., Adam, T., Hasfalina, C. M., Faridah Sfaridah, S., Zamri, I., Hashim, U., & Ariffin, S. A. B. (2013). Reviewed Immunosensor Format Using Nanomaterial for Tungro Virus Detection. *Advanced Materials Research*, *832*, 410–414. https://doi.org/10.4028/www.scientific.net/AMR.832.410

Uda, M. N. A., Hasfalina, C. M., Samsuzana, A. A., Faridah, S., R., R. A., Hashim, U., Ariffin, S. A. B., & Gopinath, S. C. B. (2017). *Determination of Set Potential Voltages for Cucumber Mosaic virus Detection using Screen Printed Carbon Electrode*. 020056. https://doi.org/10.1063/1.4975289

Vinayaka, A. C., & Thakur, M. S. (2011). Photoabsorption and Resonance Energy Transfer Phenomenon in CdTe–Protein Bioconjugates: An Insight into QD–Biomolecular Interactions. *Bioconjugate Chemistry*, *22*(5), 968–975. https://doi.org/10.1021/bc200034a

Vishwakarma, K., Upadhyay, N., Kumar, N., Tripathi, D. K., Chauhan, D. K., Sharma, S., & Sahi, S. (2018). Potential Applications and Avenues of Nanotechnology in Sustainable Agriculture. In *Nanomaterials in Plants, Algae, and Microorganisms* (pp. 473–500). Elsevier. https://doi.org/10.1016/B978-0-12-811487-2.00021-9

Wang, H., Wang, Y., Hou, X., & Xiong, B. (2020). Bioelectronic Nose Based on Single-Stranded DNA and Single-Walled Carbon Nanotube to Identify a Major Plant Volatile Organic Compound (p-Ethylphenol) Released by Phytophthora Cactorum Infected Strawberries. *Nanomaterials*, *10*(3), 479. https://doi.org/10.3390/nano10030479

Wang, Y., Jin, M., Chen, G., Cui, X., Zhang, Y., Li, M., Liao, Y., Zhang, X., Qin, G., Yan, F., Abd El-Aty, A. M., & Wang, J. (2019). Bio-barcode Detection Technology and its Research Applications: A Review. *Journal of Advanced Research*, *20*, 23–32. https://doi.org/10.1016/j.jare.2019.04.009

Wei, S., Sun, Y., Xi, G., Zhang, H., Xiao, M., & Yin, R. (2018). Development of a single-tube nested PCR-lateral flow biosensor assay for rapid and accurate detection of Alternaria panax Whetz. *PLOS ONE*, *13*(11), e0206462. https://doi.org/10.1371/journal.pone.0206462

Yao, K. S., Li, S. J., Tzeng, K. C., Cheng, T. C., Chang, C. Y., Chiu, C. Y., Liao, C. Y., Hsu, J. J., & Lin, Z. P. (2009). Fluorescence Silica Nanoprobe as a Biomarker for Rapid Detection of Plant Pathogens. *Advanced Materials Research*, *79–82*, 513–516. https://doi.org/10.4028/www.scientific.net/AMR.79-82.513

Yu, S., He, L., Yu, F., Liu, L., Qu, C., Qu, L., Liu, J., Wu, Y., & Wu, Y. (2018). A lateral flow assay for simultaneous detection of Deoxynivalenol, Fumonisin B1 and Aflatoxin B1. *Toxicon*, *156*, 23–27. https://doi.org/10.1016/j.toxicon.2018.10.305

Yüksel, S., Schwenkbier, L., Pollok, S., Weber, K., Cialla-May, D., & Popp, J. (2015). Label-free detection of Phytophthora ramorum using surface-enhanced Raman spectroscopy. *The Analyst*, *140*(21), 7254–7262. https://doi.org/10.1039/C5AN01156F

Zhan, F., Wang, T., Iradukunda, L., & Zhan, J. (2018). A gold nanoparticle-based lateral flow biosensor for sensitive visual detection of the potato late blight pathogen, Phytophthora infestans. *Analytica Chimica Acta*, *1036*, 153–161. https://doi.org/10.1016/j.aca.2018.06.083

Zhang, X., Weir, B., Wei, H., Deng, Z., Zhang, X., Zhang, Y., Xu, X., Zhao, C., Berger, J. D., Vance, W., Bell, R., Jia, Y., & Li, C. (2020). *Genome-wide identification and transcriptional analyses of MATE transporter genes in root tips of wild* Cicer *spp. Under aluminium stress* [Preprint]. Plant Biology. https://doi.org/10.1101/2020.04.27.063065

Zhao, Y., Liu, L., Kong, D., Kuang, H., Wang, L., & Xu, C. (2014). Dual Amplified Electrochemical Immunosensor for Highly Sensitive Detection of *Pantoea stewartii* sbusp. *Stewartii. ACS Applied Materials & Interfaces*, *6*(23), 21178–21183. https://doi.org/10.1021/am506104r

Index

For Product Safety Concerns and Information please contact our EU representative GPSR@taylorandfrancis.com
Taylor & Francis Verlag GmbH, Kaufingerstraße 24, 80331 München, Germany

www.ingramcontent.com/pod-product-compliance
Ingram Content Group UK Ltd.
Pitfield, Milton Keynes, MK11 3LW, UK
UKHW022216190926
475702UK00003B/395
9781032754505